SCIENTIF
KIDS in LAB COATS

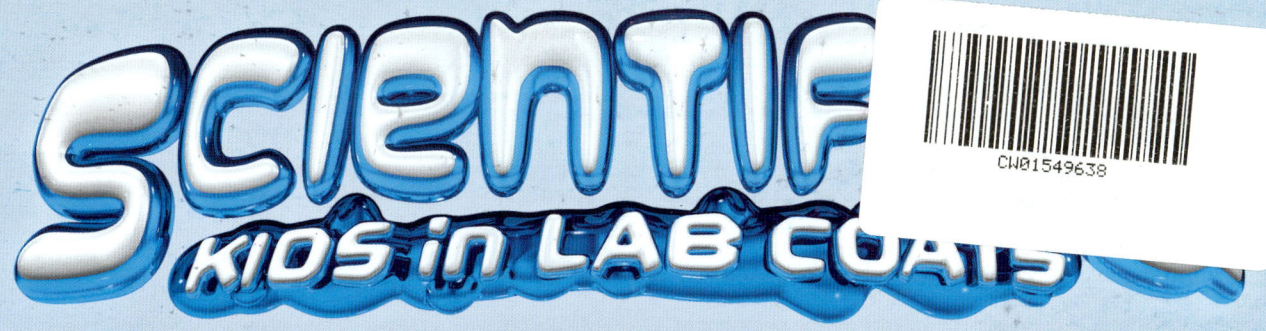

TEACHER BOOK

For Key Stage 3 Science

Peter Ellis • Phil Godding • Derek McMonagle
Louise Petheram • Lawrie Ryan
David Sang • Jane Taylor

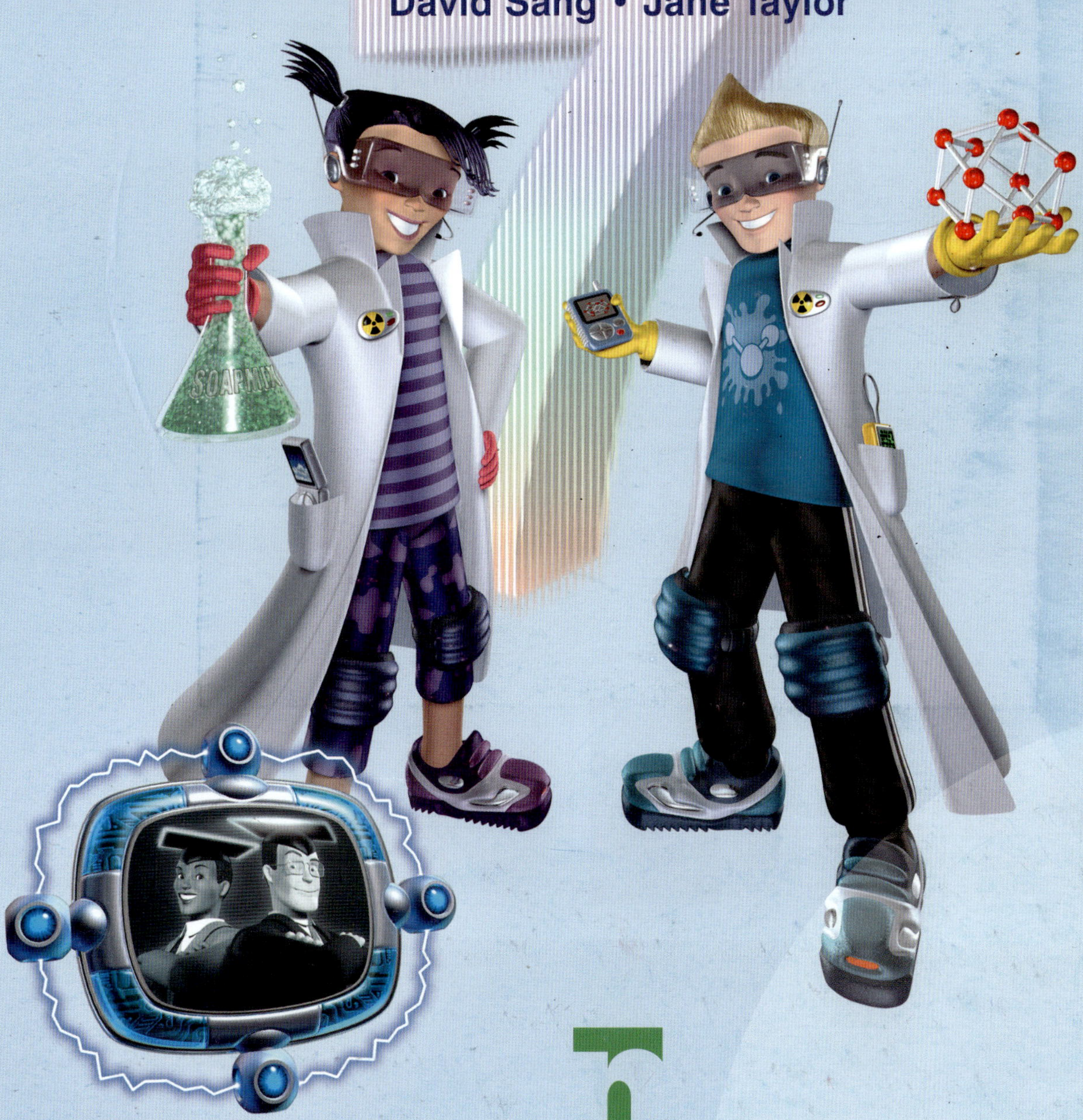

Text © Peter Ellis, Derek McMonagle, Louise Petheram 2004
Original illustrations © Nelson Thornes Ltd 2004

The right of Peter Ellis, Derek McMonagle, Louise Petheram to be identified as authors of this work has been asserted by them in accordance with the Copyright, Designs and Patents Act 1988.

All rights reserved. No part of this publication may be reproduced or transmitted in any form or by any means, electronic or mechanical, including photocopy, recording or any information storage and retrieval system, without permission in writing from the publisher or under licence from the Copyright Licensing Agency Limited, of 90 Tottenham Court Road, London W1T 4LP.

Any person who commits any unauthorised act in relation to this publication may be liable to criminal prosecution and civil claims for damages.

Published in 2004 by:
Nelson Thornes Ltd
Delta Place
27 Bath Road
CHELTENHAM
GL53 7TH
United Kingdom

04 05 06 07 / 10 9 8 7 6 5 4 3 2 1

A catalogue record for this book is available from the British Library

ISBN 0 7487 7982 5

Illustrations by Mark Draisey, Ian West, Bede Illustration
Page make-up by Wearset Ltd

Printed and bound in China by Midas Printing International Ltd.

CONTENTS

	PAGE
Introduction	
7A Cells	2
7A1 Looking at microscopes	
7A2 Plant and animal cells	
7A3 Different cells	
7A4 Tissues and organs	
7A5 Making more cells	
7A6 Pollen and ovules	
Read all about it!	
Unit review	
7B Reproduction	20
7B1 Growing up	
7B2 About girls and boys	
7B3 Menstrual cycle	
7B4 Pregnancy	
7B5 The first few months	
7B6 Healthy pregnancy	
7B7 Growth and reproduction	
Read all about it!	
Unit review	
7C Environment and feeding relationships	40
7C1 Environmental influences	
7C2 What's it like out there?	
7C3 Daily changes	
7C4 Seasons	
7C5 Food chains and webs	
7C6 Eating and being eaten	
7C7 Disturbing the web	
Read all about it!	
Unit review	
7D Variation and classification	60
7D1 Animal and plant groups	
7D2 Sorting animals	
7D3 Sorting plants	
7D4 Variation	
7D5 Passing on the information	
7D6 Making use of genes	
Read all about it!	
Unit review	
7E Acids and alkalis	78
7E1 Acids all around	
7E2 Hazard symbols	
7E3 Using indicators	
7E4 Reacting acids with alkalis	
7E5 Investigating neutralisation	
Read all about it!	
Unit review	
7F Simple chemical reactions	94
7F1 Observing reactions	
7F2 Acids and metals	
7F3 Acids and carbonates	
7F4 About combustion	
7F5 Investigating burning	
Read all about it!	
Unit review	

	PAGE
7G The particle model	110
7G1 Evidence from experiments	
7G2 Classifying materials	
7G3 The particle theory	
7G4 Applying particle theory	
Read all about it!	
Unit review	
7H Solutions	124
7H1 Separating mixtures	
7H2 Particles in solution	
7H3 Distilling mixtures	
7H4 Chromatography revealed	
7H5 Finding solubility	
Read all about it!	
Unit review	
7I Energy	140
7I1 Fuels on fire	
7I2 Fossil fuels	
7I3 Renewables	
7I4 Energy – use with care!	
7I5 Food as fuel	
Read all about it!	
Unit review	
7J Electrical circuits	156
7J1 Complete circuits	
7J2 Electric current	
7J3 Cells and batteries	
7J4 Electricity from Chemistry	
7J5 Series and parallel	
7J6 Keeping safe	
Read all about it!	
Unit review	
7K Forces and their effects	174
7K1 Go on – force yourself!	
7K2 Bend and stretch	
7K3 Mass and weight	
7K4 Floating and sinking	
7K5 All about friction	
7K6 Graphs that tell stories	
Read all about it!	
Unit review	
7L The solar system – and beyond	192
7L1 Light from the Sun	
7L2 Earth and Moon	
7L3 Four seasons – or two?	
7L4 Eclipses	
7L5 Solar system	
7L6 Beyond the solar system	
Read all about it!	
Unit review	
Best practice in Sci	210

Welcome to Scientifica

How the Teacher Book works

This Teacher Book is designed to make the planning and delivery of a great Science lesson as easy as possible.

The concept is simple. All the curriculum references, objectives, support notes and answers you need are distributed around a reduced-size facsimile of the pupil book page.

Use the colour-coding to visually locate the relevant information you require. For example, questions which appear in a green panel in the student text, will have their answers in a green panel nearby. Activity boxes are in blue, and correspondingly, their support notes appear in blue.

Additional support resources

Scientifica delivers a huge wealth of support resources for the teachers and students. A unique feature of the course is that the resources are split in three ways, according to what you need and what you can afford.

1 This **Teacher Pack** provides the support notes required for effective teaching of a *Scientifica* lesson.

2 The **Teacher Resource Pack** contains over 600 pages of further support: Homework sheets, Extension sheets, Activity sheets, Support for Misconceptions, Ideas & Evidence activities, Citizenship activities, and many more. The materials are clearly organised by individual lesson and by topic. All the materials are delivered in editable form.

3 The **ICT Power Pack** uses the Just Click solution – see below – to offer a wide range of cutting-edge resources to help deliver supercharged lessons. Alongside a portfolio of ICT skills-building resources, there are videos, animations, interactive quizzes and demonstrations.

Most activities in the pupil book are supported by a Pupil Activity Sheet in the Teacher Resource Pack – you will see the icon clearly displayed within the Activity Notes box. Many teachers like to distribute these when in a lab environment in order to provide extra support for students, and potentially to protect the textbooks from harm.

Starters and plenaries

This Teacher Book provides starters and plenaries for every lesson, in two ways . . .

1 There are 3 starter and plenary suggestions for every lesson. You will find these located in panels at the beginning and the end of the lesson, as you would expect.

2 The first of the 3 starters is delivered for you in electronic form at the beginning of the PowerPoint™ presentation. You will see this represented as an [S] in the left-hand sidebar. The first of the 3 plenaries is delivered for you at the end of the PowerPoint™ presentation. You will see this represented as a [P] in the right-hand sidebar.

To receive the second and third of the starters and plenaries in electronic form, you require the Scientifica Teacher Resource Pack, and the Scientifica ICT Power Pack respectively.

PowerPoint™ presentations

Attached to the inside front cover of this Teacher Book is a CD-ROM which contains a large series of MS PowerPoint™ presentations. You will find a presentation to correspond to every lesson in the textbook.

These presentations are designed to correlate closely to the content and sequence of each lesson. You will find them useful for whole-class teaching of the topic, with minimal preparation. Naturally, being in PowerPoint™, you will find the presentations easy to amend and customise to your own style.

Towards the outside edge of the pages, you will see sidebars with letters and numbers in boxes. These letters correspond to the starters and plenaries – see Starters and plenaries – and the numbers correspond to the relevant slide in presentation. A [2] for example, means that Slide 2 in the presentation corresponds to the pupil book content directly opposite.

Using the CD-ROM

The CD-ROM should Auto-Run when inserted in your PC or Network server. If it does not, locate and run the Setup.exe file using your File Manager. The program requires Internet Explorer and MS PowerPoint™ to run.

The CD-ROM interface is very simple to use – simply click on the lesson of which presentation you require.

Support for extension work

Towards the back of the textbook, there are a series of pages of extension work, corresponding to most of the lessons. These Phenomenal Performance activities are supported by the Extension work panels in the main lesson notes.

The Just Click solution

If you have purchased the Scientifica ICT Power Pack, you will have gained full access to a range of lesson planning tools and cutting-edge resources – the Just Click solution.

The contents of your Teacher Book Presentations CD-ROM can be seamlessly integrated into this program. Details can be found in the Just Click program.

Photos

You may find that some of the colour photos that appear in the pupil book do not appear here. This is simply due to licensing restrictions which have prevented their inclusion. With a copy of the pupil book to hand, you should not find your teaching affected in any way.

Levelometer

In the Unit review sections, a series of SAT-style questions is provided, with mark schemes. The Levelometer graphic provides an indicator of level boundaries. These indicators are not definitive, and should be used only as a guide to overall level performance.

The textbook introduction page

Scientifica is filled with the latest and best ideas we could find in Science teaching and learning today.

Pupils will find the textbook filled with them, and this introductory spread offers a brief and light-hearted explanation of some of the most visible items and features. These are more to discover throughout the textbook, and we hope that all together, they accumulate to provide a fantastic Science learning experience.

Best of luck with your teaching!

Acknowledgements

Corel 4 (NT) 106a, **Corel 12 (NT)** 43, **Corel 25 (NT)** 96, **Corel 36 (NT)** 132, **Corel 48 (NT)** 102a, **Corel 160 (NT)** 130, **Corel 243 (NT)** 94a, **Corel 403 (NT)** 66f, **Corel 417 (NT)** 104, **Corel 418 (NT)** 84, **Corel 462 (NT)** 66a-d, **Corel 573 (NT)** 14a, **Corel 602 (NT)** 34, **Corel 653 (NT)** 112, **Corel 658 (NT)** 14b, **Corel 660 (NT)** 66e, **Corel 699 (NT)** 5b, **Corel 713 (NT)** 160, **Corel 760 (NT)** 56a, **Corel 768 (NT)** 203a, **Corel 771 (NT)** 143, **Corel 799 (NT)** 144, **Corel 800 (NT)** 74; **Joe Cornish/Digital Vision LL(NT)** 47, **Stephen Frink/Digital Vision LU(NT)** 42, 185, **Digital Stock 12 (NT)** 31, **Digital Vision 2 (NT)** 68, **Digital Vision 3 (NT)** 52, **Digital Vision 5 (NT)** 35, 55, **Digital Vision 9 (NT)** 195, 202, 203b, 204b, **Digital Vision 11 (NT)** 77, 136, **Digital Vision 14 (NT)** 110, 194, **Digital Vision 15 (NT)** 126, **Digital Vision 17 (NT)** 13, **Illustrated London News V1 (NT)** 106b, **Ingram Publishing/IL V2 CD 6 (NT)** 161; **Photodisc 4 (NT)** 14c, 227b, **Photodisc 6 (NT)** 2, 48, 53, 151, **Photodisc 10 (NT)** 196, 197, **Photodisc 22 (NT)** 148a, **Photodisc 44 (NT)** 205, **Photodisc 45 (NT)** 69, **Photodisc 54 (NT)** 158, **Photodisc 61 (NT)** 36b, **Photodisc 67 (NT)** 56b, **Photodisc 72 (NT)** 114, 118a; **Royal Observatory, Edinburgh** 205bl.

Picture research by Stuart Sweatmore

Every effort has been made to trace all the copyright holders, but if any have been overlooked the publisher will be pleased to make the necessary arrangements at the first opportunity.

7A Cells

National framework/QCA SoW references

National framework – Cells: Describe a simple model for cells that recognises those features all cells have in common and the differences between animal and plant cells.

Explain that some living organisms are only one cell but that others are multi-celled.

Explain that growth means an increase in the size and number of cells.

Explain that in multi-celled organisms certain cells may become specialised, e.g. sperm and egg cells.

Explain that similar specialised cells can be grouped together to form tissues, that tissues can form organs.

QCA SoW (7A)

What are living organisms made from? (7A1)
How can using a microscope give us information about structure? (7A1)
What are cells like? (7A2)
What do cells do? (7A3, 7A4)
How are new cells made? (7A5)
What causes pollen tubes to grow? (7A6)
NC links: Sc2 1 a, b, c, d, e; Sc1 1 b, c; Sc1 2 d, e, f, g, k, m, n, o.

Learning objectives

- To know that plants and animals contain organs made up of tissues.
- To know that: plant and animal tissues are made up of cells; that cells have a cell membrane, nucleus, cytoplasm; that plant cells have, in addition, a cell wall, a vacuole and sometimes chloroplasts.
- To know that cells are specialised and to explain how the cells carry out their particular functions, such as sperm cells, red blood cells, epithelial cells, neurones, root hair cells and leaf cells.
- To describe how: cells can divide to make new cells and to grow; division begins with the nucleus.
- To know that: plants and animals have special cells that are used in reproduction; the nuclei contain information passed from one generation to the next.
- To explain the process of fertilisation in plants following pollination, including the growth of pollen tubes in the stigma.

In scientific enquiry:

- To know how observations made with a microscope helped ideas about the structure of living things to develop.
- To know how to: use a microscope safely and effectively; make observations using a microscope and record these in drawings.
- To compare and interpret information from microscopic observation.
- To draw conclusions from observations and explain these using scientific knowledge.
- To know the importance of sampling in biological investigations.
- To control relevant variables and take account of those that cannot be controlled.

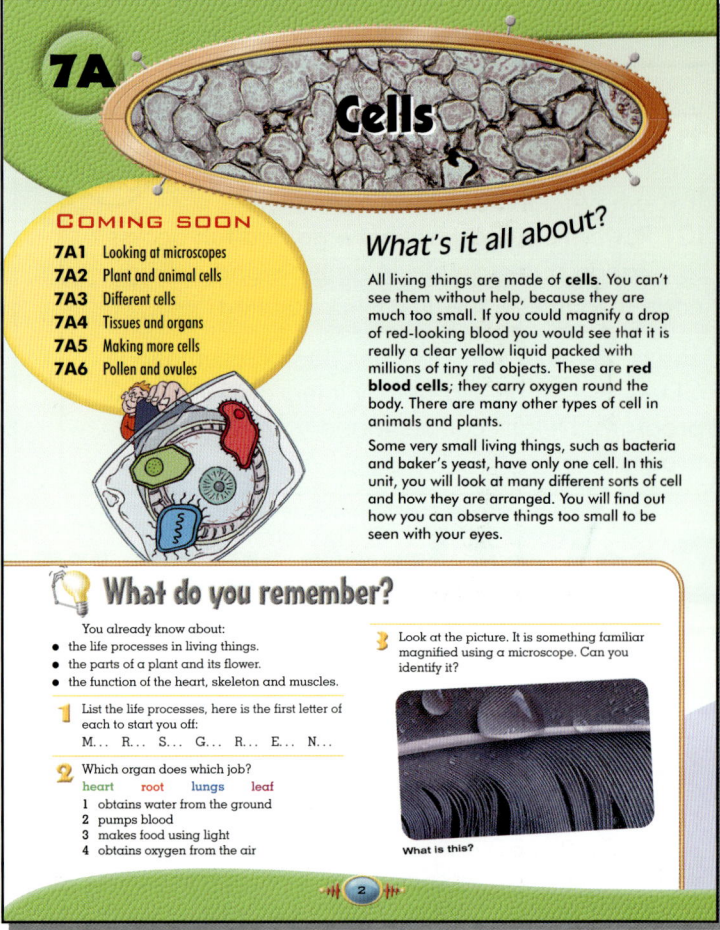

Answers to questions

What do you remember?

1. Movement, Respiration, Sensing, Growth, Reproduction, Excretion, Nutrition.
2. (1) Root.
   (2) Heart.
   (3) Leaf.
   (4) Lungs.
3. (Part of) a bird's feather.

Notes to support 'What do you remember?'

This unit builds on work from KS2. Pupils will have learned about: life processes of animals and plants (including the heart and circulation); the role of leaves and roots in plant growth and nutrition; the parts of flowering plants involved in pollination, seed formation, dispersal and germination. They will have seen individual cells in the form of micro-organisms. Before beginning this unit, it will be useful if pupils:
- Can name the processes of life.
- Can name some organs in plants and animals.
- Know the parts of the plant involved in reproduction.

Teaching strategy

This unit builds on KS2 work on reproduction in plants in unit 5B 'Life cycles' and on organs in the body in units 5A 'Keeping healthy' and 4A 'Moving and growing'. It leads into units 7B 'Reproduction', 8A 'Food and digestion', 8B 'Respiration', 8C 'Microbes and disease', 9A 'Inheritance and selection', 9B 'Fit and healthy', and 9C 'Plants and photosynthesis'.

Difficulties/Misconceptions

- Pupils may have difficulty in accepting that they and all living creatures are made up of cells that resemble the micro-organisms that they met in unit 6B.
- Pupils may confuse the processes of pollination and fertilisation in flowering plants and think that pollen and seeds are the same thing.
- Some pupils may have difficulty in focusing a microscope and in detecting growing pollen tubes.

Notes and tips

- While pupils do not need to know details of the function of parts of the cell, or of tissues formed by specialised cells, they do need to be able to distinguish between plant and animal cells. They will learn more about the functions of certain tissues in Years 8 and 9.
- This unit includes two major elements of scientific enquiry: the use of the microscope and the investigation of pollen tubes. Sufficient time should be given to enable pupils to become competent in the use of microscopes.

Learning styles

Visual
- **Obtaining** information from pictures (micrographs).

Auditory
- **Listening** and discussing.

Interpersonal
- **Debating** answers to questions in groups.

Intrapersonal
- **Reflecting** on prior knowledge.

Launch activity notes – Ideas about cells

This activity is a starting point for finding out what pupils know about cells. They can discuss the pictures and the questions. They could be asked to discuss what they think cells are. Some pupils may be unfamiliar with the word 'cells' and confuse it with 'sells'. Discussion of the questions and pictures will help pupils to understand that cells make up all organisms.

Answers

a) Pupils may know that cells are very small and can only be seen using a microscope; magnifying lenses are not powerful enough. Microscopes need a bright source of light shining through the specimen.

b) Pupils may also discuss what cells do when linked together in an organ. The liver is an organ. Other organs include: the heart, pumping blood; the lungs, exchanging gases; the skeleton, support, etc.

c) Pupils may be aware that cells divide (split in two) producing two new identical cells. Some cells (e.g. skin) are dying all the time, but being replaced by other cells that have divided. (New cells are produced in the brain of children, but not as frequently in adults.)

7A1 Looking at microscopes

National framework/QCA SoW references

National framework – Scientific enquiry: Pupils should be taught to use appropriate equipment to make observations and measurements.

QCA SoW

7A Section 2: How can using a microscope give us information about structure?
NC links: Sc1 2f, Sc1 2g

Learning objectives

▸ To use a range of equipment and materials appropriately and take action to control risks to oneself and others.
▸ To make observations and measurements.
▸ To use a microscope safely and effectively.
▸ To make observations using a microscope and to record these as drawings.

Teaching strategy

Key points to highlight

- Cells are too small to be seen clearly with the eye alone.
- Microscopes use two (or more lenses) to form a magnified image.
- Overall magnification = objective magnification × eyepiece magnification
- Use lowest magnification first.
- The microscope is focused by moving the objective away from the specimen.

Difficulties/Misconceptions

Pupils may have difficulties with these:
- Focusing the microscope.
- Drawing what is seen.
- Calculating magnification.
- Estimating the size of a specimen.

Skills

- Practical skills in manipulating microscopes.
- Writing skills in describing observations.
- Drawing skills in recording what is seen through the microscope.

Notes and tips

- Have some microscopes set up with slides focused, so that pupils can immediately have an idea of what can be seen.

Lesson starter suggestions

- **What is this? How was it produced?** Class work.
- **Using a magnifying glass:** Group work: finding the best focusing distance, estimating magnification.
- **Orders of size:** Individual work: put the following list of things in order of size. Estimate how big they are (i.e. the distance across). Use this list: (muddled) a cat, a chicken, an egg, a fingernail, hair, a red-blood cell, a virus. Which can you see clearly with your eyes?

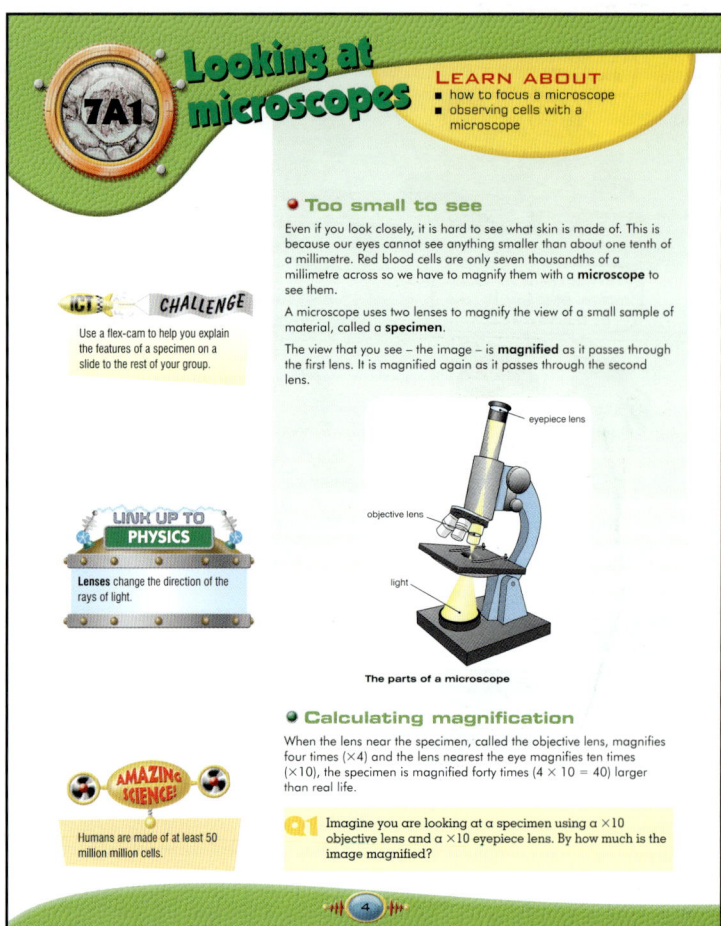

- Disabled or dyspraxic pupils should be paired with able pupils, so that they do not have to make fine adjustments or handle the slides themselves.
- Have some clear, plastic rulers available so pupils can see the size of the field of view.

Extension ideas

Measuring specimens.

Activity and technicians' notes

Using a microscope

Safety notes
- Take care not to have direct sunlight on the microscope mirror.
- Care is required with focusing to avoid cracking the slide or cover slip.
- Glass cover slips are fragile and can easily cause cuts.

Equipment required
- Microscopes
- Magnifying lenses
- Prepared slides (small insects, stem sections, small print/diagrams)
- Clear, plastic rulers

Access required to:
A TV camera mounted on a microscope

Learning styles

Visual
- **Observing** pictures of microscopes and real microscope and slides, drawing diagrams of observations.

Auditory
- **Listening** to instructions.

Kinaesthetic
- **Manipulating** a microscope.

Interpersonal
- **Collaborating** in groups to obtain observations.

Intrapersonal
- **Knowing** the learning objectives.
- **Giving feedback** on observations.
- **Reflecting** on skills learnt.

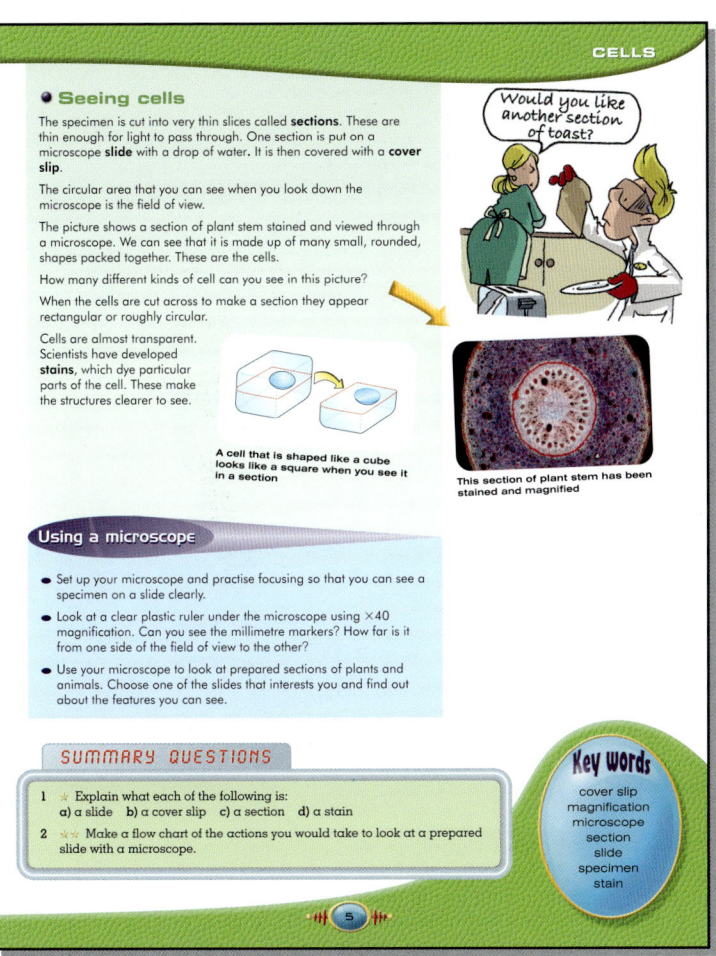

Answers to questions

Text

Q1 Magnification = 10 × 10 = 100 times

Summary

1. a) a strip of glass, on which the specimen is placed
   b) a thin piece of glass, which is placed on top of the specimen on the slide
   c) a specimen obtained by cutting across an object
   d) a dye used to make parts of the specimen coloured
2. place slide on stage → move slide into position → choose objective with lowest power → lower objective until it is just above the slide → look through the eyepiece and raise the objective until the image is in focus → repeat focusing with a higher magnification objective

Amazing science

50 million million is 50 000 000 000 000. It is also called 50 trillion or 5×10^{13}. The simplest creatures, such as bacteria and yeast, consist of just a single cell.

Learning outcomes

Following this, most pupils will:
▶ Correctly focus the microscope to view it.
▶ Describe how the objects appear under low magnification.
▶ Make careful drawings of the objects viewed.
▶ Make observations using a microscope and record them in simple drawings.

Some pupils will also:
▶ Estimate sizes of specimens viewed under the microscope.

Suggestions for plenaries

- **Common faults in using the microscope**
- **Tips for using microscopes:** What do pupils suggest? (e.g. close/cover one eye, etc.)
- **Art gallery:** Display diagrams that have been drawn of specimens.

7A2 Plant and animal cells

National framework/QCA SoW references

National framework – Cells: Describe a simple model for cells that recognises those features all cells have in common and the differences between animal and plant cells.

QCA SoW

7A Section 2: How can using a microscope give us information about structure?
7A Section 3: What are cells like?
NC links: Sc2 1b

Learning objectives

- To know: the functions of chloroplasts and cell walls in plant cells; the functions of the cell membrane, cytoplasm and nucleus in both plant and animal cells.
- To use skimming, scanning, highlighting and note taking, as appropriate, on different texts.
- To know how ideas about the structure of living things have changed.
- To know that plants and animals are made up of cells.
- To know that plant and animal cells are similar in a number of respects, but have significant differences.
- To make observations using a microscope.
- To know that plant and animal cells have a cell surface membrane that keeps the cell together and controls what enters and leaves.
- To know that cells have cytoplasm that occupies most of the cell.
- To know that cells have nuclei that control activities of the cell.

Lesson starter suggestions

- **Word search:** Individual/class: word search puzzle incorporating the words 'nucleus, cytoplasm, vacuole, cell membrane, cell wall, chloroplast'.
- **What is a cell like?** Discussion (orientation exercise).
- **Discovering cells:** Reading about what people thought plants and animals were made of (literacy exercise).

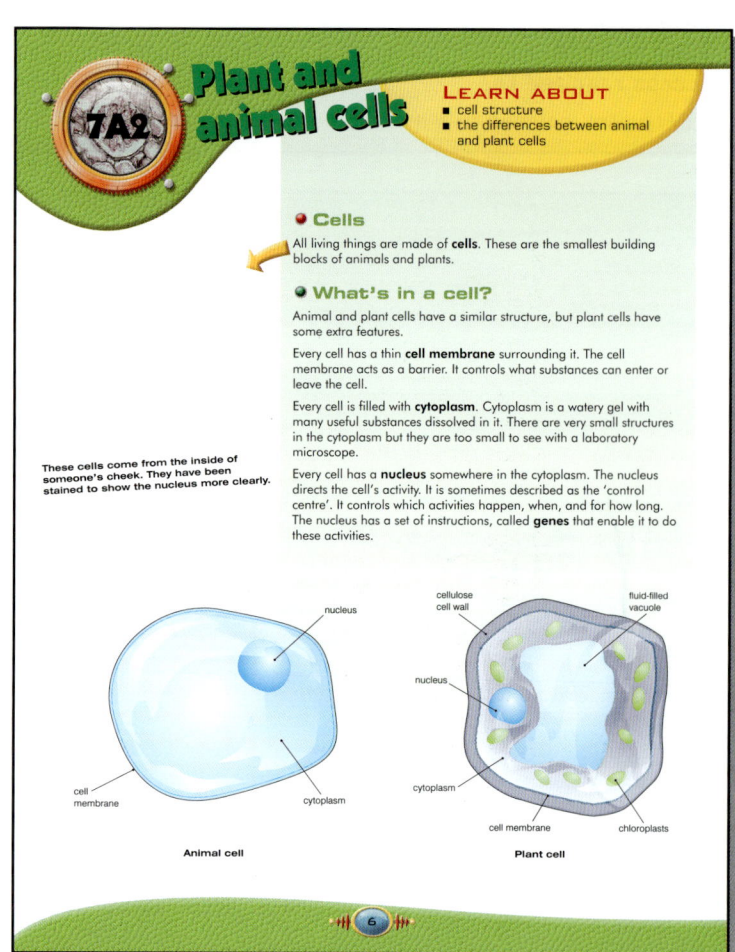

Teaching strategy

Key points to highlight

- All plant and animal cells have a cell membrane, cytoplasm and a nucleus.
- In addition, plant cells have a cell wall, chloroplasts and vacuole.
- The functions of the parts of cells.
- Preparing microscope slides.

Difficulties/Misconceptions

Pupils may have difficulties with these:
- Looking at microscope slides or pictures of cells – some pupils may not appreciate the three-dimensional nature of cells.
- In plants, most of the cell is taken up by the vacuole. The cytoplasm forms a thin layer just inside the cell wall.
- Avoiding air bubbles when preparing slides, and not confusing air bubbles with cells under the microscope.
- When focused on a cell, pupils may miss the nucleus. They need to focus up and down through a cell to try and pick out the nucleus.

Skills

- Making use of textual information.
- Preparation of microscope slides.

Notes and tips

- Demonstrate making models of plant and animal cells using jelly for cytoplasm (or allow pupils to try).
- Use stains to bring out features of cells.

Extension ideas

Measuring the size of yeast cells.

Activity and technicians' notes

Viewing plant and animal cells

Safety notes
Staining solutions – methylene blue or toluidine blue (for nuclei), iodine in potassium iodide (for chloroplasts), eosin Y (for cytoplasm and cell walls) – are all harmful.

Equipment required
- Microscopes
- Slides, cover slips, droppers, tweezers, mounted pins, (haemocytometers)
- Onion, pondweed, yeast

Access required to:
A TV camera connected to a microscope and a projector/TV

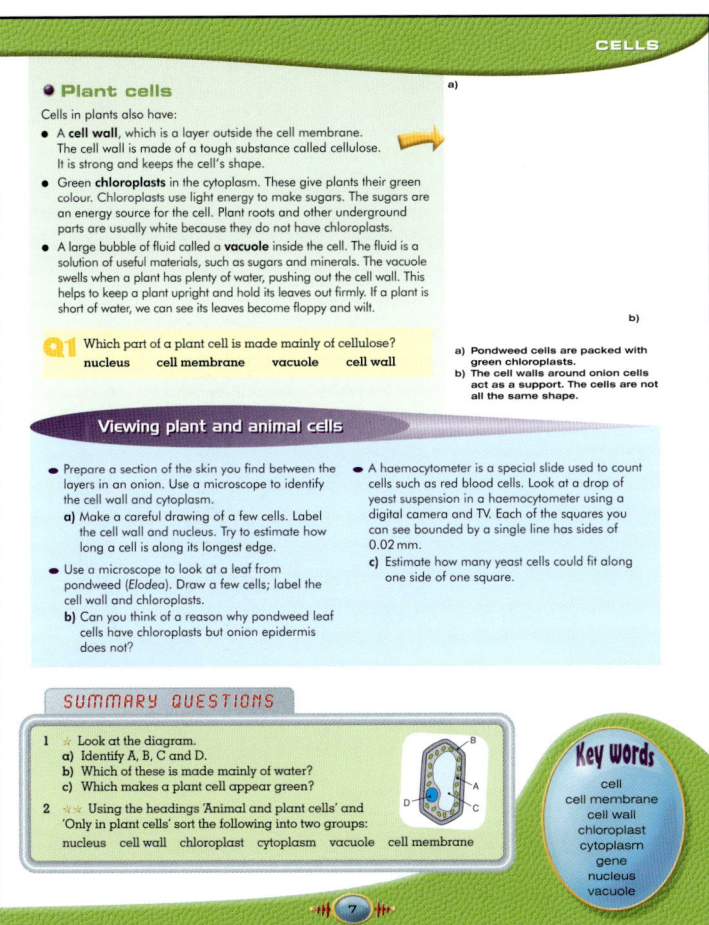

Learning outcomes

Following this, most pupils will:
▸ Describe some differences between plant and animal cells.

Some pupils will also:
▸ Describe simple cell structure and identify the differences between simple animal and plant cells.

Learning styles

Visual
- **Looking at** pictures and models of cells.
- **Studying** examples of writing.
- **Word puzzles**.

Auditory
- **Listening** to instructions (perhaps of the activity read aloud).

Kinaesthetic
- **Using microscopes** to investigate cells.
- **Building** models.

Interpersonal
- **Collaborating** on puzzles, activities and questions.

Intrapersonal
- **Reflecting** on outcomes of activities.

Answers to questions

Text

Q1 cell wall

Summary

1. a) A is a chloroplast; B is the cell wall; C is the vacuole; D is the nucleus.
   b) C
   c) A

2.

Animal and plant cells	Only in plant cells
nucleus	cell wall
cytoplasm	chloroplast
cell membrane	vacuole

Suggestions for plenaries

- **Build a cell:** Slide (or handout) shows different parts of a cell. Pupils (or teacher) move parts around screen (or cut out and stick the parts) to make an animal or plant cell.
- **Odd one out:** Pupils discuss which of the following is the odd one out:
  a) 'nucleus, cytoplasm, chloroplast, cell membrane' (The answer being 'chloroplasts are only in plant cells, the others are in all cells'.)
  b) 'vacuole, cell wall, chloroplast, nucleus' (The answer being 'all cells have a nucleus, the others are only in plant cells'.)
- **Spelling bee:** Read out this list of words, asking pupils to write them down and then check their spelling: 'nucleus, membrane, vacuole, cytoplasm, chloroplast'.

7A3 Different cells

National framework/QCA SoW references

National framework – Cells: Explain that in multi-celled organisms certain cells may become specialised, e.g. sperm cells and egg cells.

QCA SoW

7A Section 4: What do cells do?
NC links: Sc2 1c

Lesson starter suggestions

- **Sorting cells:** Class/individual work: pupils sort diagrams of various types of cell into plant and animal cells.
- **Cutlery:** Class/group work: pupils look at items from a set of cutlery (e.g. knife, fork, spoon, teaspoon) and discuss the similarities and differences in function, and the adaptations to that function.
- **Doing different things:** Individual work: pupils make a list of the different things that cells in their body do. (See Suggestions for plenaries for the conclusion to this.)

Learning objectives

▶ To know that some cells, including ciliated epithelial cells, sperm cells, and root hair cells, are adapted to their functions.
▶ To know that there are different types of cell, adapted for different functions.
▶ To know how to use secondary sources of information.

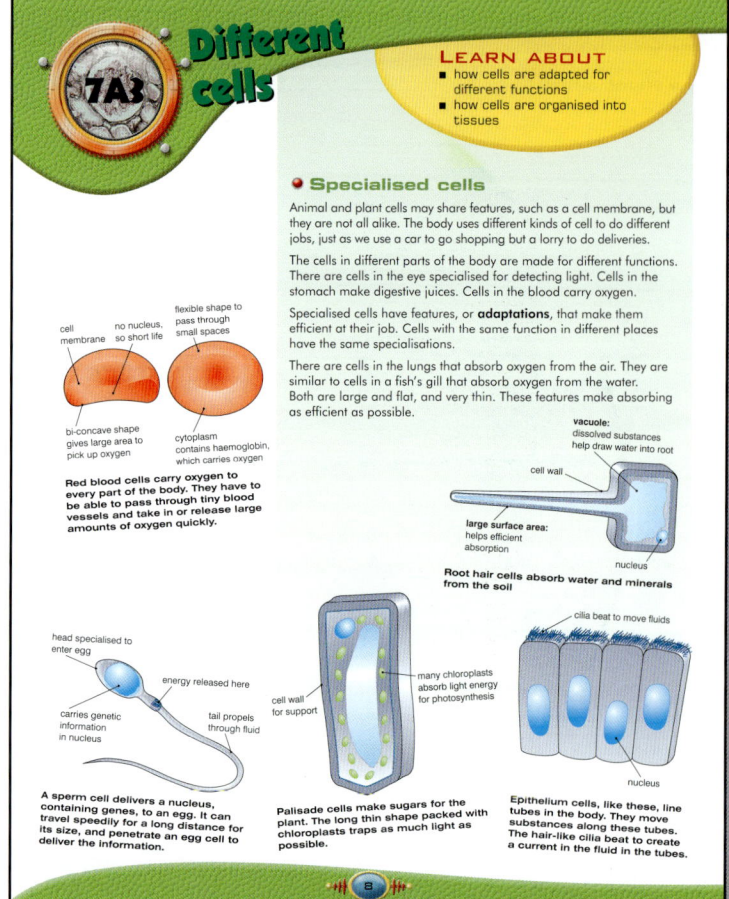

Teaching strategy

Key points to highlight

Plants and animals are made up of different types of cell that are adapted to perform various functions, for example:

- Epithelial cells have hairs (cilia) that wave to move liquids along tubes.
- Sperm cells have long tails that help them to swim.
- Root hair cells have fine tubes (hairs), which have a large surface area to absorb water and minerals from soil.
- Red blood cells contain haemoglobin and have a large surface area for absorbing oxygen.
- Palisade cells contain lots of chloroplasts for absorbing light and making sugars.

Difficulties/Misconceptions

There is a lot to learn in this topic, and pupils may have difficulty in remembering the names of the different types of cell, their adaptations and their functions. 'Average' pupils need only know that not all animal or plant cells are the same and that specialised cells do special jobs.

Skills

- Practical skills in preparing microscope slides and using the microscope.
- Presentations, where pupils will need to describe what they have done and what they have found out.

Notes and tips

- It is useful to have prepared microscope slides or PowerPoint slides of different types of cell.

- There are plenty of analogies for the way that an animal or plant is made up of different cells that work together. For example, the members of a team (e.g. football/hockey with a goalkeeper, attackers, defenders; cricket/rounders with bowlers/pitcher, wicket keeper/back-stop, batsmen/fielders; armies with infantry, artillery, reconnaissance, communications, medical workers, suppliers, drivers) each with their own equipment and job.

Extension ideas

Consider how the special features of these cells enable them to carry out their function, e.g. large surface area of root hair helps it to absorb water and minerals fast.

Activity and technicians' notes

Viewing cell adaptations

Safety notes
- Take care with scalpels, for cutting thin slices of potato, lemon rind
- Take care with stains: iodine in potassium iodide for starch; Sudan III for oil drops

Equipment required
- Microscopes
- Microscope slides, cover slips, droppers, tweezers or mounted pins
- Potatoes (small slices), lemon rind pieces
- Prepared slides of stained potato and lemon rind cells showing starch grains and oil drops respectively

Access required to:
- A TV camera, linked to a microscope and TV/computer/data projector
- Prepared slides/pictures of specialised cells, e.g. epithelial cells, red blood cells, root hair cells, sperm cells, palisade cells

 PUPIL SHEET

Learning styles

Visual
- **Looking at** pictures and slides of cells.

Auditory
- **Listening to** descriptions as part of presentations.

Kinaesthetic
- **Practical** work.

Interpersonal
- **Collaboration** on practical work and preparing presentations.

Intrapersonal
- **Answering** questions.

Answers to questions

Summary
1. a) palisade cell
   b) epithelium with cilia
   c) red blood cell or root hair cell
2. nucleus; cytoplasm; long thin rounded shape (not angular); closely packed together; cytoplasm is very dense. These are animal cells.
   The most important feature is that there is no cell wall around the cells

Gruesome science

Although skin cells are dead, they are full of nutritious goodies for creatures such as the house dust mite. Mites prefer the warm, cosy habitat found in centrally heated, carpeted houses, living in the carpets and other soft furnishings such as cushions, mattresses and duvets. Most people are unaffected by the mites, but some people become sensitive to their waste products and develop eczema and asthma.

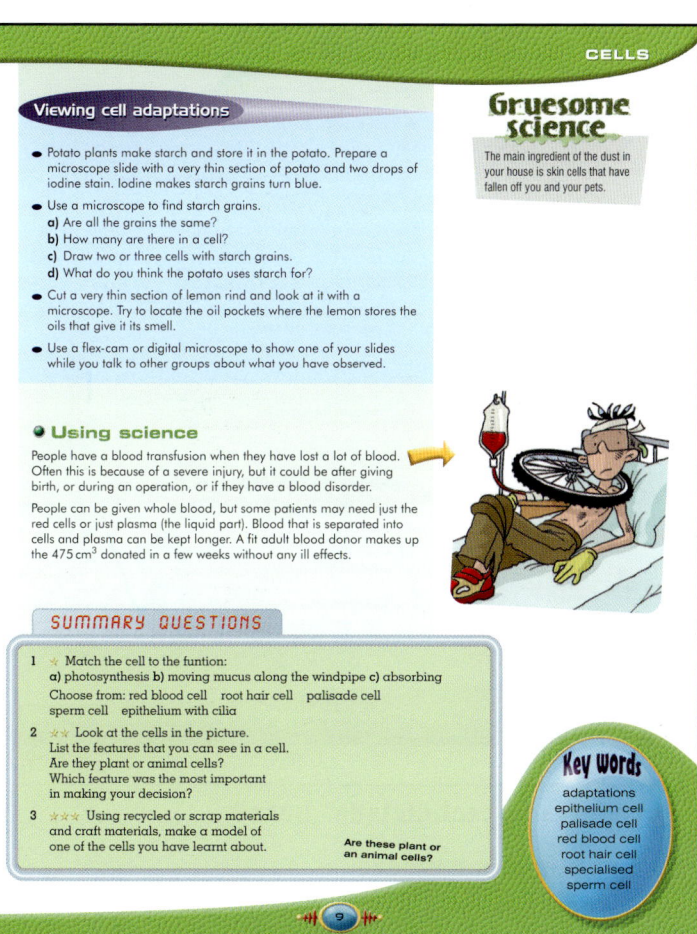

Suggestions for plenaries

- **Anagrams:** Class/individual: test pupil's ability to sort the words used in the lesson, i.e. specialised, adapted, epithelial, palisade, haemoglobin, cilia, sperm.
- **Cells mind map:** Individual/group: pupils draw a mind map for the features and functions of cells. Start with cells at the centre, with plants on one side and animals on the other.
- **Doing different things:** Individual/group: pupils look back at their list of cells in the body (from the Lesson starter suggestions) and add the particular features of the cells that allow them to do their job. They can add more cells to the list.

Learning outcomes

Following this, most pupils will:
▸ Explain that different types of cell can be found in plants and animals, and that these cells carry out specialised functions.
▸ Identify specialised features in different types of cell, and relate these to the function of a cell.
▸ Find and present relevant information on a particular cell type.

Some pupils will also:
▸ Be able to describe how some cells in an organism are specialised to carry out particular functions.
▸ Use their knowledge of cell structure to explain how cells are adapted to their functions.

7A4 Tissues and organs

National framework/QCA SoW references

National framework – Cells: Explain that similar specialised cells can be grouped together to form tissues, that tissues can form organs.

QCA SoW

7A Section 4: What do cells do?
NC links: Sc2 1a.

Learning objectives

- To know that plants and animals contain organs.
- To know that tissues make up organs.
- To draw inferences from data.
- To know that cells form tissues, and tissues form organs.
- To name some important tissues in plants and humans.
- To explain the organisation of tissues using a model.

Teaching strategy

Key points to highlight

- Tissues are groups of similar cells.
- Tissues carry out a special function.
- Organs contain a number of tissues that work together.
- Organs are linked together to form organ systems.

Difficulties/Misconceptions

- Pupils may not realise that plants are made up of organs.
- Pupils need to be able to distinguish between a tissue and an organ.

Skills

Visual recognition of organs and organ systems in the human body.

Lesson starter suggestions

- **House building:** Class/individual work: pupils make a list of the materials and parts of a house.
- **Bicycle:** Group work: pupils exchange cards until they have assembled all the parts needed to put a bicycle together.
- **Beetle drive:** Group work: pupils play the game in which a dice is thrown to assemble a beetle.

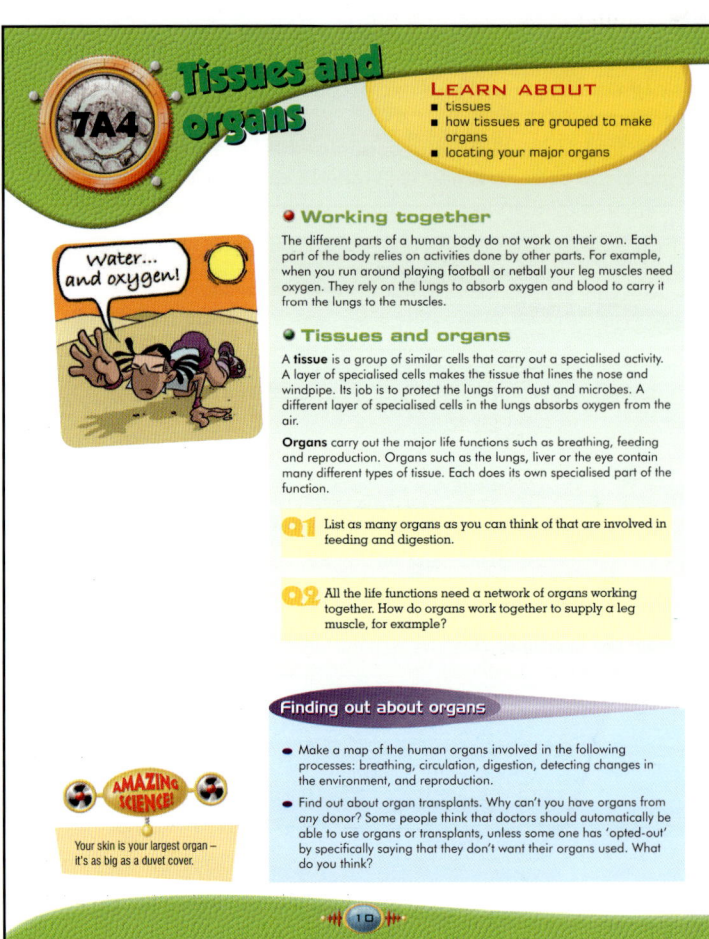

Activity and technicians' notes

Finding out about organs

Access required to:
- a data projector

Teaching strategy contd.

Notes and tips
- There are many analogies for organ systems, e.g. houses, cars, bicycles, computers. In these inanimate systems, the materials used represent the tissues (e.g. the rubber of a tyre, the steel of a wheel rim). It is difficult for the analogies to represent the full hierarchy of cell – tissue – organ – system.
- Other models could be the members of a team or a business (or school), where different people have different functions but work together.
- Each pupil can touch the parts of their body where various organs are located.

Extension ideas
- Dissecting an organ.
- Examining plant tissues.

Learning styles

Visual
- **Seeing** images of tissues, organs and systems.
- **Mind-mapping**.

Auditory
- **Hearing** names of tissues and organs.
- **Spelling**.

Kinaesthetic
- **Touching** the part of the body where an organ is located.

Interpersonal
- **Collaborating** with other pupils in groups, or as a class.

Intrapersonal
- **Reviewing** learning.

Answers to questions

Text
Q1 mouth, gullet (oesophagus), stomach, intestines (large and small), rectum, anus, liver, pancreas, blood

Q2 The lungs absorb oxygen into the blood. The stomach and intestines absorb food into the blood.
The blood vessels carry food and oxygen to the cells and carry away waste materials.
Nerves provide a signal to make the muscles contract and relax.

Summary
1. lungs
2. A, brain – interpreting sensory data and control;
   B, heart – pump;
   C, stomach – digestion;
   D, uterus – nurtures embryo

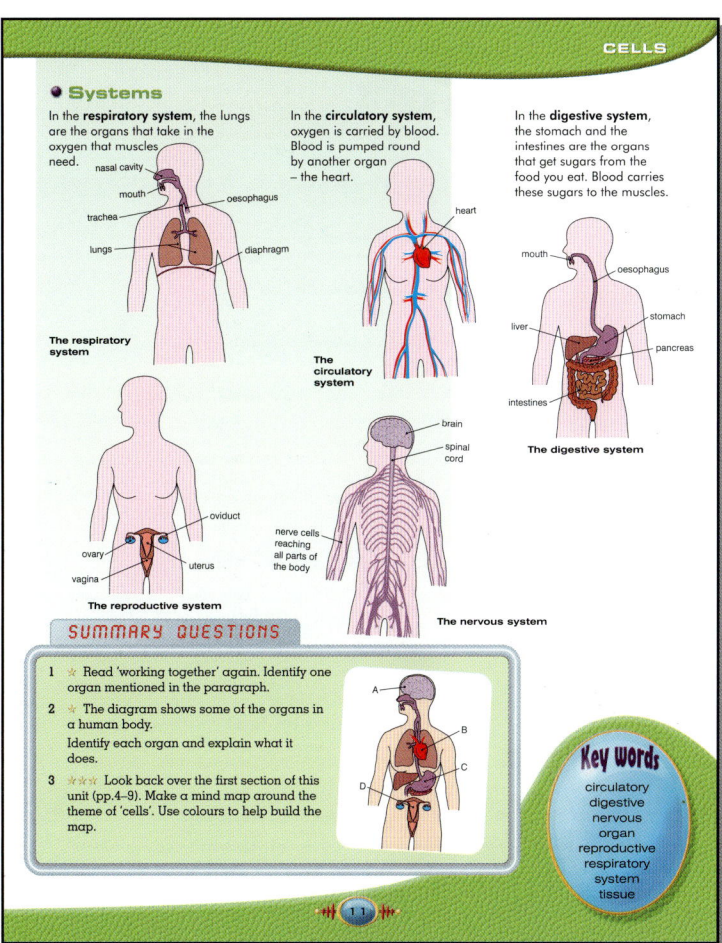

Learning outcomes

Following this, most pupils will:
- Name some examples of tissues from plants and humans.
- Relate the different parts of a model to the cells and tissues making up an organ in a living organism.
- Describe how cells are grouped to form tissues.
- Use scientific names for the major organs of body systems and identify organs of plants.
- Describe the main functions of organs of the human body and plants.

Some pupils will also:
- Relate the cellular structure of organs to life processes.

Amazing science

Skin is more than just a waterproof covering for the body. It has many functions, including: sense and control of temperature, sensing touch. The skin is made up of a number of different types of tissue and cell. Pupils can find out more about the tissues in the skin, i.e. the epidermis, dermis, glands, hairs and nerves.

Suggestions for plenaries

- **Name that organ!** Divide the class into teams, competing to name which organ system different organs belong to.
- **Mind-mapping:** Individual work: make a mind map around the theme of 'cells' (see Summary question 3).
- **What we have learned:** Individual/group: Pupils to make a list of the things they have learned during the lesson.

7A5 Making more cells

National framework/QCA SoW references

National framework – Cells: Explain that growth means an increase in the size and number of cells.

QCA SoW

7A Section 5: How are new cells made?
NC links: Sc2 1e.

Learning objectives

- To relate cells and cell functions to life processes.
- To know that cells can make new cells by dividing.
- To know that growth occurs when new cells are made and increase in size.
- To know that cell division begins with division of the nucleus.
- To know that cells have nuclei containing the information that is transferred from one generation to the next.

Teaching strategy

Key points to highlight

- Growth in plants and animals is a result of cells multiplying.
- Cells multiply by asexual reproduction involving binary fission.
- Chromosomes, which carry the genes in the nucleus, are duplicated.
- The nucleus divides first, then the cytoplasm. A new membrane, and in plants a cell wall, is formed. The new cells are initially small, but soon grow to full size.
- Bacteria are single-celled creatures, which reproduce by binary fission.
- Cell division occurs quickly and frequently when there is a good supply of nutrients.

Difficulties/Misconceptions

- Cells go through a period of growth after each division. They do not divide up into smaller and smaller bits.
- Full details of how the chromosomes duplicate and separate is not required at this stage.
- Pupils may be concerned about how cell division starts and stops. This depends on the availability of nutrients and, in plants and animals, chemical messengers which trigger cell division.

Lesson starter suggestions

- **How does a plant grow?** Group: pupils discuss how a seedling turns into a fully grown plant.
- **Multiples of 2:** Individual: pupils are given a sheet of paper and told to fold it in half and then tear it along the fold. They repeat this as many times as they can, and count the number of pieces they have made.
- **Good/bad/find out more:** Groups discuss the statement 'cells can make more cells very quickly'. Ask the groups to think of a reason for this being good (positive), bad (negative) and a question they want an answer to.

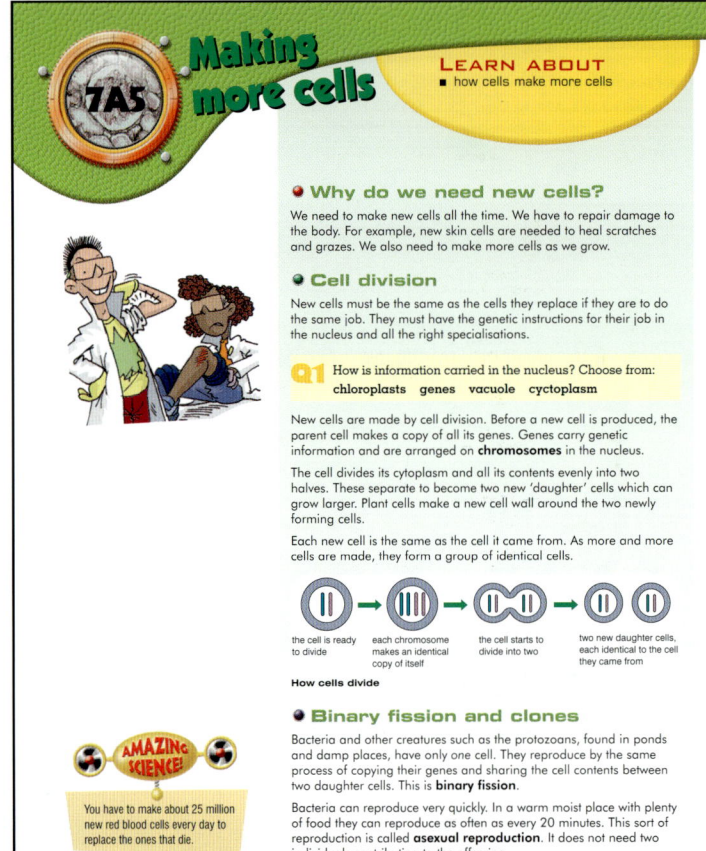

Amazing science

Red blood cells live for just 120 days, as they do not have a nucleus and hence are unable to divide or regenerate. They are produced in the marrow of long bones. A blood donor gives about 10% of their blood; this is replaced in a few days.

Teaching strategy contd.

Skills
Reading and writing; vocabulary.

Notes and tips
- Pupils may have heard the term 'clone' elsewhere and it may have acquired a negative connotation. Explain that we are all made up of cloned cells; we started life as a single cell which has produced many copies of itself.
- Make sure that pupils realise that cell division occurs in all multi-cellular creatures (i.e. plants and animals) as normal growth and in bacteria.

Extension ideas
Viruses.

Learning styles

Visual
- **Studying** images and video clips of cell division.
- **Looking** at slides of cells dividing.

Auditory
- **Repeating** words to learn how they are spelt, i.e. cell division, binary fission, asexual reproduction, chromosome, clone.

Kinaesthetic
- **Sorting** diagrams.

Interpersonal
- **Sharing** work.

Intrapersonal
- **Feedback** on completion of tasks.

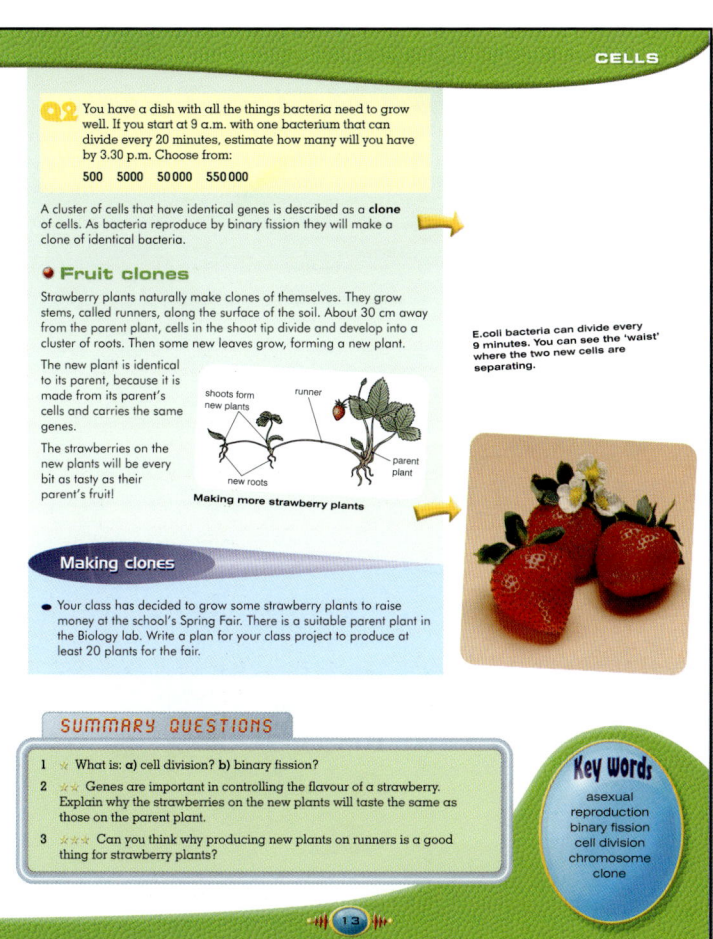

Answers to questions

Text
Q1 genes
Q2 550 000

Activity (Making clones)
Take as many cuttings as possible from the parent plant. Plant them in a growing medium. When they have rooted and grown sufficiently, take more cuttings.

Summary
1. **a)** cell division – parent cell copies its genes and divides into two identical new cells
   b) binary fission – process by which single celled creatures divide into two identical daughter cells
2. The runners have the same genes as the parent plant, so the fruits will be similar.
3. It allows the plant to spread into new habitats, where conditions may be more favourable for growth. It also improves the chance of survival of the plant, by having multiple versions of itself.

Learning outcomes

Following this, most pupils will:
- Explain that growth of living things occurs by cells dividing to make new cells, and these cells increasing in size.
- Represent the process of cell division as a sequence that begins with division of the nucleus.
- Demonstrate an increasing knowledge and understanding of life processes.

Some pupils will also:
- Use knowledge and understanding to describe life processes. They will be able to use appropriate scientific terminology.

Suggestions for plenaries

- **The cell division game:** Groups: pupils play the game where throws of the dice move them around the cell division cycle. The one with the highest number of times round in the time available is the winner.
- **Same or different:** Group/class: pupils discuss the statement 'cell division is like making photocopies' and make a list of similarities and differences between the two processes.
- **What have I learned?** Individual: pupils write down a summary or key points of their learning.

7A6 Pollen and ovules

National framework/QCA SoW references

National framework – Cells: Describe fertilisation as the joining of the nucleus of a male sex cell to the nucleus of a female sex cell.

QCA SoW

7A Section 6: What causes pollen tubes to grow?
NC links: Sc2 1d.

Learning objectives

- To know that fertilisation in flowering plants is the fusion of a male and a female cell.
- To know that in plants, pollen and ovule are specialised cells, which enable information to be transferred from one generation to the next.
- To know that in plants, at fertilisation, nuclei from pollen and ovule fuse to make a new and unique individual.
- To be able to frame a question that can be investigated.
- To realise the importance of sample size and the number of observations in biological investigations.
- To identify trends shown in graphs.
- To evaluate the strength of the evidence.

Teaching strategy

Key points to highlight

- Pollen formed in the stamen of a flower is carried to the stigma, where a tube grows to carry the pollen nucleus to the ovule.
- Pollen and ovules are the specialised sex cells of the plant (pollen: male; ovule: female).
- Fertilisation is the joining together of the pollen and ovule nucleus to form a seed of a new plant.

Difficulties/Misconceptions

- Pollination and fertilisation are often confused. Pollination only involves the transport of the male sex cells, while fertilisation is the joining of the male and female gametes.
- There are many variables that can affect the success of the pollen tube investigation in 'Viewing pollen'. It can often take over an hour for the pollen tubes to start growing, so it is a good idea to have 'some I made earlier'.

Skills

Practical.

Lesson starter suggestions

- **Flower power:** Group/individual: pupils label a diagram of a flower.
- **Good, bad, find out more – pollen:** Divide pupils into groups as they arrive, and give them the statement 'every flower produces millions of pollen grains'. Ask the groups to think of a reason for this being good (positive), bad (negative) and a question they want an answer to.
- **Pollination:** Group: pupils respond to the term 'pollination' with a list of what they know and what they want to find out.

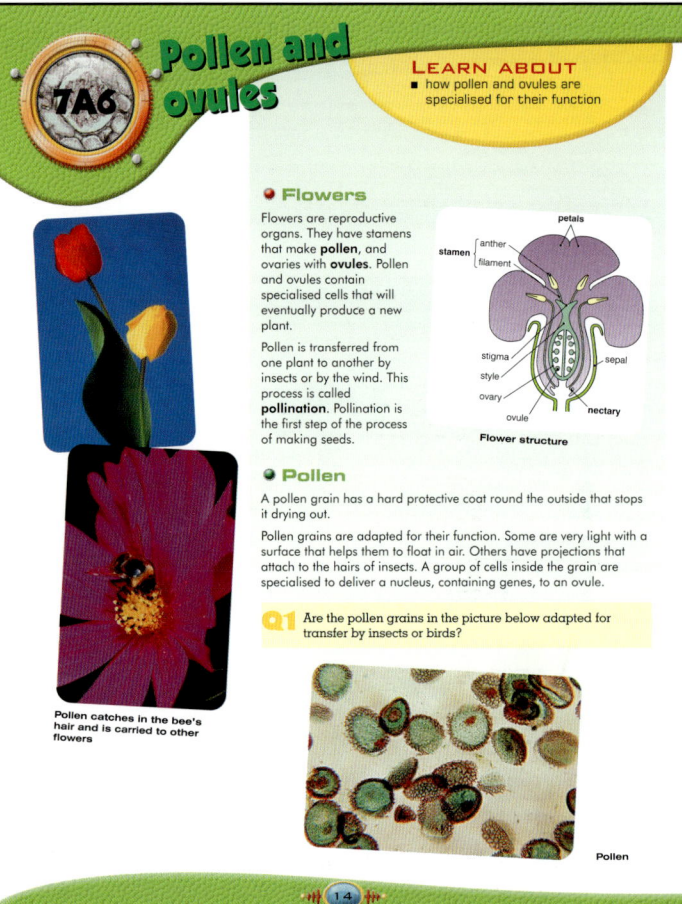

Activity and technicians' notes

Viewing pollen

Equipment and materials required

- Microscopes (with eyepiece graticules or graduated slides), slides, cover slips, tweezers, mounted pins, droppers, sugar solution (5%, 10%, 15% and 20% – all with 0.01% sodium borate) glass rod, Vaseline, watch glasses, mature flowers (at least two types) e.g. *Osteospermum* (hooky coat), London pride (teddy bear ears), lilies, deadnettle.
- Prepare some slides about an hour before the lesson.

Safety notes

- Risk of breakage with glass slides and cover slips.

Access required to:

- Microscope with digital camera and TV/projector/computer
- Prepared slides of pollen grains, some with pollen tubes

Teaching strategy contd.

Notes and tips
It would be helpful to show pupils images of pollen and pollen tubes before they embark on the activity, so that they know what to look for.

Extension ideas
Carry out a further pollen tube experiment changing a different variable.

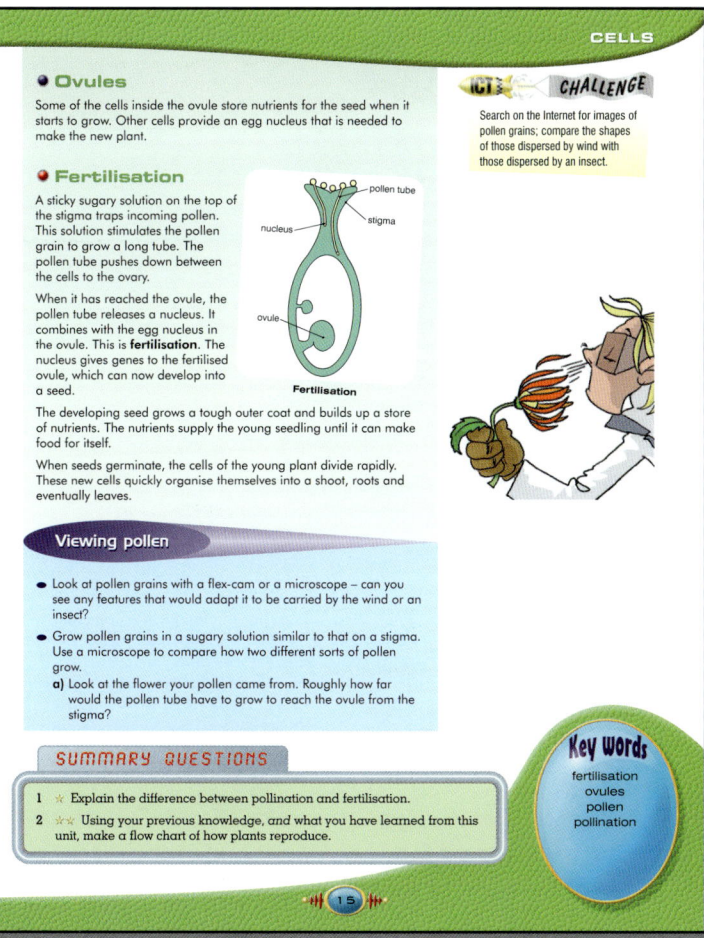

Learning styles

Visual
- **Observing** pictures and slides of pollen germinating.

Auditory
- **Discussing** investigative technique with partner.

Kinaesthetic
- **Practical** work.

Interpersonal
- **Collaborating** with partner on the investigation.

Intrapersonal
- **Considering** the results of the investigation.
- **Drawing conclusions**.
- **Evaluating**.

Answers to questions

Text

Q1 Wind

Summary

1. Pollination is the transfer of the male sex cells from the stamen of a plant to the stigma.
   Fertilisation is the joining together of the male and female sex cells (the pollen and the ovule).
2. Stamens make pollen → pollen carried from stamen to stigma → pollen tube grows down to the ovaries → nucleus of pollen travels down tube → pollen nucleus joins with ovule → seed formed

Suggestions for plenaries

- **From pollination to fertilisation:** Class/group: pupils sort out events in the correct order.
- **Mop up:** Class: pupils pool the results of their investigations and come to a general conclusion.
- **What do we know now?** Group: pupils look back at the list of things they wanted to know (see Lesson starter 'Pollination') and review what they have learnt.

Learning outcomes

Following this, most pupils will:
▶ Explain the process of fertilisation in flowering plants as the transfer of information within nuclei from parents to offspring.
▶ Distinguish between the processes of pollination and fertilisation in flowering plants.
▶ Make drawings to illustrate the sequence of events during fertilisation in plants.
▶ Explain why they needed to use a particular number of pollen grains in the investigation.
▶ Make accurate observations and record these appropriately.
▶ Draw an appropriate graph of data collected.
▶ Use the graph to identify trends and make generalisations.
▶ Compare graphs produced by different groups, and use these to evaluate the strength of evidence.
▶ Suggest a question about pollen tubes that can be investigated and use an appropriate sample; present results in an appropriate graph, explaining what these show.
▶ Identify organs (stamen and stigma) of plants.
▶ Describe the functions of organs of the plant and explain how they are essential to the plant.

Some pupils will also:
▶ Explain how some cells are adapted to their function.
▶ Justify the sample chosen in an investigation of pollen tubes.

7A Read all about it!

Teaching strategy

Antony van Leeuwenhoek

- Pupils should read the passage, out loud perhaps.
- Then they should discuss the questions listed below in groups and deliver their answers to the whole class. If necessary explain the answers as described below.
- Other tasks: Imagine you are Leeuwenhoek's illustrator. Use his description of *Vorticella* to draw a diagram. Display the diagrams around the room and vote on which you think is the most accurate.

Difficulties/Misconceptions

- Pupils may have difficulty in conceiving of the originality of Leeuwenhoek's work and may find his language difficult. They may not realise that there were not even words to describe some of his observations.
- Pupils may think that a large lens has a greater magnification than the tiny lenses that Leeuwenhoek used. This is incorrect; the magnifying power is in inverse proportion to the radius of curvature of the lens. Compound microscopes (such as Hooke's) can give greater magnifications, but the high quality lenses were not available in Leeuwenhoek's time.

Skills

- Literacy – reading (in private or to a group)
- Discussion – putting forward arguments
- Visual – drawing diagrams

Notes and tips

- Leeuwenhoek's microscope was simpler than Hooke's since it used just a single lens. The power came from the minute beads of glass that he used. This made the field of view very small and the use of the microscope very tiring. Leeuwenhoek was a gifted amateur as his main job was running his drapery business. While Hooke gave cells their name, it was Leeuwenhoek who actually saw living cells, particularly bacteria and sperm, that he called 'animalcules'.
- This was the first time that such minute organisms had been observed. The observations had a great effect on scientists and the public and lead to a new view of how life was formed.
- There are many pictures of Leeuwenhoek's drawings on internet sites. The Oxford Museum for the History of Science has examples of Leeuwenhoek-style microscopes.

IDEAS AND EVIDENCE

Antony van Leeuwenhoek (1632–1723)

Antony van Leeuwenhoek, born in Holland in 1632, was an early microscopist. He knew about the microscope used by Robert Hooke to observe animal and plant materials. Hooke's microscope could magnify objects to about 20 or 30 times natural size. Hooke gave us the word 'cell'.

Leeuwenhoek took this newly invented technology, developed it and used it to discover new things. He knew how to make and use lenses and also that he needed good lighting to see clearly. His skills enabled him to build microscopes that magnified over 200 times, with much clearer and brighter images. Some of his 500 simple microscopes still survive.

As Leeuwenhoek had not studied science at University, he did not have pre-existing ideas about how a scientist should work. He worked in a different way to others and developed new techniques to find out about living things. Some of these are part of the scientific method we use today. He was willing to alter his ideas when his observations did not fit with what he had expected.

Leeuwenhoek followed his interest in the natural world to discover bacteria, single-celled pond creatures, sperm cells and red blood cells. He wrote careful descriptions of his observations and used an illustrator to prepare the drawings. These are so good that anyone today could recognise his specimens. Leeuwenhoek regularly sent his findings to the Royal Society in London. The Royal Society published his work, making microscopic life available to other scientists.

A letter, dated December 25, 1702, gives descriptions of many single-celled pond creatures including *Vorticella*:

'In structure these little animals were fashioned like a bell, and at the round opening they made such a stir, that the particles in the water thereabout were set in motion thereby. And though I must have seen quite 20 of these little animals on their long tails alongside one another very gently moving, with outstretched bodies and straightened-out tails; yet in an instant, as it were, they pulled their bodies and their tails together, and no sooner had they contracted their bodies and tails, than they began to stick their tails out again very leisurely, and stayed thus some time continuing their gentle motion: which sight I found mightily diverting.'

Extension ideas

Make a Leeuwenhoek microscope

Safety: Pupils must be supervised while carrying out this experiment. Take care with hot glass. When using the lens it is impossible to wear goggles. Take great care.

Heat and draw out a glass rod. Break the glass fibre and heat the tip to form a droplet. Let it cool and snap the bead off. Hold the bead between the teeth of a clothes peg. Hold the bead to the eye and look at a specimen in bright light.

Note the difficulties that Leeuwenhoek had in making his microscopes.

Questions

1. What does a magnification of 200 mean? How big would something 1 mm across appear to be?
2. Leeuwenhoek's lenses were less than 1 mm in diameter. What problems would this have caused him?
3. Leeuewenhoek didn't draw all his own diagrams. What skills did he and his illustrator need to have?
4. Why was it important for Leeuwenhoek that the Royal Society published his letters?
5. Some people didn't believe that Leeuwenhoek's observations were real. Why was it difficult to believe his work?

Danger – common errors

Pupils may have difficulty in accepting that all organisms, including themselves, are made up of cells which are too small to be seen except with a microscope. Pictures and microscope slides of cells may give the wrong two-dimensional impression of cells. It is important to use models to give a three-dimensional representation.

Pupils may also think that cells are passive as their observations do not reveal the activity that is constantly taking place in live cells and the transport of materials across the cell membrane.

Pupils may be confused by the terms cell 'division' leading to a 'multiplication' of cells. Explain that dividing something by two results in two parts where there was originally one. A crucial point in cell reproduction is the period of growth prior to cell division.

Learning outcomes

Life processes and living things

Following this unit, most pupils will:
- Identify and name features of cells and describe some differences between plant and animal cells.
- Explain that growth occurs when cells divide and increase in size.
- Describe how cells are grouped to form tissues.

Some pupils will also:
- Recognise that viruses are not cells.
- Describe how some cells in an organism are specialised to carry out particular functions.

In scientific enquiry:

Following this, most pupils will:
- Describe some earlier ideas about the structure of living things and relate these to evidence from microscope observations.
- Make observations using a microscope and record them in simple drawings.
- Suggest a question about pollen tubes that can be investigated and use an appropriate sample.
- Present results in an appropriate graph, explaining what these show.

Some pupils will also:
- Explain how evidence from microscope observations changed ideas about the structure of living things.
- Estimate sizes of specimens viewed under the microscope.
- Justify the sample chosen in an investigation of pollen tubes.

Teaching strategy contd.

Answers

1. The length of something appears 200 times bigger through the microscope:

   200 mm = 20 cm

2. difficult to handle and fit into the frame; very small field of view; tiring to look through. (The microscope has to be held very close to the eye.)

3. Leeuwenhoek – ability to describe all the features clearly. Illustrator – skilled artist, able to understand the terms Leeuwenhoek used (as this was all new there were no accepted scientific terms for Leeuwenhoek to use, hence the language he used was quite flowery).

4. It allowed other people to repeat his work and show that he was correct. It showed that his work was respected. (The Royal Society had the King's patronage; members included the most famous scientists of the time – Hooke, Newton, Halley – as well as many wealthy and eminent gentlemen. Note that even at this early time, scientific work was being shared across national boundaries.)

5. no one had seen these things before; the microscope was difficult to use so others found it difficult to repeat Leeuwenhoek's work. They thought he may have been imagining his observations or making them up. (The same was said of Galileo's observations with the telescope early in the 17th century.)

7A Unit review

Answers to review questions

1 microscope, section, cover slip, magnification, nucleus, cytoplasm, cell wall, cells, cell division

2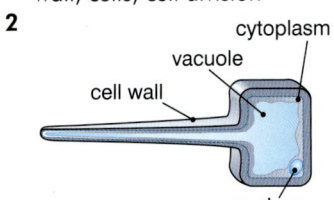
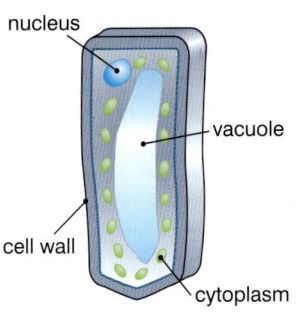

3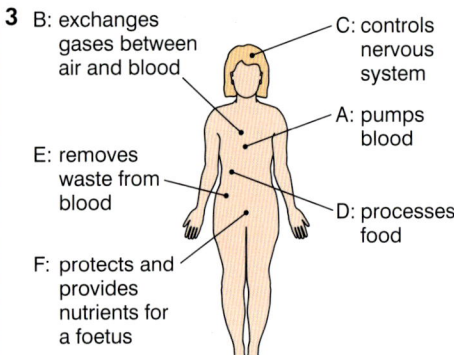
- B: exchanges gases between air and blood
- C: controls nervous system
- A: pumps blood
- E: removes waste from blood
- D: processes food
- F: protects and provides nutrients for a foetus

5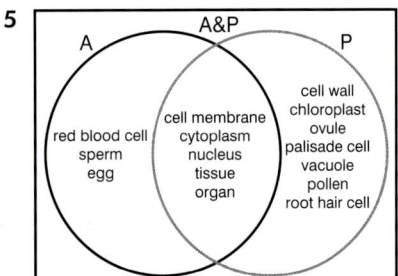

A: red blood cell, sperm, egg
A&P: cell membrane, cytoplasm, nucleus, tissue, organ
P: cell wall, chloroplast, ovule, palisade cell, vacuole, pollen, root hair cell

6 For example: All plant and animal cells contain a round structure called the nucleus, which controls the cell's activity.

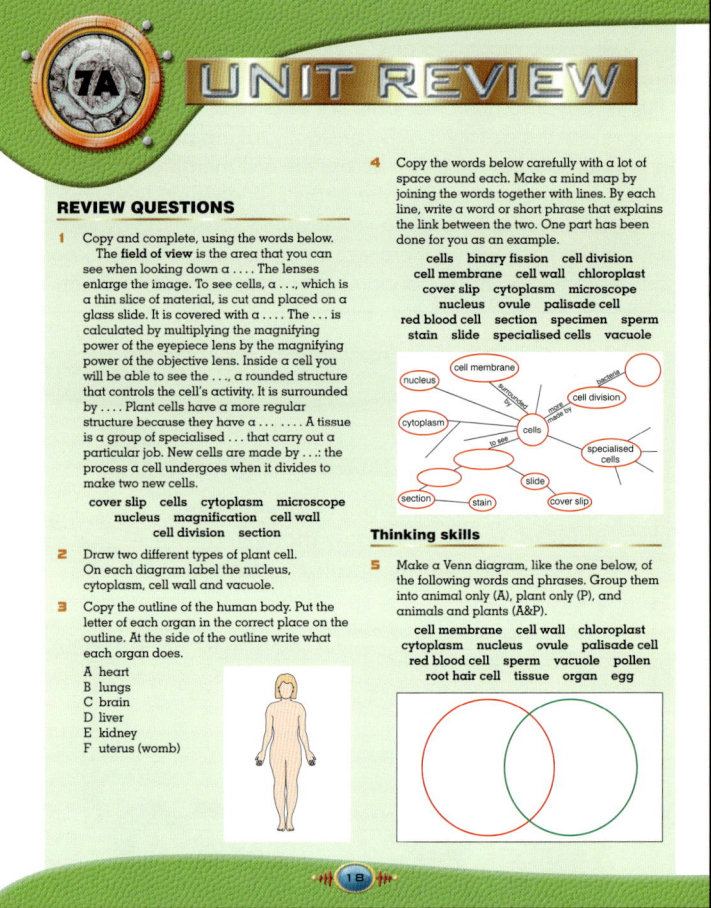

UNIT REVIEW QUESTIONS

Ways with words

6 Physicists, chemists and biologists use the word 'nucleus' in different ways. Write a sentence using 'nucleus' as a biologist studying cells would use it.

SAT-STYLE QUESTIONS

1 The diagram shows a cell from a leaf.

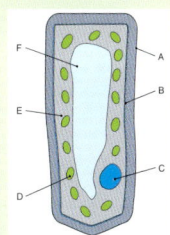

a Match the parts of the cell with a letter on the diagram.
cell wall   cytoplasm   nucleus
vacuole   chloroplast   (5)
b Which parts are also found in animal cells? (2)

2 The diagram shows a cell from the lining of the windpipe.

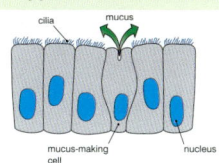

a What is the function of the nucleus in this cell? (1)
b Give one way in which the cell is adapted for its function of keeping the lungs free of dust and bacteria. (1)
c What word describes a group of similar cells that work together? (1)

3 Josh and Arun had been researching bees for a project. Bees see blue, violet and ultra-violet light best. They wondered if bees collecting nectar and pollen were more likely to visit flowers with these colours. They made model flowers out of filter paper, coloured with food dye, and dropped sugar solution on the centre. They made some blue and some pink flowers. They used a security marker pen that shines under UV light to mark dots on some of the pink and blue flowers.
Arun and Josh placed the model flowers near a flowerbed in the school grounds. They recorded the number of bees visiting the flowers. Arun predicted that bees would visit blue flowers more frequently than pink.

Flower colour	Number of visits by bees
blue	6
blue with UV spots	8
pink	4
pink with UV spots	7

a Why did they put sugar on the flowers? (1)
b Josh and Arun wanted to make this a 'fair test'. What factors do they need to think about to make it a fair test? (2)
c Does the data support Arun's prediction? (1)
d Is there another explanation for these results? (1)
e Does the data support the statement that bees can detect UV reflections from flowers? (1)

Key words
Unscramble these:
reclean emblem
toy clamps
specci moor
nice mens
suites

Answers to SAT-style questions

1 a A – cell wall
     B – cell membrane
     C – nucleus
     D – chloroplast
     E – cytoplasm
     F – vacuole (5)
  b B, C, E (2)
2 a It controls the activity of the cell. (1)
  b It has cilia which move dust and bacteria away from the lungs. (1)
  c tissue (1)
3 a to take the place of nectar to attract the bees (1)
  b Use the same amount of sugar; place flowers in the same place with the same lighting, wind, etc. (2)
  c Yes (1)
  d Bees are attracted to the UV colours. (1)
  e Yes (1)

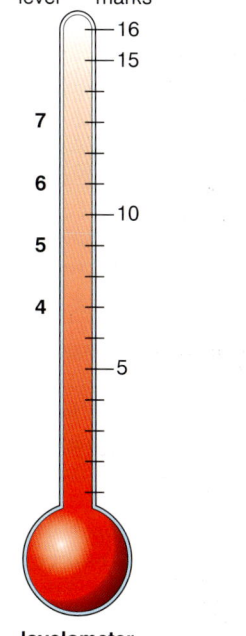

levelometer

7B Reproduction

National framework/QCA SoW references

National framework – Cells: Explain that growth means an increase in size and number of cells. Explain that, in multi-celled organisms, certain cells may become specialised, e.g. sperm and egg cells, and that not all organs grow at the same time; use this to explain why and how some organisms care for and protect their offspring.
Describe fertilisation as the joining of the nucleus of a male sex cell to the nucleus of a female sex cell, and use this knowledge to explain that resulting offspring are always similar to their parents but never identical.

QCA SoW (7B)

How does a new life start? (7B2, 7B7)
When can human fertilisation take place? (7B3)
How is the human foetus supported as it develops? (7B4, 7B6)
What do new-born babies need? (7B5)
How do humans change as they grow? (7B1)
NC links: Sc2 1d, g; Sc2 2 f, g, h; Sc1 2e, j, k, m.

Learning objectives

- To know that patterns of reproduction vary in animals and that human babies are more dependent on parental support than most young animals.
- To name, locate and describe the main human reproductive organs, including the ovary, oviduct, uterus, vagina, penis, testis, sperm duct.
- To describe the functions of the sperm and egg cells and the changes that occur before and after ovulation.
- To describe fertilisation as the fusion of sperm with the egg.
- To explain that the sperm and egg cells contain hereditary material from the father and mother.
- To describe the menstrual cycle and to be able to calculate when fertilisation is most likely to happen.
- To describe how the foetus grows and is nourished during pregnancy, and how it may be harmed by substances crossing the placenta.
- To explain the process of birth, and describe how the new-born baby gets nourishment from the mother.
- To describe growth as a result of cell division.
- To describe the development of humans, in particular the development of the reproductive organs and the effects of hormones at puberty.
- To explain the differences between individuals.

In scientific enquiry:

- To consider sample size in biological investigations.
- To present data in bar charts and graphs.
- To interpret data collected and data from secondary sources.

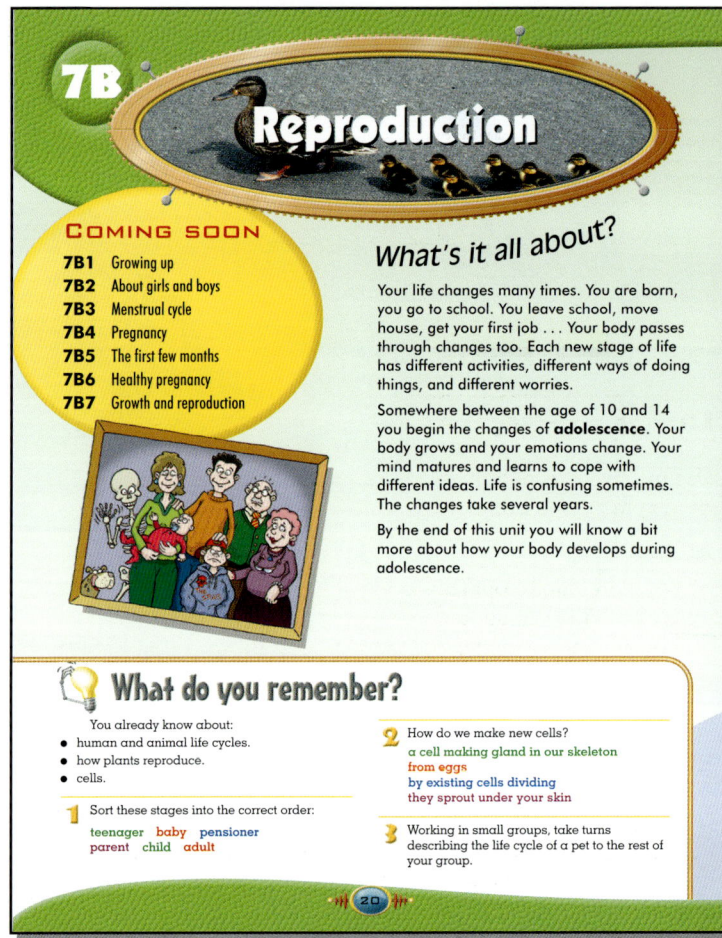

Notes to support 'What do you remember?'

At KS2 pupils will have learnt about the life cycles of animals and plants; in the previous unit '7A Cells' they will have met sperm and egg cells. They will also have learnt about fertilisation in plants. Before beginning this unit, it would be useful if pupils can:
- Describe the stages in the life of humans and animals.
- Know that cells, including sperm and egg cells, contain a nucleus.
- Know that in fertilisation (in plants) two sex cells join together.

Launch activity notes – Ideas about growing up

Use the cartoons and questions to initiate discussion about development and pregnancy.

Answers

a) What are pupils' opinions about their own and other people's bodies? They could look at pictures from fashion magazines and catalogues and cut out photos to make a collage to show how bodies develop from infancy to adulthood. Pupils may discuss the models' figures and their own self-esteem and body image. Look out for subtle signs of bullying.

b) Discuss the physical and emotional changes that occur, as children become young adults. What changes are pupils expecting to happen? How might their relationships with parents, siblings, and friends change? Pupils could use role-play to portray scenes in the life of a teenager.

c) Periods can last from 3 to 7 days. Some pupils, especially boys, may be unaware of what a period is, and both boys and girls may be embarrassed to talk about it. Try to get pupils to see a period as a normal body function. Ask groups to list what they have heard about periods and the treatment of the subject by society, including jokes about PMT and adverts for tampons and pads. How does society expect women having a period to behave?

d) Ask groups to list all the help and support that a baby needs, such as adequate food, balanced diet, clean air, warmth, protection from dangers, stimulation. Why is it easier to look after a kitten? Most young animals are less dependent on parental support than humans and can fend for themselves to some extent.

Teaching strategy

This unit builds on KS3 unit 7A 'Cells' and also KS2 units 5B 'Life cycles' and 4A 'Moving and growing'. It leads into unit 7D 'Variation and classification', 9A 'Inheritance and selection' and 9B 'Fit and healthy'.

Difficulties/Misconceptions

- Be aware of pupils' religious and cultural backgrounds and attitudes to sexual development and conception.
- Pupils themselves will be at varying stages of their development and will be self-conscious of their own developing sexuality. This needs to be considered when comparing pupils in the class.
- Pupils may have various misunderstandings concerning conception, e.g. 'a virgin cannot conceive on her first experience of intercourse,' and about puberty, e.g. 'acne is caused by not washing regularly.'

Notes and tips

- Pupils can draw on their own experiences of growing up from infancy, and they may have older siblings that they have observed going through puberty. Some pupils, particularly girls, will be going through puberty in Year 7 and may have a different perspective on the unit than those for whom the changes lie in the future.
- This unit adopts a slightly different pattern to the QCA scheme of work.
- While the 'mechanics' of puberty, conception, pregnancy and birth are all that need to be studied, remember that emotional development and moral and ethical issues play a large part in this topic.

Learning styles

Visual
- **Using cartoons and text** to stimulate ideas for discussion.

Auditory
- **Discussing** answers to questions.

Kinaesthetic
- **Collecting photos** from magazines to make into a collage.

Answers to questions

What do you remember?

1. baby, child, teenager, adult, parent, pensioner
2. by existing cells dividing
3. Discussion: pupils may describe the appearance of newly born, or very young animals (kittens, puppies, rabbits, hamsters), and the changes that take place as they grow older.

7B1 Growing up

National framework/QCA SoW references

National framework – Cells: Explain that growth means an increase in the size and number of cells; that organs do not all develop at the same time.

QCA SoW

7B Section 5: How do humans change as they grow?
NC links: Sc2 2f.

Learning objectives

- To know that periods of rapid growth occur during the human life cycle.
- To decide what sort of graph is appropriate to represent different data.
- To know that cell division and increased cell size lead to growth of the body.
- To know about the importance of sample size in obtaining reliable evidence.
- To decide on an appropriate graph to display data.
- To interpret class data and compare it with national data.
- To know that changes in hormone concentrations result in the development of secondary sexual characteristics and emotional changes at puberty.
- To collaborate with others, sharing information and ideas and solving problems.
- To answer questions using relevant evidence or reasons.

Lesson starter suggestions

- **True or false:** Pupils analyse statements about puberty and adolescence.
- **Differences:** Pupils list ways in which others in older year groups (e.g. Year 10) are different to them.
- **Match the slang:** Pupils to pair up the slang words with more scientific words (e.g. boobs/breasts, dick/penis, spunk/semen, pubes/pubic hair, the curse/periods).

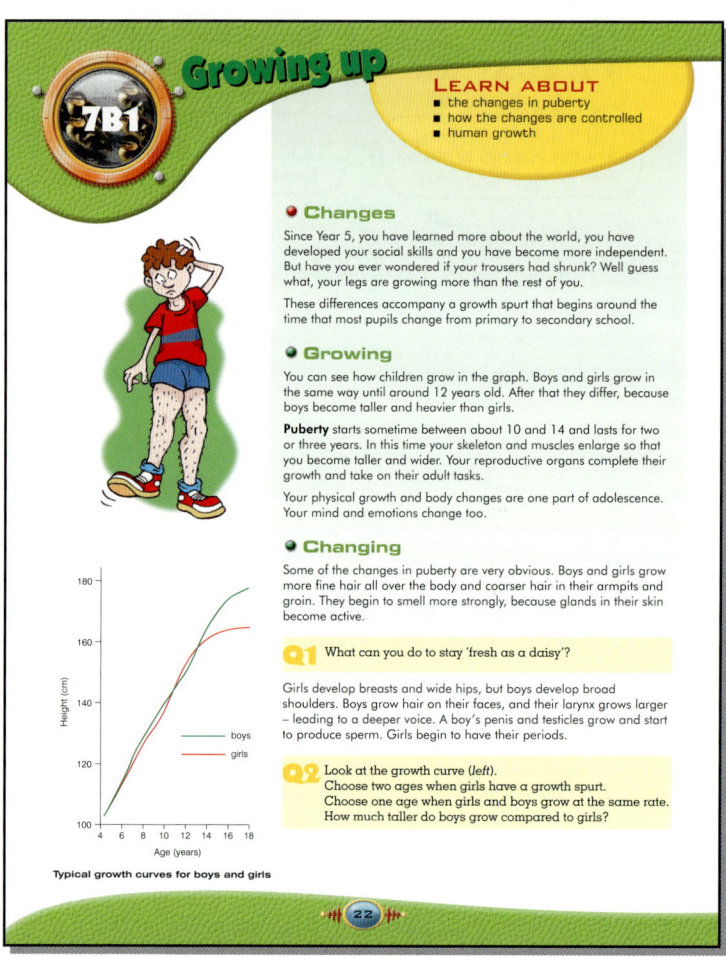

Teaching strategy

Key points to highlight

- Puberty usually occurs between the ages 11 and 14 (adolescence encompasses these ages, but lasts until growth ceases at ages 14–20). However a significant number of boys and girls experience puberty outside this range, with the age for girls gradually getting earlier in developed countries.
- The main physical changes that occur during puberty are growth spurts, maturation of reproductive organs, growth of body hair.
- Spots, body odour and emotional instability are by-products of puberty.

Difficulties/Misconceptions

- Some pupils will be worried by the changes taking place, or about to take place, in their bodies. Some may be too embarrassed to discuss sexual or body matters.
- Pupils may have difficulty understanding the role of hormones in triggering changes in puberty.
- Pupils may have many confused ideas about what happens at puberty and the causes of the changes. For example, they may not realise that body odour is a result of the changes in skin secretions, is a normal consequence of puberty and that personal hygiene can solve the problem. Spots and acne are also a natural occurrence, but are not always so easy to deal with and are not the individual's 'fault'.
- Pupils may not understand that once a boy or girl commences puberty, they are capable of having sexual intercourse leading to pregnancy.

Skills

- Selecting a sample in a statistical survey.
- Graph plotting.

Teaching strategy contd.

Notes and tips
This is a very personal topic: you will have to be aware of pupil's sensibilities and misconceptions.

Extension ideas
Draw and analyse line graphs for continuous variables and compare gradients.

Activity and technicians' notes

PUPIL SHEET

Carrying out surveys
Equipment required
- Metre rules or tape measures
- Sticky tape to fix rulers to walls

Learning styles

Visual
- **Using drawings** to show changes at puberty.

Auditory
- **Discussion** of issues.

Kinaesthetic
- **Measuring** heights.

Interpersonal
- **Working together**.

Intrapersonal
- **Awareness** of learning objectives.
- **Reflection**.

Answers to questions

Text

Q1 Regular washing with a cleanser (soap) and water removes the substances that produce smells as they decay. There are many products on the market, e.g. antiperspirants, deodorants, etc. that slow the processes that produce smells.

Q2 4–6, 10–12; 4–6; about 15 cm.

Summary

1

Boys	Girls	Both sexes
sex organs get bigger	breasts get larger	hair grows in armpits and between legs
hair grows on face and chest	ovaries start to make hormones	
voice gets deeper	periods start	

2 The pituitary gland in the brain produces a hormone, which activates the sex organs to produce other hormones that stimulate growth and the other changes that take place.

3 Yes

Learning outcomes

Following this, most pupils will:
▶ Present and interpret data about growth in bar charts and graphs, indicating whether increasing the sample they used would have improved the work.

Some pupils will also:
▶ Explain whether the sample size in their investigation of growth was sufficient for comparisons to be made with national data.

Suggestions for plenaries

- **Label the diagram:** Pupils to label copies of child/adult pictures noting the changes.
- **Concept mapping:** Place 'puberty' at the centre.
- **Advertising puberty:** Pupils to think up slogans and advertising copy to make the changes of puberty the 'in thing' (nothing too rude!).

7B2 About girls and boys

National framework/QCA SoW references

National framework – Cells: Explain that, in multi-celled organisms, certain cells become specialised e.g. sperm and egg cells. Describe fertilisation as the joining of the nucleus of a male sex cell (e.g. sperm) to the nucleus of a female sex cell (e.g. egg).

QCA SoW

7B Section 1: How does a new life start?
NC links: Sc2 1d and Sc2 2g.

Learning objectives

- To know the structure and function of the human male and female reproductive organs.
- To know that fertilisation involves the fusion of the nuclei of sperm and egg cells.
- To know that the fertilised egg divides into 2, 4, 8, etc. cells as it passes down the oviduct.
- To know that sperm and egg cells are specially adapted for their functions.
- To know that male and female nuclei contain the characteristics of male and female parents respectively.

Teaching strategy

Key points to highlight

- The basic anatomy of the genital organs:
  male – penis, testes, scrotum, sperm tube, seminal fluid glands, urethra;
  female – vagina, ovaries, oviduct, uterus, cervix, vulva, clitoris.
- The functions and specialisations of the egg cell (large, food supply), and sperm (tail for swimming, head carrying nucleus).
- The process of fertilisation: penetration and ejaculation, movement of egg down oviduct meeting sperm, sperm enters egg, nuclei join.
- The fusion of genetic material from male and female to produce a unique embryo.

Difficulties/Misconceptions

Pupils' backgrounds may have given them preconceptions that have to be overcome for them to understand this topic, e.g. the facts that at least partial penetration is necessary for sperm to enter the uterus (kissing or oral sex won't work), that a boy or girl who have entered puberty and hence are producing sperm and eggs can achieve fertilisation. The navel has nothing to do with sexual intercourse!

Skills

Literacy: spelling and use of scientific terms.

Lesson starter suggestions

- **Remembering changes:** Pupils are asked to recall the changes that occur to the genital organs during puberty.
- **Know/want to know:** What do pupils know about sexual reproduction and what do they want to learn?
- **Birds and the bees:** What odd stories have pupils heard about how babies are made?

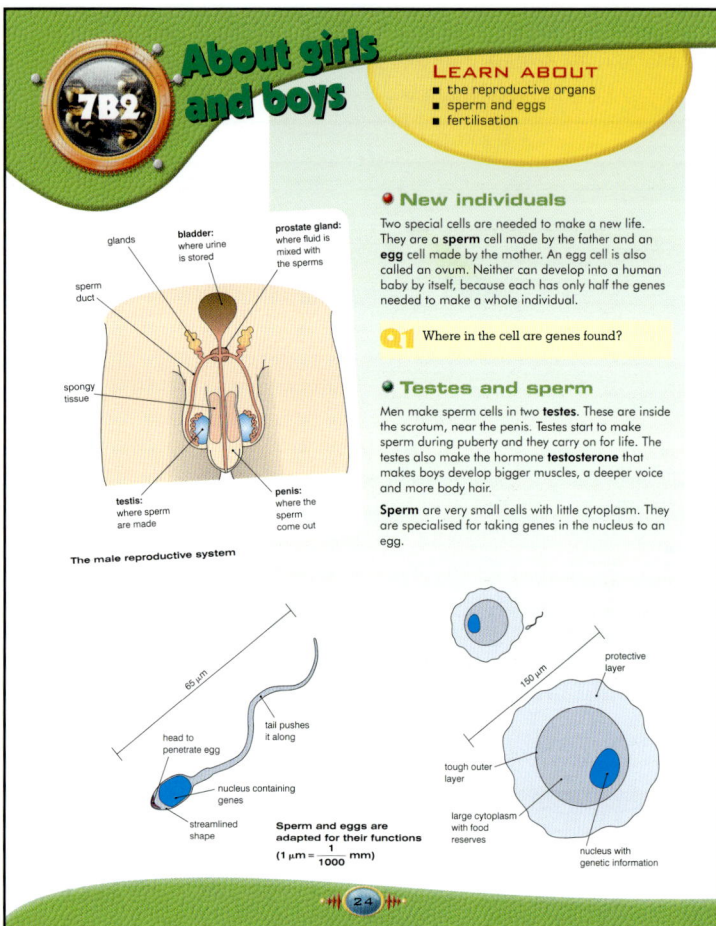

Answers to questions

Text

Q1 In the nucleus

Summary

1. Sperm: tail to 'swim', streamlined head, small size so easier to move, head contains chemicals that allow sperm to enter egg cell.
   Egg: large size with plenty of food reserves.
2. Men: if there are only a few sperm, the chance that one will make contact with the egg is small.
   Women: if the oviduct is blocked, the egg cannot travel down it to meet the sperm. Low pituitary hormone levels mean that the ovaries do not receive the trigger to release an egg.

Teaching strategy contd.

Notes and tips

- Pupils may be embarrassed by this topic and afraid to discuss it. Try and put them at their ease by using the terms in a relaxed fashion.
- Some pupils may try to embarrass you by using slang terms and making crude comments. Asking them to explain their statements using the correct terms may give them a pause for thought.
- Although the emphasis here is in the 'mechanics' of intercourse and fertilisation, try not to ignore the emotional side. Most people find intercourse a most pleasurable experience and this should not be denied. Similarly the possible consequences (emotional attachment, pregnancy, disease) should not be downplayed, being part of the whole topic of sexual intercourse. These ideas will be covered in more depth, perhaps, in PSHE.

Extension ideas

Sex cells and inheritance, extending understanding of specialisation of cells and the merging of genetic information in the offspring.

Learning styles

Visual
- **Using** diagrams to convey understanding.

Auditory
- **Hearing** the words spoken.
- **Discussion** of the processes.

Kinaesthetic
- **Using drag and drop** in ICT.
- **Designing** a flow chart (optional plenary activity).

Interpersonal
- **Sharing ideas** and understanding about sexual intercourse.

Intrapersonal
- **Reflecting** on ideas learnt.

Amazing science

The tube is not many times wider than the diameter of the egg cell. This means that the tube can easily become blocked, stopping the egg from travelling to the uterus or meeting the sperm. Diseases and defects that block the tubes are a major cause of infertility.

Learning outcomes

Following this, most pupils will:
- Identify and name the main reproductive organs and describe their functions.
- Describe fertilisation as the fusion of two cell nuclei.
- Describe egg and sperm cells.

Some pupils will also:
- Explain how egg and sperm cells are specialised.
- Describe how they carry the information for development of a new life.

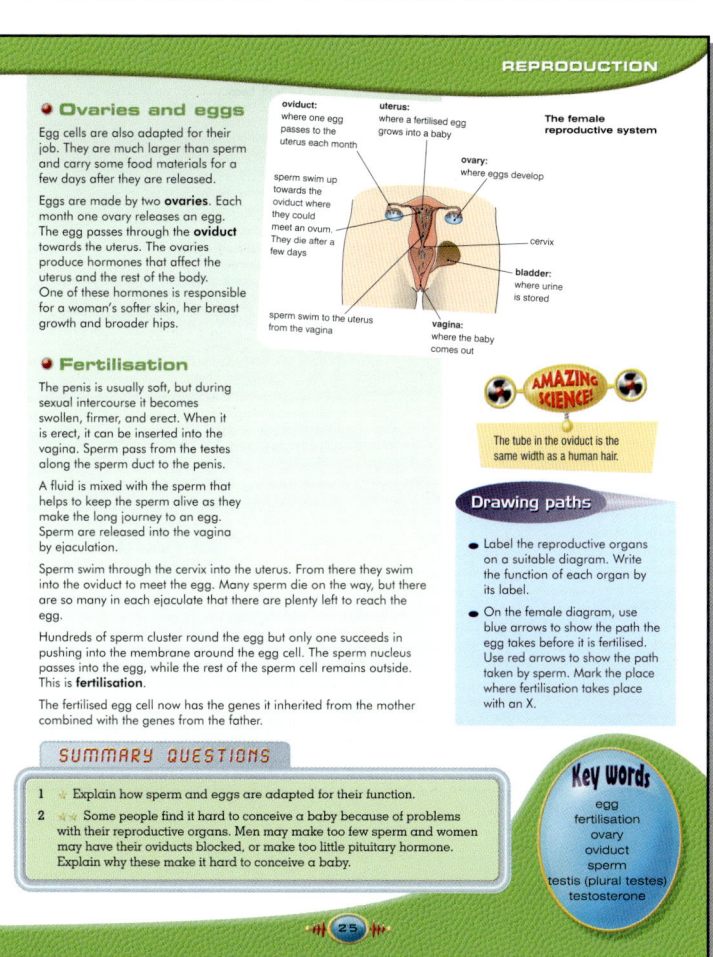

Suggestions for plenaries

- **Flow chart:** Pupils to draw a flow chart of the stages of sexual intercourse and fertilisation.
- **What have I learnt?** Pupils to list the new facts and ideas that they have learnt in the lesson (see Lesson starter 'Know/want to know').
- **Was it useful?** Pupils to discuss the importance of knowing about sexual intercourse and fertilisation.

7B3 Menstrual cycle

National framework/QCA SoW references

National framework – Cells: Specialised cells can be grouped together to form tissues, and that tissues form organs. These do not develop and grow at the same time.

QCA SoW
7B Section 2: When can human fertilisation take place?
NC links: Sc2 2g.

Learning objectives

- To know that egg cells are released from the ovaries at regular (approximately monthly) intervals.
- To know that menstruation is a monthly cycle, which stops during pregnancy.
- To know that the stages in the menstrual cycle are controlled by hormones.

Amazing science

In IVF treatment, women are given a hormone that stimulates their ovaries to make a lot of eggs mature at once. A suggestion has been made to 'harvest' the egg cells of aborted foetuses for use in stem cell research. It was even suggested that one of these cells could be fertilised and then implanted in a surrogate mother. The baby would be the child of a mother who never lived.

Lesson starter suggestions

- **Rites of passage:** Pupils read and discuss the passage.
- **What do you remember?** Pupils discuss what they remember about female puberty and fertilisation.
- **Good, bad, want to know more?** Pupils discuss the statement, 'during her life a woman may have about 500 periods'.

Teaching strategy

Key points to highlight
- The main stages of the menstrual cycle: ovulation, uterus thickening, menstruation, ripening of next egg.
- That ovulation and menstruation are governed by hormones.
- That women's menstrual periods vary, but average 28 days.
- By monitoring her menstrual cycle, a woman can work out approximately when she will ovulate and when an egg is available for fertilisation.

Difficulties/Misconceptions
Some pupils may have difficulty in calculating the timing of events in the menstrual cycle from a previous cycle's dates.

Skills
Listening and giving advice.

Notes and tips
- Boys may be reluctant to discuss girls' periods, but they should be persuaded that as they get older (and have girl friends) the matter will become relevant.
- Girls who have not reached puberty may be apprehensive about pain and mess of a period and will need reassuring.

Extension ideas
Get more-able pupils to probe more deeply into the questions that arise about ovulation and menstruation and the need for carefully considered advice.

Learning styles

Visual
- **Examining, drawing and labelling** diagrams/cycles.

Auditory
- **Listening** to reading (starter activity).
- **Discussion**.

Kinaesthetic
- **Moving** into work groups.
- **Manipulating** labels on diagrams.

Interpersonal
- **Discussing** things to think about (starter), problems (plenary).

Intrapersonal
- **Reviewing** knowledge and understanding.

Answers to questions

Text

Q1 The cilia move in waves (like a Mexican wave) and this pushes the fluid along making a current that carries the egg.

Q2 The women could be given hormones to help the eggs mature.

Summary

1 Ovulation is the process of the release of the egg from the ovary. Menstruation is the discharge of the uterus lining through the vagina. Menstruation occurs after ovulation, but only if the egg has not been fertilised.
2 She stops menstruating.
3 Every month girls lose a considerable amount of blood at menstruation. Replacing these blood cells requires an intake of iron (as well as proteins and carbohydrates). Beef and spinach would be good for the girl having heavy periods.

Activity (Monthly cycles)

1 about 28th July
2 approx. 17–21 June, 12–16 July
3 Her period should be finished by the first or second of August. She will not have a period in the fortnight she is away.

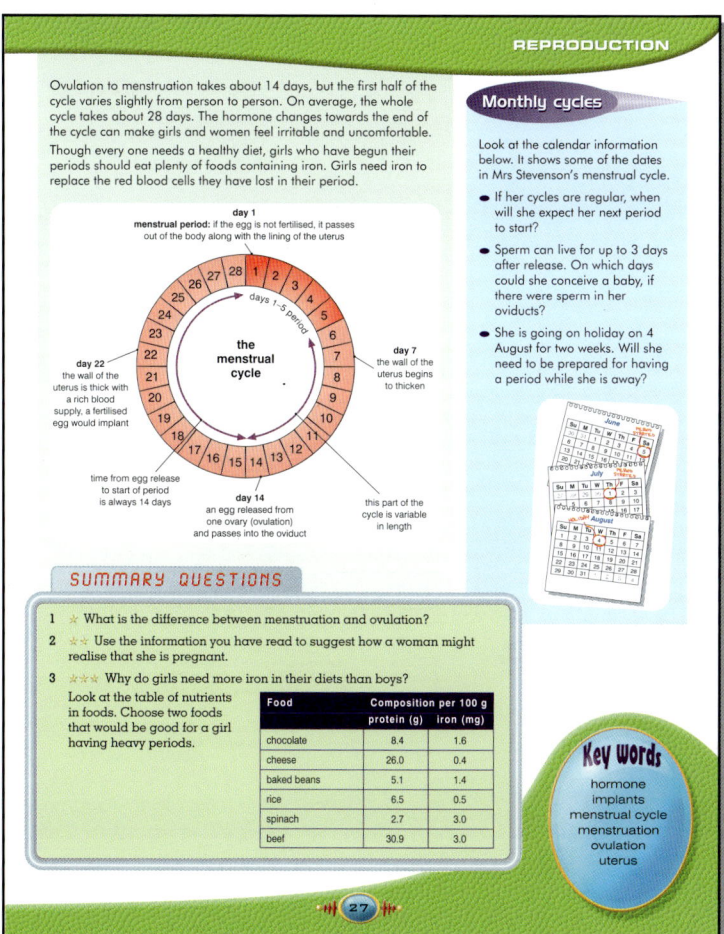

Suggestions for plenaries

- **The journey of the egg:** Pupils label a diagram of female genital organs with times and stages in the menstrual cycle.
- **Brainstorm problem solving:** State the problem, 'a man and woman are not achieving fertilisation'. What are the possible causes? Pupils discuss as many reasons as they can think of.
- **Information pack:** Pupils plan a poster/pamphlet to explain puberty and menstruation to young girls and boys.

Learning outcomes

Following this, most pupils will:
▶ Describe the menstrual cycle.

7B4 Pregnancy

National framework/QCA SoW references

National framework – Cells: Explain why and how some organisms care for and protect their offspring.

QCA SoW

7B Section 3: How is the human foetus supported as it develops?
NC links: Sc2 2h.

Learning objectives

- To know that the foetus develops within a membranous bag and is supported and cushioned by amniotic fluid.
- To know that the placenta supplies nutrients and oxygen to the foetus via the umbilical cord, and removes carbon dioxide and other waste products.

Teaching strategy

Key points to highlight

- An embryo implants into the lining of the uterus after the fertilised egg has divided a number of times.
- The amnion and amniotic fluid protect and cushion the foetus.
- The placenta and umbilical cord connect the foetus to the uterus. The placenta carries blood vessels from the mother and from the foetus.
- Nutrients, oxygen and antibodies diffuse from the mother's blood to the foetus' blood; waste products diffuse in the opposite direction.

Difficulties/Misconceptions

- Pupils may think that the mother's and the foetus' blood mixes. It does not.
- They may also think that the foetus breathes in the womb. Although the lungs of the foetus may move, it gets all its oxygen via the placental blood supply.
- Diffusion occurs when substances move from a zone of high concentration to a zone of lower concentration. Pupils may not know or understand this; they will study diffusion in more detail in unit 7G 'The particle model'.

Skills

Drawing and interpreting bar charts.

Lesson starter suggestions

- **Pregnant women:** Pupils discuss the changes that happen to pregnant women – both those that they know about and have heard mentioned.
- **After fertilisation:** Pupils answer questions to remind them of what must happen before a woman becomes pregnant.
- **What does the embryo need?** Pupils consider all the requirements of a growing embryo.

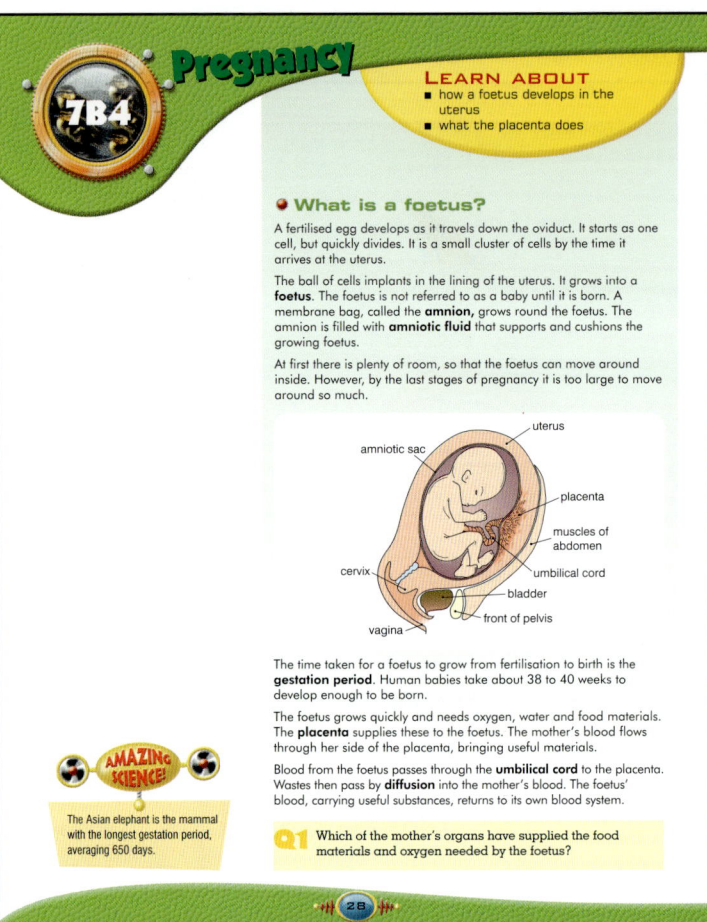

Notes and tips

If you choose to do the model womb activity, the clear plastic bag can be filled with water to represent the amniotic fluid, or to avoid mess, with just air. The rubber tube umbilicus must pass into the amnion to connect the 'foetus' with the 'placenta'.

Extension ideas

Pupils could take a more detailed look at the development of the foetus, exploring the monthly changes and growth.

Amazing science

Mammals show a wide range of gestation periods that roughly correlates with the size of the animal. Primates, including humans, have somewhat longer gestation periods.

Activity and technicians' notes

Making a model
Safety notes
Pupils should not put plastic bags over their heads.

Equipment required
Materials for building a model womb, e.g. clear plastic bags with wire seals, rubber tube, dolls or stuffed toys or balls (as the foetus), foam rubber or other spongy material, something to act as the uterus (e.g. swimming caps, strong supermarket bags, small holdalls/shopping bag, etc.) sticky tape, scissors.

Access required to:
A model womb would be useful.

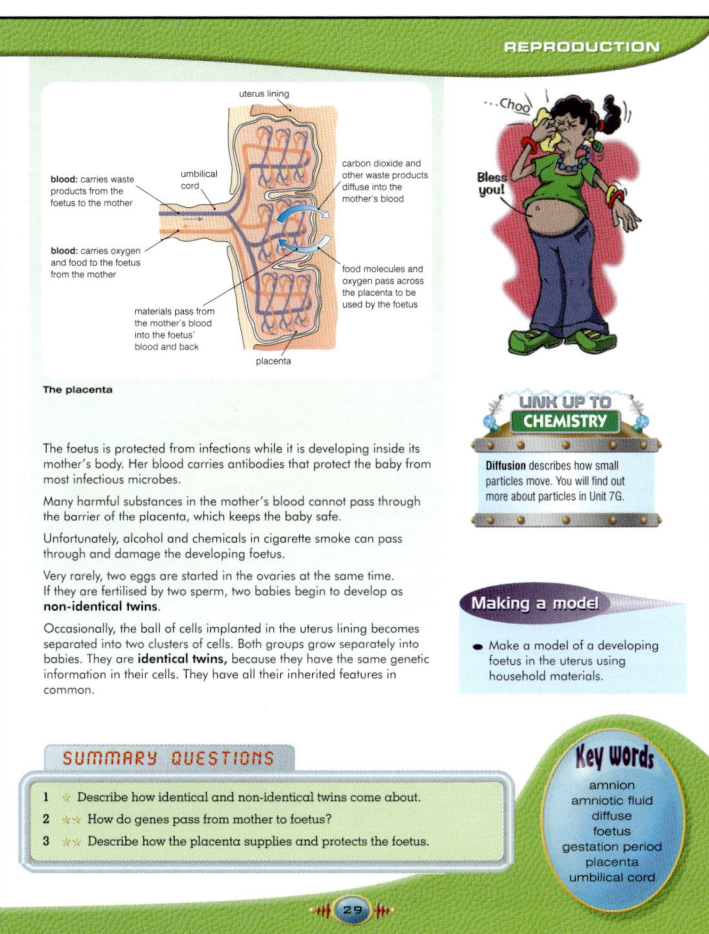

Learning outcomes

Following this, most pupils will:
▸ Explain how the foetus obtains the materials it needs for growth.

Learning styles

Visual
- **Examining** or sorting diagrams.
- **Drawing** a mind map or completing a word square.

Auditory
- **Hearing** the important words spoken (for example in a spelling bee).
- **Discussing** pregnancy.

Kinaesthetic
- **Manipulating** models (2D or 3D).

Interpersonal
- **Collaborating** on various activities.

Intrapersonal
- **Giving** feedback from activities to raise self-esteem.

Answers to questions

Text
Q1 Nutrients enter the mother's blood from the digestive system (stomach and intestines) and the liver. Oxygen enters the mother's blood in the lungs.

Summary
1. Identical twins: a single embryo divides to form two identical foetuses.
   Non-identical: two eggs are fertilised by two sperm, each fertilised egg forms an embryo that becomes a foetus.
2. in the nucleus of the egg
3. Nutrients and oxygen diffuse from the mother's blood into the foetus' blood in the placenta, while harmful waste materials diffuse out of the foetus' blood into the mother's. The mother's blood also delivers substances such as antibodies that protect the foetus from diseases.

Suggestions for plenaries

- **Word square:** A puzzle containing many of the words met in the last few lessons.
- **Embryo to baby:** Pupils draw a memory map to remind them of the changes during pregnancy.
- **Spelling bee:** Pupils test their spelling of words, i.e. sperm, foetus, fertilisation, embryo, uterus, placenta, pregnancy, amnion, umbilical cord, diffusion.

7B5 The first few months

National framework/QCA SoW references

National framework – Cells: Explain why and how some organisms care for and protect their offspring.

QCA SoW
7B Section 4: What do new-born babies need?
NC links: Sc3 2.

Learning objectives

▸ To know that uterine muscle contracts during birth, expelling the foetus and placenta through the vagina.
▸ To know that the baby is nourished by milk from mammary glands; the milk provides nutrients and protects from infection.

Teaching strategy

Key points to highlight
- At birth, the cervix dilates and the muscles in the uterus push the baby out, head first, followed by the placenta.
- Breast milk provides nutrients appropriate for a baby's digestive system, and antibodies.
- New-born babies are not inert, but have certain skills and abilities.

Difficulties/Misconceptions
- Pupils may find it difficult to relate to the struggle of birth, the force required to push the baby from the womb and the degree of dilation of the cervix and vagina required.
- Be aware that some pupils may know of babies that have died at or soon after birth, or have been born prematurely.
- Pupils may not fully understand the reason why babies need breast milk, as they have not studied digestion in detail.

Notes and tips
A full-sized model baby would be a useful resource, also a large balloon (or condom) to demonstrate how the cervix and vagina expands from its normal size to allow the baby to pass through.

Extension ideas
Take a more detailed look at the capabilities of babies.

Lesson starter suggestions

- **The foetus is ready!** A diagram of the fully developed foetus in the womb triggers pupils to think about what happens next.
- **The baby's coming!** Pupils plan what to do if a woman is about to give birth.
- **Good, bad, want to know more?** Pupils comment on the statement 'at the end of a normal pregnancy, a foetus has a mass of 3–4 kg and a length of about 0.5 m from head to toe'.

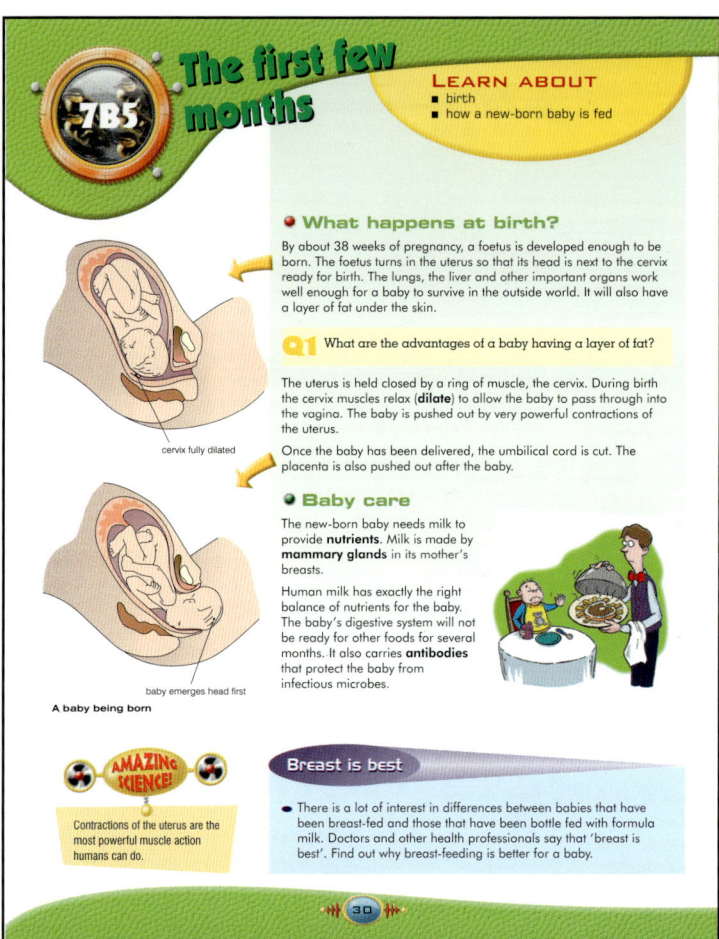

Activity and technicians' notes

Class activity
Equipment required
- Scissors
- Paste

Access required to:
- Birth models

Learning styles

Visual
- **Using diagrams and models** to visualise the birth process.

Auditory
- **Discussing or reading aloud** passages in the text or activities.

Kinaesthetic
- **Cutting and pasting** (or re-writing) sentences in the correct order.

Interpersonal
- **Collaborating** on activities.

Intrapersonal
- **Reflecting** on their learning.

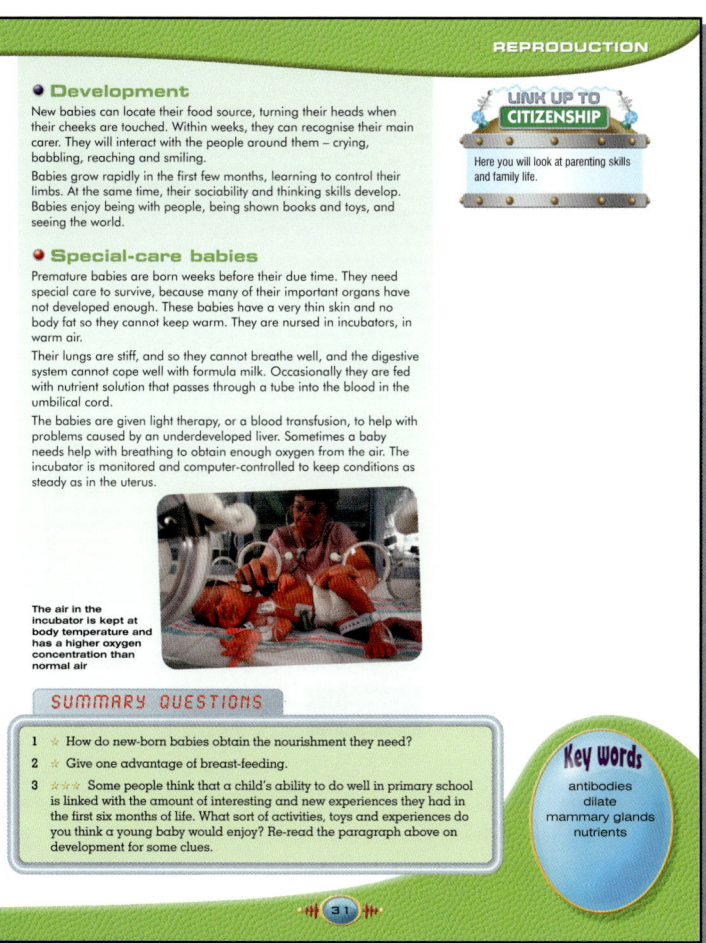

Answers to questions

Text

Q1 A layer of fat insulates the baby from cold (the temperature in the womb is warmer than most places). It also protects the baby from knocks, and provides the baby with a source of energy if food doesn't arrive from the usual source.

Summary

1. usually from their mother's breast milk
2. It provides the baby with nutrients it can digest and other useful substances (antibodies).
3. Being touched, cuddled, moved. Toys that can be gripped, that have various shapes, colours, movements, make sounds; picture books; interaction with people who make sounds and movements directed at the baby.

Amazing science

During normal life, the muscles in the uterus do not do a lot. A muscle in the jaw that helps us chew food is probably the strongest normally. However, during pregnancy the muscles in the uterus get bigger and stronger for the task of pushing the baby out. During birth, the uterus works like heart muscle contracting and relaxing at intervals. The heart has to work hard too, perhaps increasing blood flow to the muscles in the uterus and pelvic muscles by a third.

Suggestions for plenaries

- **Keyword puzzle:** Clues to words used in this unit. Keyword is 'foetus'.
- **Just imagine!** Pupils imagine that they are a baby being born. What would happen? What would they feel? What changes would they notice?
- **Babies survive:** Pupils list the things about babies and birth that contribute to their survival.

Learning outcomes

Following this, most pupils will:
▶ Describe the independence of the young of humans.

7B6 Healthy pregnancy

National framework/QCA SoW references

National framework – Cells: Explain why and how some organisms care for and protect their offspring.

QCA SoW

7B Section 3: How is the human foetus supported as it develops?
NC links: Sc2 2h.

Learning objectives

▸ To learn more about how the foetus develops in the uterus and the role of the placenta.
▸ To know that harmful substances and viruses can cross the placenta into the foetus and affect development.

Teaching strategy

Key points to highlight

- Poor diet (e.g. lack of vitamins) can damage the foetus.
- Some viruses (e.g. rubella) can pass across the placenta and damage the foetus.
- Smoking, alcohol and drugs can affect the growth of the foetus and cause small, weak babies to be born.

Difficulties/Misconceptions

- Pupils may have difficulty interpreting the data on smoking and pregnancy, as it is taken from a variety of sources with inconsistent variables.
- Pupils may lack knowledge of what constitutes a healthy diet, or the importance of vitamins, although they should be aware that growth requires a healthy diet.
- Pupils may become disturbed about the risks of smoking and alcohol if their own parents are smokers and/or drinkers.

Notes and tips

- Remind pupils that in developed countries the vast majority of babies are born healthy and that the dire effects mentioned in this spread only affect a small minority of births.

Lesson starter suggestions

- **Making babies – what do you remember?** Pupils recall what they have learnt in previous lessons.
- **Lifestyles:** Pupils consider diet and use of tobacco, alcohol and drugs in people's lives.
- **What does a baby need?** Pupils consider the requirements for a healthy baby.

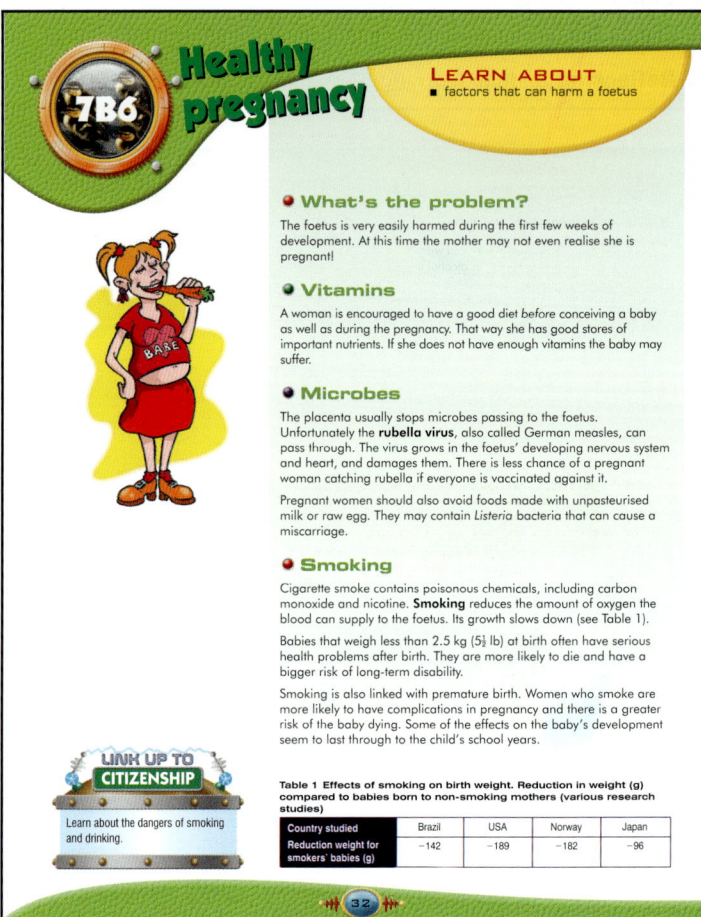

- There are other factors (e.g. genetic, age of parents) that can affect the foetus.
- Pupils need access to food labels showing composition (especially vitamins) and leaflets produced by medical organisations on the effects of smoking and alcohol.

Extension ideas

'The AIDS babies of Africa' explores the effects of the HIV on babies in southern Africa.

Activity and technicians' notes

PUPIL SHEET

Designing posters

Equipment required
- Scissors
- Paste
- Coloured pens/pencils
- Colour magazines with pictures/ads on smoking/alcohol/food
- Labels from food packets

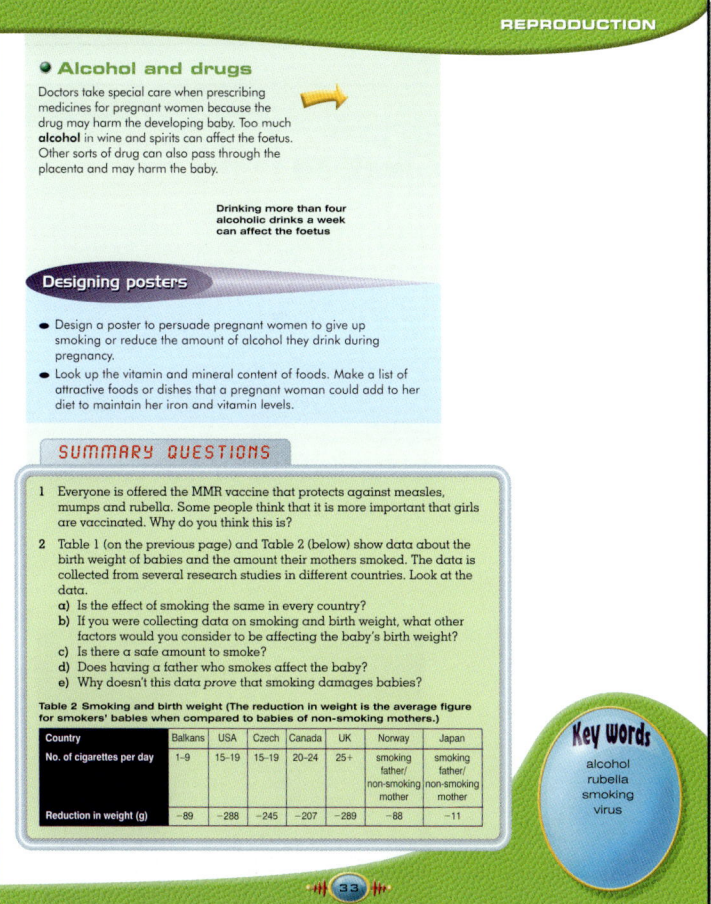

Learning outcomes

Following this, most pupils will:
▶ Recognise that the foetus can be harmed by substances in the mother's blood.

Some pupils will also:
▶ Understand the possible effects of smoking on the foetus and of viruses such as rubella.

Learning styles

Visual
- **Using mind maps** or games to reinforce learning.

Auditory
- **Discussing** or reading passages aloud.

Kinaesthetic
- **Matching** phrases.

Interpersonal
- **Discussing** ideas.

Intrapersonal
- **Reflecting** on what they have learnt.

Answers to questions

Summary

1. Rubella (German measles) is a minor disease for boys; only in girls will it possibly affect the foetus. Thus the small but finite risks of vaccination are only justified for girls.
2. a) All countries listed show the same effect, although the magnitude varies.
   b) Mother's diet and health, environmental factors (pollution).
   c) No, even 1 cigarette a day seems to have an effect.
   d) Yes, passive smoking.
   e) While the data suggests a link between low birth weight and smoking, a causal effect could only be proved by showing a mechanism of how smoking affects the foetus.

Suggestions for plenaries

- **The pregnancy game:** Pupils play the game modelled on snakes and ladders.
- **Odd one out:** Pupils discuss the odd one out in the trio: rubella, folic acid, tobacco.
- **Mind map:** Pupils draw up a mind map on pregnancy, summarising the do's and don'ts for a healthy baby.

7B7 Growth and reproduction

National framework/QCA SoW references

National framework – Cells: Explain why and how some organisms care for and protect their offspring.

QCA SoW
7B Section 1: How does a new life start?
NC links: Sc3 2.

Learning objectives

- To relate cells and cell functions to life processes in a variety of organisms.
- To know that animals have different patterns of reproduction and development.

Teaching strategy

Key points to highlight

- Animals adopt various methods of reproduction.
- External fertilisation in fish.
- Internal fertilisation in insects, reptiles, birds and mammals.
- External development (lay eggs) in amphibians, insects, reptiles and birds.
- Internal development in mammals.

Difficulties/Misconceptions

- Pupils may have problems understanding the difference between the (unfertilised) egg laying of fish and the (fertilised) egg laying of reptiles, birds, etc.
- It may be easy to see the advantages of internal fertilisation and development (fewer eggs needed) but not the disadvantages (commitment in energy and time of mother in raising the young).

Extension ideas

- **Comparing patterns of reproduction:** A more detailed examination of the advantages and disadvantages of internal and external fertilisation and development.
- **Parents' joy:** A discussion of the part that human parents play in the emotional and intellectual development of their children.

Lesson starter suggestions

- **Sexual reproduction:** Pupils sort words into order and write sentences to describe the process of sexual reproduction.
- **Species:** Pupils match animal species to branches of the animal kingdom.
- **Parents and offspring:** Pupils match adult animals to their infant offspring.

Learning styles

Visual
- **Using pictures** to assist learning and looking at patterns of words.

Auditory
- **Discussing** reproduction and talking about research.

Kinaesthetic
- **Moving** and picking cards.

Interpersonal
- **Collaborating** in research and other activities.

Intrapersonal
- Reviewing understanding.

Answers to questions

Q1 The piglets are sucking at the sow's mammary glands; only mammals produce milk.

Summary

1 **Internal fertilisation:** Female only has to produce a small number of eggs as there is a greater chance that they will be fertilised. It saves the female's energy and resources.
Internal development: Only a small number of embryos need to develop as they are protected.

2

Venn diagram:
- Lays eggs: salmon, goldfish
- Both: chicken, cobra, spider, ladybird, robin, turtle, crocodile
- Internal fertilisation: leopard, deer, cat, human

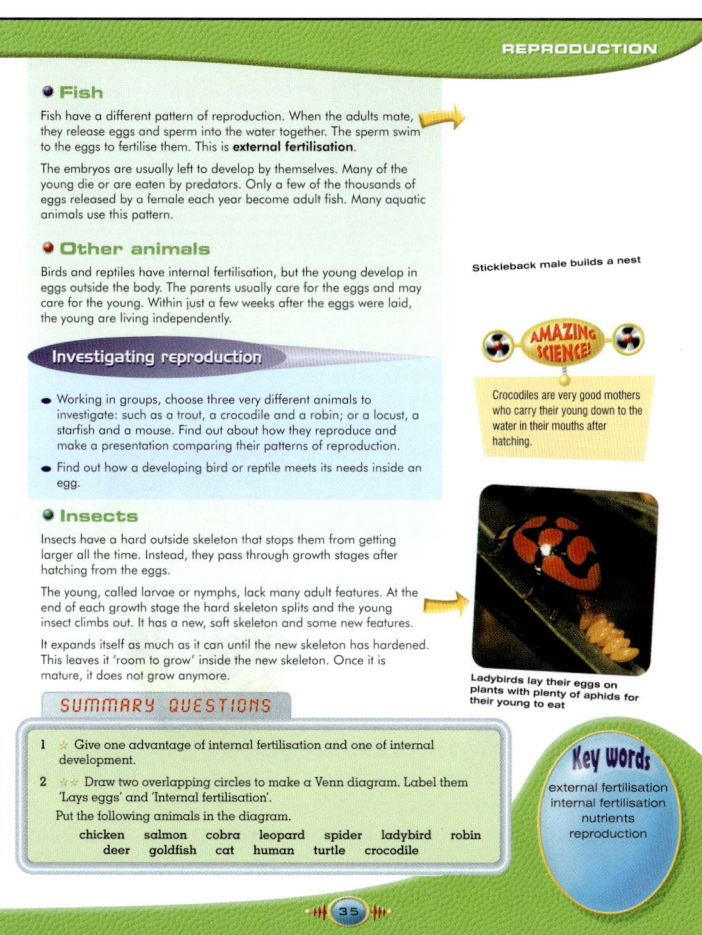

Activity and technicians' notes

Investigating reproduction

Access required to:
Internet, encyclopaedia, books on animals

Amazing science

The behaviour of crocodiles in some ways counteracts the disadvantage of external development, by protecting their young from predators. Most reptiles have no parenthood duties after the eggs are laid, and the young have to fend for themselves from the moment that thy hatch. Other animals do have parenting strategies that encourage more offspring to survive, e.g. birds nurture and endeavour to protect their young, sticklebacks build nests.

Suggestions for plenaries

- **Writing frame:** Pupils write a summary of fertilisation and development.
- **Parents and offspring:** Pupils identify the mode of fertilisation and development used by the pairs of animals in Lesson starter 'Parents and offspring'.
- **Same but different:** Pupils consider explanations for the statement 'the appearance of the sperm and of the eggs of all animals are very similar'.

Learning outcomes

Following this, most pupils will:
▶ Describe the differences in the independence of the young of mammals and other animals.

Some pupils will also:
▶ Be able to compare the advantages and disadvantages of internal and external fertilisation.

7B Read all about it!

Teaching strategy

Spina bifida

- Pupils read the passage (perhaps out loud). Give examples of correlations they have met.
- Put pupils into groups to discuss these options for testing the link between vitamins in vegetables and incidence of spina bifida:
  1. Select a large group of pregnant women. Give half a normal diet of vitamins and the rest none. Count how many babies are born with spina bifida.
  2. Find an animal that also develops spina bifida (e.g. monkeys). Select a large group of pregnant animals. Give half a normal diet of vitamins and the others none. Count how many offspring are born with spina bifida.

  Groups should decide the advantages and disadvantages of each method and whether they think either should be allowed.

Difficulties/Misconceptions

- Pupils may think that any correlation implies a causal link. In this case the time of year of the birth is not the cause of the birth defect but is itself linked to the lack of vitamins in the diet.

Skills

Understanding the terms correlation and cause. Discussion of medical ethics.

Notes and tips

- **Correlation and cause.** Pupils will have met the idea that two factors or sets of observations may be linked. For example in unit 7A the correlation between sugar concentration and the rate of growth of pollen tubes; in this unit the correlation between incidence of smoking and low birth weight. While it is relatively easy to show pupils that a correlation implies a shared pattern between two factors, it is more difficult to explain that correlation and cause are different things. A causal link implies that one factor has a direct impact on the second. Correlations may be coincidental. For instance, the low birth weight of babies born to smoking mothers may be because smokers eat less than non-smokers, so the babies of the former are not getting all the nutrients they need. The article shows that there is correlation between the number of babies born with spina bifida and the time of year. The time of year isn't the direct cause but it could be linked to the cause, i.e. availability of fresh vegetables.
- Medical ethics: Normally in science an experiment can be set up to check a correlation by changing just one variable. In medical experiments, the outcome of an experiment may cause harm to a human.

For instance in the case described here, the prediction is that lack of vegetables will increase the number of cases of spina bifida. As medical ethics demands that no harm be done to a patient, this sort of experiment cannot be carried out. Instead a similar experiment could be carried out with animals, but this is increasingly proving impossible with animal welfare groups against animal testing. Tests could be done on embryos, but this also has ethical considerations.
- Spina bifida is a birth defect in which the spine does not enclose the spinal column completely, resulting in, at worst, the inability to control the lower limbs.

Questions

1. What could harm babies?
2. State two possible causes of spina bifida suggested in the article.
3. How might the researchers be responsible for harming babies?

Answers

1. smoking cigarettes, lack of vitamins
2. lack of vitamins normally found in fresh vegetables; low temperatures
3. The researchers may have stopped the mothers from receiving the correct vitamins.

Infertility and IVF

- Read the article.
- Discuss the following points:
  1. Should IVF treatment be available free to families who need it to have children?
  2. Should a man and woman be able to choose which embryo they use (it may be possible to choose the sex or characteristic of the baby)?

Danger – common errors

- 'The sperm and egg are similar in size.' The egg is actually many times bigger than sperm.
- 'Eggs move towards sperm.' In fact the egg cannot move itself. Sperm are active.
- 'Male and female twins can be identical.' They cannot, as they must have formed from separate eggs. Identical twins have the same genes including the sex genes.
- 'The mother's blood flows through the foetus.' It does not. There is no contact between the mother's and the foetus' blood.
- 'A woman can get pregnant at any time that she is not having her period.' In fact the window for fertilisation is just a day or two after she ovulates. Sperm can remain active for up to three days after intercourse.

Learning outcomes

Following this unit, most pupils will:
- Identify and name the main reproductive organs and describe their functions.
- Describe fertilisation as the fusion of two cell nuclei.
- Describe egg and sperm cells.
- Explain how the foetus obtains the materials it needs for growth.
- Describe differences between the gestation periods and the independence of the young of humans and other mammals.
- Describe the menstrual cycle.

Some pupils will also:
- Explain how egg and sperm cells are specialised, and describe how they carry the information for development of a new life.

In scientific enquiry:

Following this most pupils will:
- Select information about reproduction from secondary sources.
- Present and interpret data about growth in bar charts and graphs, indicating whether increasing the sample they used would have improved the work.

Some pupils will also:
- Explain whether the sample size in their investigation of growth was sufficient for comparisons to be made with national data.
- Describe how reproduction was explained before the role of cells was understood.

3 hormone changes at puberty
4 warm, clear water
5 no (at least not in excess)

Spots
- Read the article.
- Ask pupils to design a leaflet for teenagers to explain what causes acne and how it can be treated.

Notes
- IVF is available free to some families but most have to pay a large sum of money.

Difficulties/Misconceptions
- Pupils may think that the foetus develops outside the womb, but only fertilisation and the first stages of cell division take place on a glass dish.

Questions
1 What may cause a woman to seek IVF?
2 How is a woman made to release eggs?
3 Where does fertilisation take place in IVF?
4 How are embryos 'banked'?
5 What is the age of the oldest baby born by IVF?

Answers
1 blocked oviduct; egg not released because of low hormone levels; father does not produce enough active sperm
2 She is given hormones.
3 in a glass dish
4 by freezing them
5 Louise Brown is 26 years old in 2004.

Teaching strategy contd.

Note
Acne is not caused by insufficient washing; harsh soaps and over-energetic rubbing can make things worse.

Difficulties/Misconceptions
Pupils may have a many ideas about how spots form from hygiene habits to diet. Those that have acne may be very sensitive about it.

Questions
1 Where do spots form?
2 What substances make a spot form?
3 What is linked to the cause of acne?
4 What should be used to rinse your face?
5 Does eating chocolate cause spots?

Answers
1 in the openings of hair follicles
2 oil and dead skin cells

7B Unit review

Answers to review questions

1. sperm, eggs, testes, ovaries, oviduct, implants, uterus
2. Reproductive organs develop, body hair thickens, hormones cause more skin secretions, mood changes occur, and girls start having periods.
3. a sperm
   b carries the male genetic information to the egg
   c long tail to aid swimming; streamlined shape; enzymes in head to penetrate egg
4. Oxygen and nutrients move from the mother's blood stream into the foetus' blood in the placenta. Carbon dioxide is passed from foetus to mother in the same way.
5. Internal fertilisation takes place inside the body of the female, external fertilisation takes place outside the body.

	Advantage	Disadvantage
Internal	good chance that the egg will be fertilised	male and female may be in danger while they are copulating
External	female does not have to remain close to male	lots of eggs and sperm must be produced to ensure that fertilisation takes place

6. a sperm – a cell (other two are organs)
      or ovary – in female (other two are male)
   b growth – a process (the other two are stages in the process)
   c uterus – an organ (the other two are processes)
      or ovulation – occurs in the ovaries (menstruation occurs in the uterus)
   d uterus – a female organ (the other two are to do with the foetus)
      or amniotic sack – part of foetus (placenta connects uterus to foetus)
   e hormone – a chemical (the other two are products of the action hormones)
7. 20–22 October

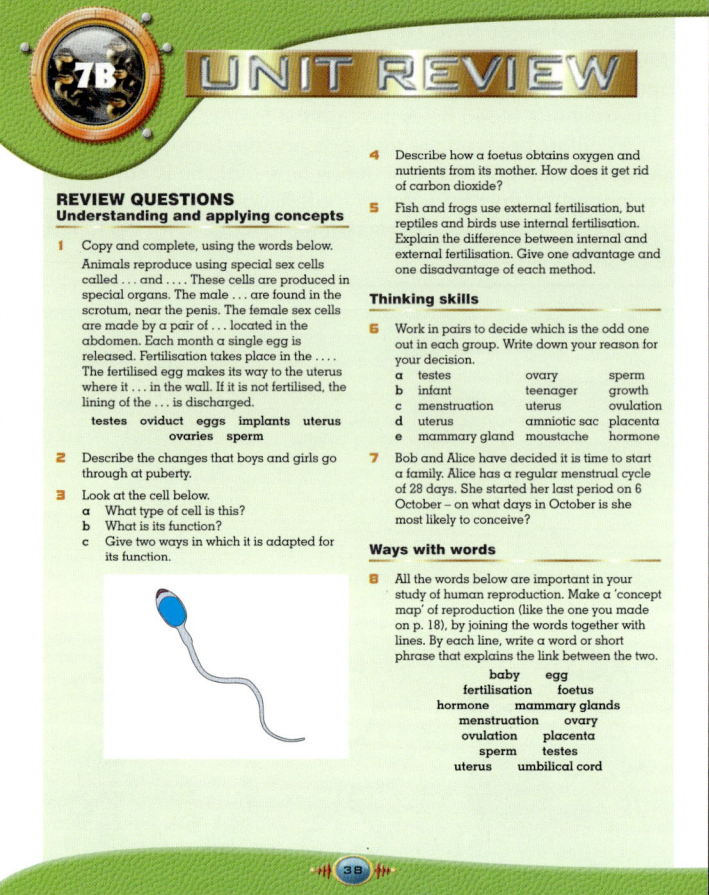

UNIT REVIEW QUESTIONS

SAT-STYLE QUESTIONS

1 This diagram shows part of the female reproductive system:

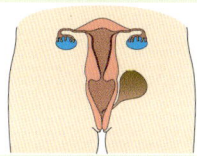

a Where does fertilisation take place? (1)
b Where are the sperm deposited? (1)
c Where is the ovum released? (1)
d Where does the fertilised egg implant? (1)
e What happens at fertilisation? (1)

2 Look at the diagram of a developing foetus:

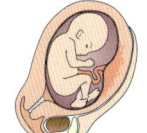

a What links the foetus to the placenta? (1)
b How does the foetus get the food and oxygen that it needs? (2)
c How is the baby pushed out during birth? (2)
d Describe how a baby obtains its food after it is born. (1)

3 Lizzie kept pet rats. She bought two young female rats when they were six weeks old. She measured Millie and Coco from nose to tail tip every Sunday.

Week	Millie's length (cm)	Coco's length (cm)
1	24.0	22.5
2	25.5	24.0
3	27.5	25.5
4	29.0	26.5
5	—	—
6	32.5	29.5
7	34.5	31.0
8	36.0	32.0

a Which rat was the longest? (1)
b Which rat grew most quickly in the first four weeks? (1)
c Use the information in the table to plot a graph.
 • Put the number of weeks on the horizontal axis – this is the **independent variable**.
 • Put the length of rat on the vertical axis – this is the **dependent variable**.
 • Label each axis and write the units on each.

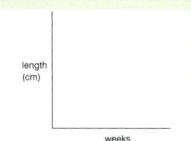

(5)

d Lizzie had to spend one weekend with her Grandma and she could not measure her rats.
Use your graph to predict the length of each rat in Week 5. (1)
e How old were the rats in Week 6? (1)
f Give one reason why Lizzie's measurements might not be accurate. (1)

Key words
Unscramble these:
stoufe
cantaple
minicoat lidfu
stumer nation
pry tube

Answers to SAT-style questions

1 a–d

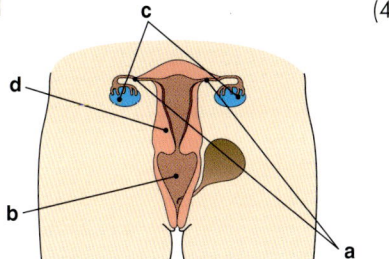
(4)

e The sperm and egg cell join together. (1)

2 a the umbilical cord (1)
b Food and oxygen diffuse from the mother's blood into the foetus' blood in the placenta. (2)
c The cervix dilates and muscular contractions in the uterus push the baby out head first. (2)
d The mother's mammary glands produce milk to feed the baby. (1)

3 a Millie (1)
b Millie (1)
c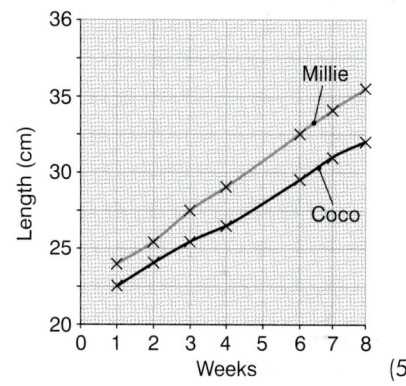
(5)

d Millie: 30.5–31.0; Coco: 28.0 (1)
e 12 weeks (1)
f The rats may move so that they do not stay the same length. (1)

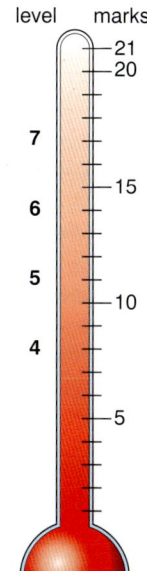

levelometer

7C Environment and feeding relationships

National framework/QCA SoW references

National framework – Interdependence:
Explain how food chains within a habitat can be combined into food webs.
Describe ways in which organisms are adapted to daily or seasonal changes in their environment and to their mode of feeding; use this idea to explain why some organisms can live more successfully than others in different habitats.

QCA SoW (7C)
How does the environment influence the animals and plants living in a habitat? (7C1)
How do environments vary? (7C2, 7C3, 7C4)
What is a feeding relationship? (7C5, 7C6)
What do food webs tell us? (7C7)
NC links: Sc2 5 b, c, d, e; Sc1 2 d, e, f, g, k.

Learning objectives

- To be able to identify features that can vary in habitats, and know that different habitats support different forms of life.
- To understand that the distribution of plants and animals depends on environmental factors.
- To know how animals and plants are adapted to daily and seasonal changes in the environment of their habitats.
- To know some features of predator and prey animals.
- To understand how feeding relationships can be linked to form food webs.
- To determine the feeding relationships of a habitat from observations.
- To understand how changes in populations affect feeding relationships.

In scientific enquiry:

- To consider the importance of sample size.
- To make measurements of environmental changes and interpret these.
- To survey the variety of living things within a habitat.
- To investigate the activity of a small invertebrate, taking into account variables pupils cannot control.

Answers to questions

What do you remember?

1. lettuce → slug → bird → cat
2. desert = camel
   freshwater = common frog
   woodland = squirrel
   sea = shark
   arctic = polar bear
3. For example: **Carnivores:** cat, eagle, owl, wolf, killer whale, ladybird
   Herbivores: cow, elephant, giraffe, sperm whale
   Predators: cat, eagle, owl, wolf, killer whale, spider
   Prey: mice, aphids, wildebeest, pigeons

Notes to support 'What do you remember?'

At KS2 pupils will have learned about living things in their environment, the different plants and animals found in different habitats and how animals and plants in two different habitats are suited to their environment. They will have learned about how food chains show feeding relationships in a habitat, and that nearly all food chains start with a green plant. Before beginning this unit, it would be useful if pupils can:

- Identify the habitat that some animals and plants belong in.
- Write a food chain for a given group of producers and consumers.
- Understand the terms 'carnivore', 'herbivore', 'predator' and 'prey'.

Launch activity notes – Who's eating whom?

This activity can be used to elicit pupils' understanding of feeding relationships and the adaptations that plants and animals have for living in their habitat.

Answers

a) The song is an example of a rather long food chain. Pupils may like to perform the song if they know the words in full:
   fly → spider → bird → cat → dog → cow → horse
   Comment on what is wrong with the latter parts of the food chain given in the song, i.e. the cow and horse are herbivores.

b) In groups, pupils can list the requirements for a successful predator and why these are needed, e.g. teeth to grab and tear the prey, claws to grab and hold the prey, powerful muscles to chase the prey. Is size important? Not all successful predators are as big as T.Rex. Pupils can draw their own 'super-predator'.

c) Discuss the ways prey animals can evade capture. Some pupils may be anglers and will have experienced the ways in which fish try to get away. The slime coating a fish is an adaptation that enables it to move swiftly through water, not to slip out of anglers' hands.

d) Discuss how animals and plants are adapted to extreme climates. In the desert, most animals are nocturnal only emerging from underground when it has cooled. In this way, they conserve water which would evaporate rapidly in the heat of the Sun. Humans must adopt the same strategy if they want to survive in the desert.

Teaching strategy

This unit builds on KS2 units 6A 'Interdependence and adaptation' and 4B 'Habitats'. It leads into units 7D 'Variation and classification', 8D 'Ecological relationships' and 9D 'Plants for food'.

Difficulties/Misconceptions

- Pupils sometimes get the directions of arrows incorrect in food chains. Energy transfer diagrams are introduced in a later unit, but here it is sufficient to suggest that the arrows show the direction that energy moves through the food chain, starting with a plant.
- This unit does not mention decomposers, so food chains are assumed to end with the top predator.
- Some pupils may not have visited the coast and will be unfamiliar with tidal changes.

Notes and tips

- A lot of the practical work in this unit involves fieldwork. If the school has any grounds, however small, these can be sufficient. Even a small garden or an overgrown wall can provide sufficient habitats and organisms to study. Work outside the classroom and any trips off the school site must comply with school and/or LEA regulations, including completion of appropriate risk assessment forms. Fieldwork is time consuming and requires organisation, but gives pupils confidence in using the instruments and enables them to see the ideas discussed in the classroom at work.
- In addition to the pictures of organisms and habitats included in this pack, it would be useful to have a variety of other pictures of different habitats.

Learning styles

Visual
- **Using cartoons** and diagrams to stimulate discussion.
- **Studying** food chains.

Auditory
- **Pronouncing** words correctly.
- **Discussing** questions.

Interpersonal
- **Discussion**.

Intrapersonal
- **Considering objectives**.
- **Recollecting** work done previously.

7C1 Environmental influences

National framework/QCA SoW references

National framework – Interdependence: Explain why some organisms can live more successfully than others in different habitats.

QCA SoW

7C Section 1: How does the environment influence the animals and plants living in a habitat?
NC links: Sc2 5b.

Learning objectives

- To know that different habitats have different features.
- To know that different habitats support different organisms.
- To know that the distribution of organisms in different habitats is affected by environmental factors, e.g. light, nutrients or water availability.
- To organise, sequence and link what is said, so listeners can follow it.

Lesson starter suggestions

- **Home sweet home:** Pupils look at pictures of habitats and describe the conditions they may find there.
- **Pick an animal:** Pupils name an animal, and others say the kind of habitat where it may be found and what its food may be.
- **Odd one out:** Given a trio of animals, pupils pick the odd one and explain their reasoning, e.g. cod, herring trout.

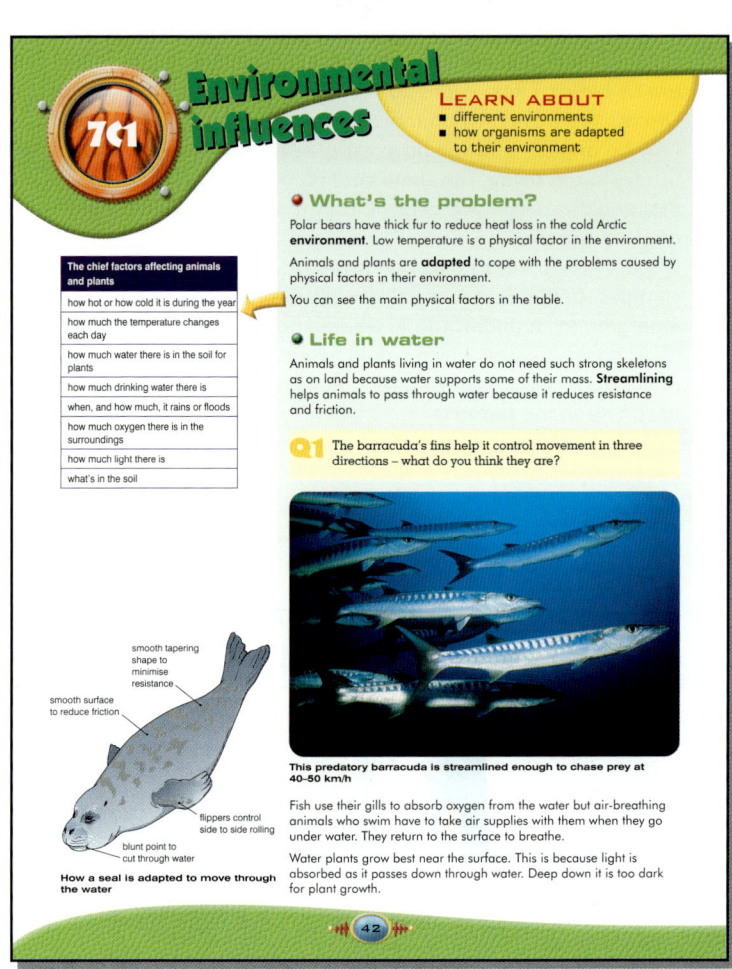

Teaching strategy

Key points to highlight

- Some of the physical factors in habitats that influence animals and plants are the amount of light, the temperature, availability of water and oxygen (and for animals, food).
- The conditions vary from one habitat to another.
- Animals and plants are adapted to the habitats in which they live.
- Animal adaptations include body shape (for swimming, flying, burrowing, running), methods of maintaining stable temperature (insulation, radiators e.g. perspiraton, panting, tongue hanging out, big ears), and means of getting food (teeth, claws).

Difficulties/Misconceptions

- This lesson relies on pupils having some knowledge of a range of animals from earlier years. It may be useful to have plenty of pictures of animals in habitats available for pupils to refer to and to identify.
- Evolution is not the subject of this unit, but pupils may have the misconception that animals and plants choose to live in a particular habitat because it suits them, rather than having evolved, as a result of competition in the habitat, from species less adapted to the conditions.

Skills

Planning and carrying out a short investigation.

Extension ideas

Exploration of more habitats and their occupants.

Answers to questions

Text

Q1 Any description of the three degrees of movement, which in aeronautical terms are:
- pitch (nose-to-tail angle) (rotation about y axis),
- yaw (turn left/right) (rotation about z axis),
- roll (rotation about x axis).

Summary

1. streamlined shape and powerful muscles for swimming fast; eyes and other senses to detect prey; sharp teeth to grab and tear prey
2. **physical factors:** conditions affecting plants and animals caused by the natural features of the habitat, e.g. light/shade, hot/cold, wet/dry;
   adaptation: a feature of the plant or animal that helps it to survive in its habitat, e.g. body shape, size of eyes, amount of fur.
3. Water plants are near the surface to collect light, so consumers must be near the surface too. Predators can lurk lower down, only rising to take prey. Scavengers can remain at the bottom, because bits of plants and dead animals will sink.
4. They lose less water by transpiration (less surface area).
5. Small birds lose heat faster than larger birds so need more insulation.

Learning styles

Visual
- **Looking at** pictures of habitats and animals.

Auditory
- **Discussing** the topic.

Kinaesthetic
- **Experimenting.**

Interpersonal
- **Collaborating.**
- **Discussing.**

Intrapersonal
- **Recalling** prior knowledge.

Activity and technicians' notes

Smooth movers

Safety notes
- Tall, glass measuring cylinders can be unstable when full of water. Watch for broken glass if any topple over.

Equipment required
- Plasticine
- Large measuring cylinders and/or tall glass/plastic containers or tubes
- Stopwatches

Access required to:
- Water

Suggestions for plenaries

- **Spot the difference:** Pupils to compare a pair of habitats, noting the different conditions and suggesting the type of creatures that may be found.
- **Good, bad, find out more:** Give pupils the statement 'Global warming is likely to make Britain hotter and drier'. Encourage pupils to discuss something good and something bad about this statement, and to pose a question.
- **Our home:** Pupils to describe the conditions in their own habitat (local community) and the types of animal and plant that they encounter.

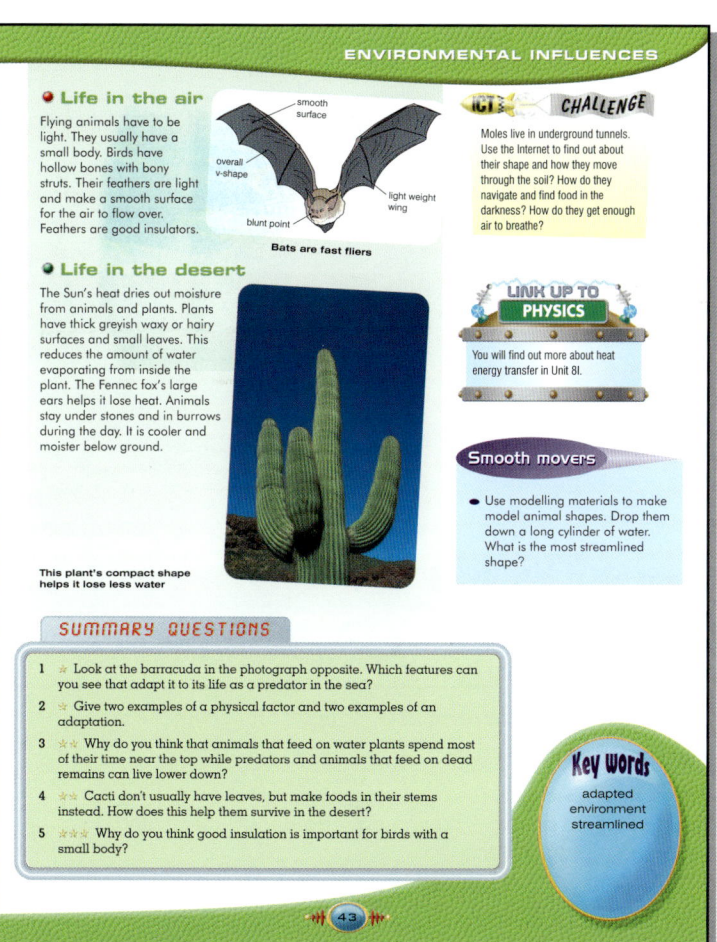

Learning outcomes

Following this, most pupils will:
- Identify differences between different habitats and relate these to the organisms found in them.

Some pupils will also:
- Have knowledge of a wider range of habitats.

7C2 What's it like out there?

National framework/QCA SoW references

National framework – Interdependence: Explain why some organisms can live more successfully than others in different habitats.

QCA SoW

7C Section 2: How do environments vary?
NC links: Sc1 2.

Lesson starter suggestions

- **Describing conditions:** Describe the conditions shown on the slide.
- **Suit you?** Describe the weather today and agree on what words mean.
- **Odd one out:** Pick the odd one out of the trios and explain why they are different: warm/mild/scorching; wet/damp/moist; dim/bright/shady

Learning objectives

▸ To know how to measure and record changes in environmental factors.
▸ To know how to interpret patterns in data.
▸ To know how to frame a question to be investigated.
▸ To know how to decide what factors are relevant to a question.
▸ To know about the importance of sample size.
▸ To consider results in relation to the sample used.

Teaching strategy

Key points to highlight

There is a lot of work in this section; time must be allowed to complete the activities. There are three parts to the work:

- Using instruments and observations to record the conditions in a chosen habitat. This habitat will be returned to in later lessons. The initial stage involves getting used to the instruments and interpreting readings and observations. This is an opportunity to develop skills using data logging equipment, but it can also be done with simple apparatus and observation.
- An investigation of the living conditions preferred by a specific small animal. This is an opportunity for pupils to plan their own investigation.
- A comparison of the plants found in various habitats. This demonstrates that different species occupy different environmental niches.

Difficulties/Misconceptions

- A variety of habitats should be studied to illustrate different conditions, and to stop pupils crowding into a small area with detrimental effects on the organisms present.
- Habitats can be small (e.g. under a bush, the side of an old stone wall) or large (e.g. a lawn or pond). Pupils need to think carefully before choosing the habitat they wish to study.
- When doing the choice chamber study, it is important that only one variable is being changed, e.g. wet cotton wool can provide moisture and shade for woodlice; a lamp, light and heat.

- Have plant guides available, or access to the Internet.

Skills

- Scientific enquiry (see Learning objectives).
- Setting up and interpreting data from instruments and data logging equipment.
- Making and recording observations.
- Looking after living creatures.

Notes and tips

- Check suitable habitats to investigate and make sure there are no hazards.
- Set up some data logging apparatus several days before the lesson, in order to collect data for more than one day.
- Identify the plants in the different habitats before the lesson.

Extension ideas

Some pupils can use more advanced sensors and data logging packages (e.g. pH meters, oxygen sensors).

Activity and technicians' notes

Safety notes

Here are some hazards involved in fieldwork:
- Broken glass/pieces of metal on ground
- Wasp/bee stings (note pupils who are allergic to stings)
- Plants that have thorns/stings/or are poisonous (yew, foxgloves, etc.)
- Uneven ground
- Unstable trees/walls.

Pupils should be warned of hazards and report any cuts or inflammation at once.

Equipment required

For the physical factors and habitat investigations:
- Instruments, or data logging equipment, for measuring light level, temperature, humidity, oxygen level (in water), pH, wind speed
- A rain gauge
- Card, plastic cups, Sellotape, dowelling (or pencils) for making wind direction indicators
- Wind strength indicators

For the choice chamber investigation:
- Petri dishes
- Black card
- Filter paper
- Droppers
- Lamps
- Thermometers
- A supply of woodlice (a lot needed)
- Aquaria of brine shrimps

Access required to:
- Information on plants (guides or Internet)

Learning styles

Visual
- **Making** and interpreting observations.
- **Reading** instruments.

Auditory
- **Discussing** plans.
- **Choosing** words in recording observations.

Kinaesthetic
- **Exploring** habitats.
- **Carrying out** an investigation.

Interpersonal
- **Setting up** equipment together.
- **Sharing** data.
- **Collaborating** on investigation.
- **Sharing** research.

Intrapersonal
- **Reviewing** vocabulary.
- **Reflecting** on discoveries.

Answers to questions

Text

Q1 light/shade; wet/dry; mineral abundance; temperature; wind strength; trampling

Q2 wind; insects/birds/animals; stuck to people's clothes/shoes

Summary

1. community
2. a) rain gauge
   b) anemometer
   c) thermometer

Learning outcomes

Following this, most pupils will:
- Make a series of measurements of environmental variables appropriate to the task.
- Identify a question to investigate about the activity of an invertebrate, suggesting a suitable approach and sample size.

Some pupils will also:
- Describe, in terms of approach and sample size, how strongly any patterns or associations identified are supported by the evidence.

Suggestions for plenaries

- **Habitat conditions:** Pupils to summarise observations of habitats on the conditions scale.
- **Changing conditions:** Pupils to consider the changes that may occur in the habitats studied and the reasons for the changes.
- **Reflections:** Pupils to reflect on the skills learnt in this lesson, the vocabulary used and the ideas discussed.

7C3 Daily changes

National framework/QCA SoW references

National framework– Interdependence: Describe ways in which organisms are adapted to daily changes in their environment.

QCA SoW:
7C Section 2: How do environments vary?
NC links: Sc2 5c.

Learning objectives

▶ To know that some animals are adapted to daily changes in their habitat.

Teaching strategy

Key points to highlight

- Plants and animals are adapted to the daily cycle of day and night, light and dark, warm and cold.
- Nocturnal animals make more use of their senses of smell and hearing than sight.
- Seashore plants and animals are also adapted to the tidal cycle.
- Pitfall traps, pooters, nets and branch shaking are ways of collecting specimens of small animals.

Difficulties/Misconceptions

- The main problem with this sub-topic is the length of time required for a worthwhile examination of a habitat, particularly to see diurnal variation.
- Pupils (and teachers!) may lack the necessary knowledge of plants and animals required for identification of specimens. A supply of good field guides is essential.
- Pupils who live some distance from a coast may have no experience of tides.

Lesson starter suggestions

- **Daily routine:** Pupils record the times when they might feel hungry, tired, cold, and hot, and when they do things.
- **Think of a reason:** Ask pupils to consider all the reasons why crows and other birds may be seen in the school grounds in the early morning and evening, but not during the middle of the day or at night.
- **Senses:** Pupils discuss the senses that animals use to detect food and predators.

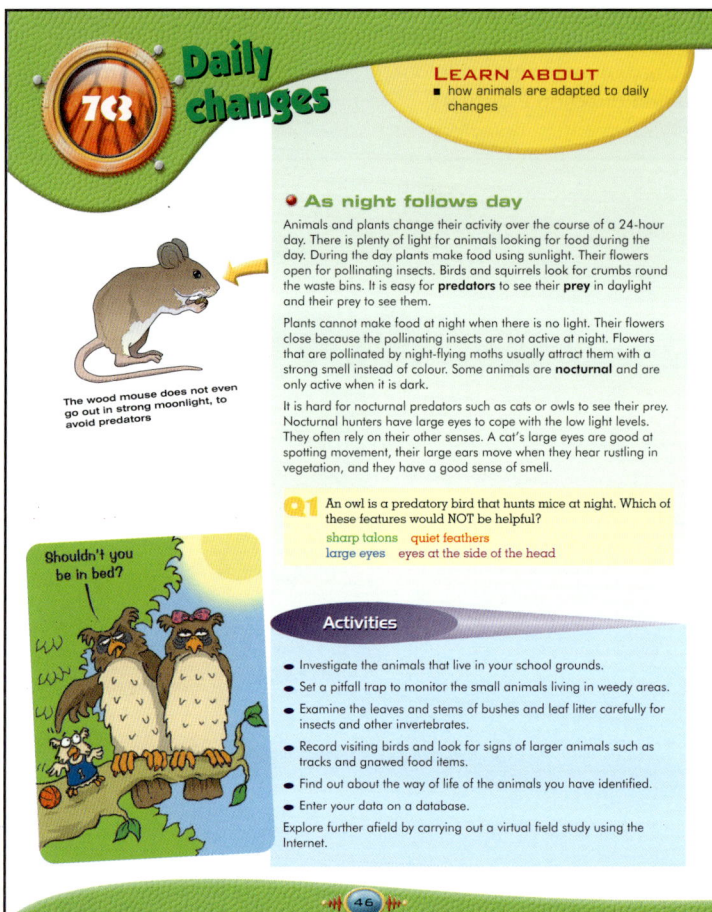

Activity and technicians' notes

Safety notes

- Pupils must be supervised while working in the field.
- Make sure that pooters have gauze fixed securely over one tube, and that pupils know how to use them carefully.
- Pupils should wash their hands before and after handling animals.

Equipment required:

- Clean yoghurt pots or jars for pitfall traps
- A trowel
- Pooters
- Magnifiers
- Large nets
- Plastic sheets (opened out plastic bags will do)

Access required to:

Field guides of:
- Plants
- Trees
- Insects and other invertebrates
- Birds

ICT:

- Database
- Internet for virtual field study

Teaching strategy contd.

Skills
Observations of birds and animal tracks.

Notes and tips
Look for suitable habitats before the lesson. Choose somewhere with a variety of plant growth and micro-habitats (walls and trees to provide shade, ponds or places that are frequently wet), as far as possible from traffic, noisy children etc.

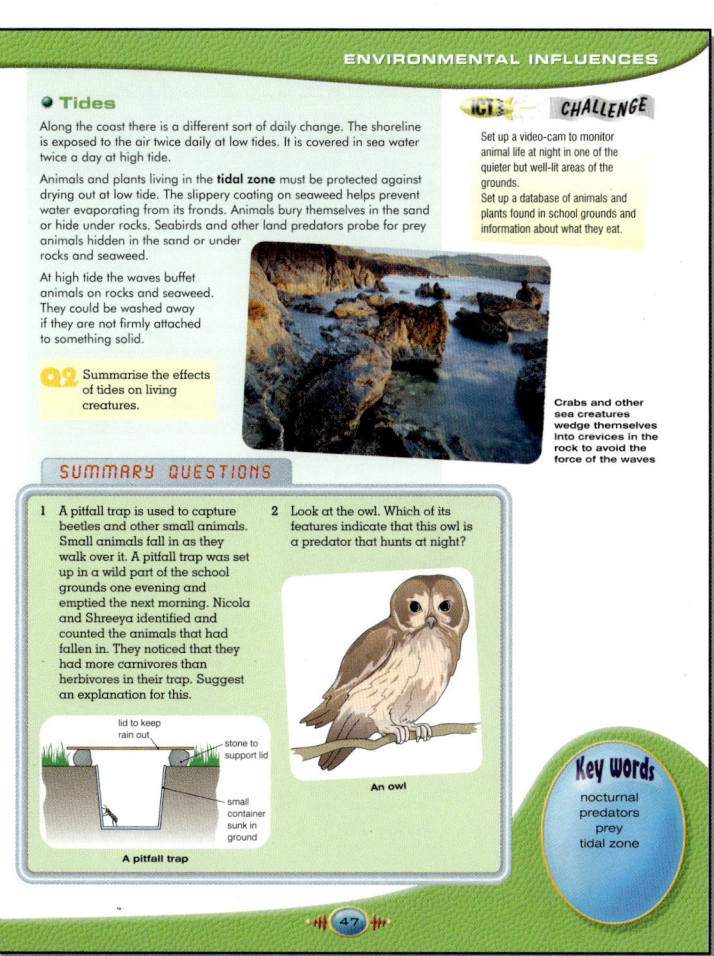

Learning styles

Visual
- **Looking at** pictures and graphs to get an idea of daily cycles.

Auditory
- **Discussing** observations and thoughts.

Kinaesthetic
- **Collecting and examining** animals and their habitats.

Interpersonal
- **Sharing** observations amongst the class.

Intrapersonal
- **Reflecting** on daily changes in their lives.

Answers to questions

Text
- **Q1** eyes at the side of the head
- **Q2** Organisms are submerged and exposed twice a day. They are battered by waves and currents.

Summary
1. Carnivores have to move to find their prey. Herbivores tend to stay where their plant food is.
2. predator because of strong claws and forward facing eyes; night-time because of large eyes

Suggestions for plenaries

- **Day feeders/night feeders:** Pupils sort pictures of animals into those that feed during the day and those that feed at night.
- **Advantage/disadvantage:** Pupils discuss the pros and cons of some animals being nocturnal.
- **Cycles:** Pupils summarise the daily cycles.

Learning outcomes

Following this, most pupils will:
▸ Describe ways in which organisms are adapted to daily changes in their environment.

7C4 Seasons

National framework/QCA SoW references

National framework – Interdependence: Describe ways in which organisms are adapted to seasonal changes in their environment.

QCA SoW

7C Section 2: How do environments vary?
NC links: Sc2 5c.

Learning objectives

- To know how some animals are adapted to seasonal changes in their habitats.
- To know that adaptations may be to avoid climatic stress.

Teaching strategy

Key points to highlight

- In winter, low temperatures (and less light) reduce plant growth. Some plants prepare for winter by storing food in roots and by becoming dormant.
- Animals cope with the lack of food in the winter by migrating or hibernating. They may grow thicker fur or feathers and increase the layer of fat under the skin for insulation.

Difficulties/Misconceptions

- In winter, migrating animals move away from the poles. Thus some animals move from polar regions (e.g. the Arctic) to temperate areas (e.g. the UK) while some move from temperate areas to tropical regions (e.g. North Africa). The ability of some animals to over-winter in areas that others find inhospitable is because of their different feeding habits, e.g. geese eat grass and tubers (e.g. potatoes), while swallows rely on flying insects.
- Hibernation does not necessarily mean that the animal is asleep. Some animals remain awake, but do not move from their burrows or nests and they reduce their level of activity.

Skills

Observing and measuring (using thermometers).

Notes and tips

Investigating insulation is a good opportunity for pupils to plan an investigation and to learn how to use and read thermometers correctly.

Extension ideas

Under stress: a literacy and comprehension exercise on the effects of climatic stress.

Lesson starter suggestions

- **Seasonal changes:** Pupils look at two pictures (summer and winter) and identify the differences in environmental conditions.
- **Adapting for change:** Pupils are given a set of objects, asked to identify them and suggest how they relate to the changing seasons.
- **Where are the animals?** Ask pupils why we see fewer birds in winter than in summer and why there are much fewer animals.

Learning styles

Visual
- **Studying pictures** to help respond to discussion questions.

Auditory
- **Listening** to descriptions and discussion.

Kinaesthetic
- **Carrying out** an investigation.

Interpersonal
- **Discussing and collaborating** on investigations.

Intrapersonal
- **Evaluating** investigations.
- **Reviewing** what's been learnt.

Activity and technicians' notes

Safety notes
Here are some hazards involved in fieldwork:
- Broken glass/pieces of metal on ground
- Wasp/bee stings (note pupils who are allergic to stings)
- Plants that have thorns/stings/or are poisonous (yew, foxgloves, etc.)
- Uneven ground
- Unstable trees/walls.

Pupils should be warned of hazards and report any cuts or inflammation at once.

Equipment required
For the physical factors and habitat investigations:
- Instruments, or data logging equipment, for measuring light level, temperature, humidity, oxygen level (in water), pH, wind speed
- A rain gauge
- Card, plastic cups, Sellotape, dowelling (or pencils) for making wind direction indicators
- Wind strength indicators

For the choice chamber investigation:
- Petri dishes
- Black card
- Filter paper
- Droppers
- Lamps
- Thermometers
- A supply of woodlice (a lot needed)
- Aquaria of brine shrimps

Access required to:
- Information on plants (guides or Internet)

Learning styles

Visual
- **Making** and interpreting observations.
- **Reading** instruments.

Auditory
- **Discussing** plans.
- **Choosing** words in recording observations.

Kinaesthetic
- **Exploring** habitats.
- **Carrying out** an investigation.

Interpersonal
- **Setting up** equipment together.
- **Sharing** data.
- **Collaborating** on investigation.
- **Sharing** research.

Intrapersonal
- **Reviewing** vocabulary.
- **Reflecting** on discoveries.

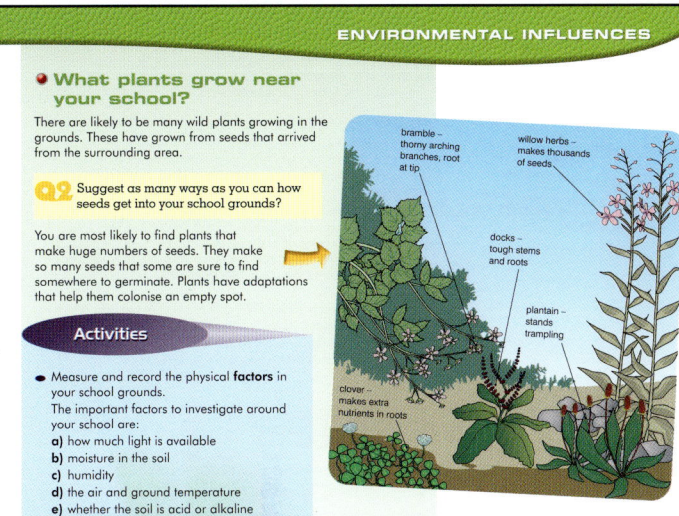

Answers to questions

Text
- **Q1** light/shade; wet/dry; mineral abundance; temperature; wind strength; trampling
- **Q2** wind; insects/birds/animals; stuck to people's clothes/shoes

Summary
1. community
2. a) rain gauge
   b) anemometer
   c) thermometer

Learning outcomes
Following this, most pupils will:
- Make a series of measurements of environmental variables appropriate to the task.
- Identify a question to investigate about the activity of an invertebrate, suggesting a suitable approach and sample size.

Some pupils will also:
- Describe, in terms of approach and sample size, how strongly any patterns or associations identified are supported by the evidence.

Suggestions for plenaries
- **Habitat conditions:** Pupils to summarise observations of habitats on the conditions scale.
- **Changing conditions:** Pupils to consider the changes that may occur in the habitats studied and the reasons for the changes.
- **Reflections:** Pupils to reflect on the skills learnt in this lesson, the vocabulary used and the ideas discussed.

7C3 Daily changes

National framework/QCA SoW references

National framework– Interdependence: Describe ways in which organisms are adapted to daily changes in their environment.

QCA SoW:
7C Section 2: How do environments vary?
NC links: Sc2 5c.

Learning objectives

▸ To know that some animals are adapted to daily changes in their habitat.

Teaching strategy

Key points to highlight

- Plants and animals are adapted to the daily cycle of day and night, light and dark, warm and cold.
- Nocturnal animals make more use of their senses of smell and hearing than sight.
- Seashore plants and animals are also adapted to the tidal cycle.
- Pitfall traps, pooters, nets and branch shaking are ways of collecting specimens of small animals.

Difficulties/Misconceptions

- The main problem with this sub-topic is the length of time required for a worthwhile examination of a habitat, particularly to see diurnal variation.
- Pupils (and teachers!) may lack the necessary knowledge of plants and animals required for identification of specimens. A supply of good field guides is essential.
- Pupils who live some distance from a coast may have no experience of tides.

Lesson starter suggestions

- **Daily routine:** Pupils record the times when they might feel hungry, tired, cold, and hot, and when they do things.
- **Think of a reason:** Ask pupils to consider all the reasons why crows and other birds may be seen in the school grounds in the early morning and evening, but not during the middle of the day or at night.
- **Senses:** Pupils discuss the senses that animals use to detect food and predators.

Activity and technicians' notes

Safety notes

- Pupils must be supervised while working in the field.
- Make sure that pooters have gauze fixed securely over one tube, and that pupils know how to use them carefully.
- Pupils should wash their hands before and after handling animals.

Equipment required:

- Clean yoghurt pots or jars for pitfall traps
- A trowel
- Pooters
- Magnifiers
- Large nets
- Plastic sheets (opened out plastic bags will do)

Access required to:

Field guides of:
- Plants
- Trees
- Insects and other invertebrates
- Birds

ICT:

- Database
- Internet for virtual field study

Activity and technicians' notes

Surviving the seasons

Safety notes

Pupils should take care in pouring boiling water into small beakers and should not carry the beakers around the room.

Equipment required

For the surviving winter investigation:
- Carrots, onions, insect pupae, winter twig with buds, dog or cat fur (enough sets for each group of 3 or 4 pupils in the class)

For the insulation investigation:
- Wool (knitted or raw), fur (fake or real), feathers (down from a pillow or duvet), (enough of each material for pairs of pupils to wrap around small beakers)
- Cling film, card, scissors, small beakers, thermometers, stopwatches

Access required to:
- Boiling water (kettles)

Answers to questions

Text

Q1 various species of geese (not Canada geese)

Summary

1. The north of Scotland is snow-covered in winter. The ermine coat is an adaptation that camouflages the stoat. White fur would not be an advantage in the snow-free Midlands.
2. When the animals become active, they use up their fat stores quicker. This means that they may not have enough to last them until food becomes available again.

Amazing science

Male grizzly bears weigh about 500 kg (half a tonne!) on average, with the females slightly lighter. Most of their diet is vegetarian: grass, leaves, berries and nuts; but they supplement their eating with insects, fish (they are especially partial to salmon) and mammals (small and large). They will eat carrion and kill large animals like caribou. As they do not have the special digestive system that cattle have, they have to eat a huge amount of food during the six months that they are awake. The females give birth in February, before they leave their winter home, and they provide milk for their young for most of the year.

Suggestions for plenaries

- **Adapt and survive:** Drag and drop puzzle matching the survival technique to the conditions.
- **I will survive:** Ask the pupils: 'You have been stranded in the Antarctic with winter approaching. What would you do to survive?'
- **Memory-map 'Seasonal changes':** Pupils to put the title in the centre and add branches, to summarise the changes that occur in summer and winter and the ways in which animals and plants adapt to the changes.

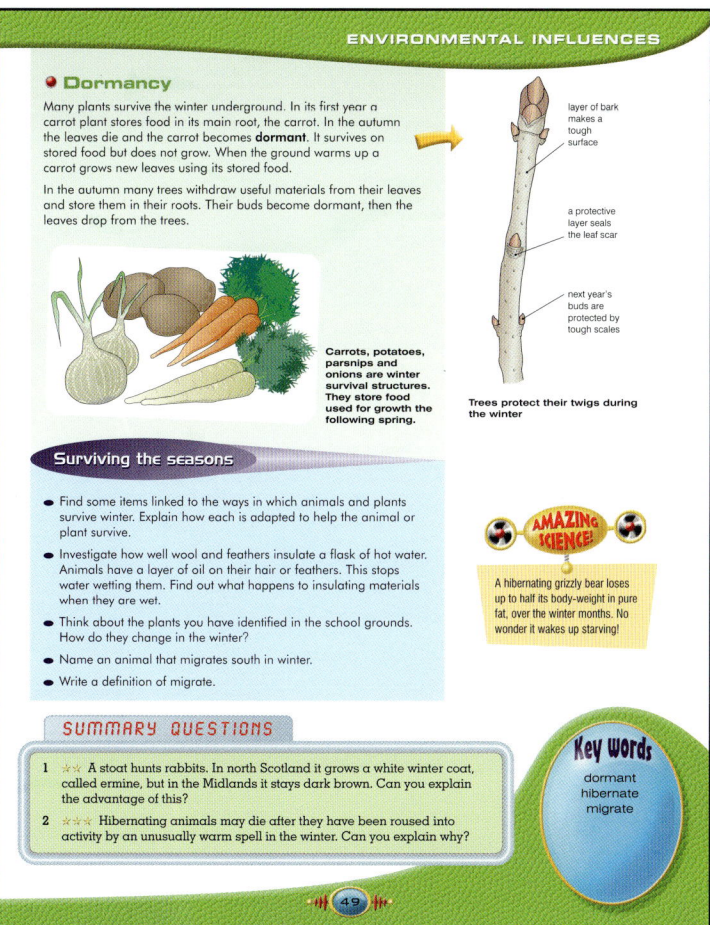

Learning outcomes

Following this, most pupils will:
▶ Describe ways in which organisms are adapted to seasonal changes in their environment.

Some pupils will also:
▶ Have a more detailed understanding of climatic stress.

7C5 Food chains and webs

National framework/QCA SoW references

National framework – Interdependence: Explain how food chains within a habitat can be combined into food webs.

QCA SoW

7C Section 3: What is a feeding relationship?
7C Section 4: What do food webs tell us?
NC links: Sc2 5e.

Lesson starter suggestions

- **Who eats what?** Pupils use their prior knowledge to identify the food source of the animals in the picture.
- **Good, bad, want to know more?** Pupils discuss the statement 'Animal life on Earth depends on the Sun shining and plants growing'.
- **Do you remember food chains?** Ask pupils to explain what they understand by the term 'food chain' and to give a simple example.

Learning objectives

- To know that all the organisms in a habitat can be linked together in food webs.
- To know that food webs are made up of a number of food chains, which start with plants.
- To know that arrows in a food chain represent energy transfer.
- To make careful observations of plants and animals and sources of evidence about animals' foods.
- To link organisms together in food webs.

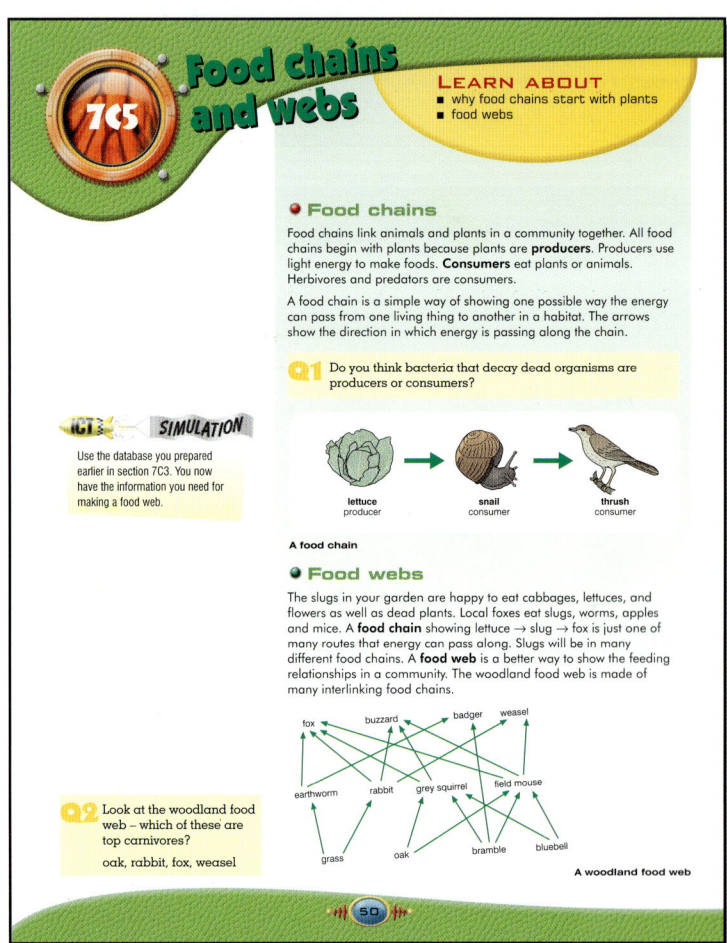

Teaching strategy

Key points to highlight

- The Sun is the source of energy for (nearly) all life on Earth.
- Food chains show how the energy taken in by plants is passed from producers to consumers.
- Food webs show the more complex feeding relationships that occur in any habitat, where most consumers have more than one food source and are often the prey for more than one predator.
- The top predator is at the top of the food web: it is a consumer, but not prey to any other animal.

Difficulties/Misconceptions

- Make sure that pupils write the arrows in food chains and webs in the correct direction. They should show energy (food) passing from producers to consumers.
- More-able pupils should learn that some of the energy taken in by plants is lost at every stage (animals use the energy for moving muscles and making new substances). This is why the number (and total mass) of animals decreases up the food web. There are very few top predators compared to the number of primary consumers.

Skills

Observation (animal watching) and analytical (interpreting observations and determining feeding relationships).

Notes and tips

- This lesson is an opportunity to continue the observations begun in animal watching in unit 7C3.
- In many food webs, humans are the top predator or principal consumer.

Extension ideas

Further work relating food chains to energy consumption.

Activity and technicians' notes

PUPIL SHEET

Safety notes
- Pupils must be supervised while working in the field.
- Make sure that pooters have gauze fixed securely over one tube, and that pupils know how to use them carefully.
- Pupils should wash their hands before and after handling animals.

Equipment required
- Clean yoghurt pots or jars for pitfall traps
- A trowel
- Pooters
- Magnifiers
- Large nets
- Plastic sheets (opened out plastic bags will do)

Access required to:
Posters of various habitats showing producers, consumers, predators.

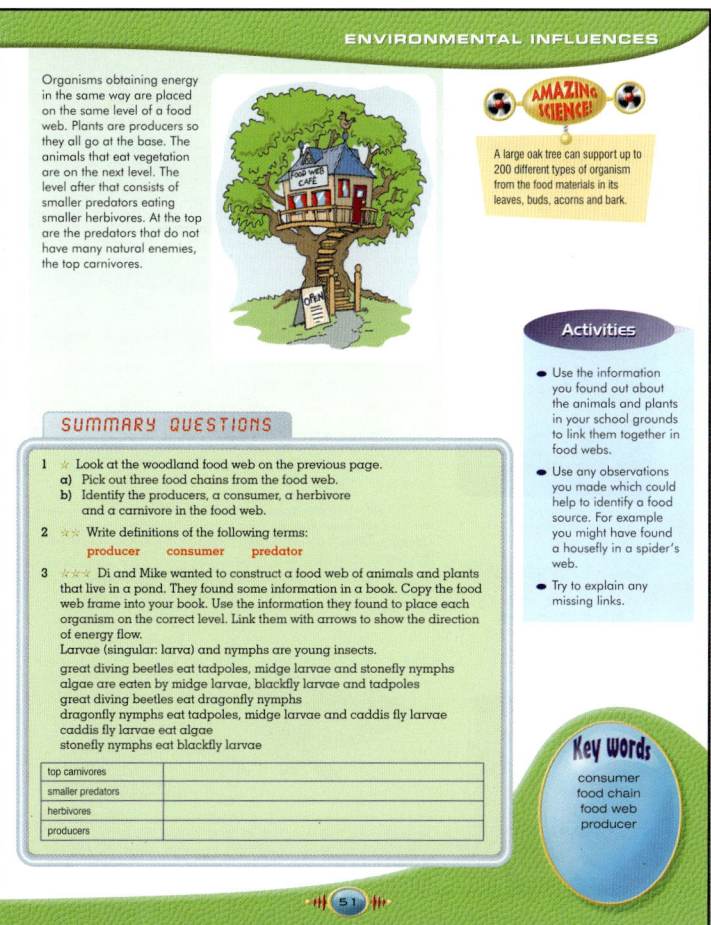

Learning styles

Visual
- **Observing** in fieldwork.

Auditory
- **Discussing** answers to questions.

Kinaesthetic
- **Exploring** habitats.

Interpersonal
- **Sharing data**

Intrapersonal
- **Considering** own position in food webs (see Plenary 'Where do we fit in?').

Amazing science

Oak and other deciduous trees support many more species than conifers. The galls formed by oak trees in response to insects burrowing into acorns were once ground up to make black ink.

Answers to questions

Text

Q1 consumers   Q2 fox, weasel

Summary

1 a) For example:
   grass → earthworm → badger
   oak → grey squirrel → buzzard
   bramble → field mouse → buzzard
   bluebell → grey squirrel → fox

   b) **producers:** grass, oak, bramble, bluebell
   consumers: earthworm, rabbit, grey squirrel, field mouse, fox, buzzard, badger, weasel
   herbivores: earthworm, rabbit, grey squirrel, field mouse
   carnivores: fox, buzzard, badger, weasel

2 Plants are **producers** because they use light energy to make foods.
   A **consumer** is an animal that gains its energy by eating other animals or plants.
   A **predator** is an animal that gains energy by catching and eating other animals.

3

Top carnivores	great diving beetles,
Smaller predators	stonefly and dragonfly nymphs
Herbivores	blackfly, midge and caddis fly larvae, tadpoles
Producers	algae

Learning outcomes

Following this, most pupils will:
▶ Describe food chains and combine these into food webs.

Some pupils will also:
▶ Describe how different organisms contribute to the community in which they are found and relate food chains to energy transfer.

Suggestions for plenaries

- **Word matching:** Pupils match the words to definitions and explanations.
- **Where do we fit in?** Pupils to think about the food that humans eat and discuss where humans fit into food webs.
- **Glossary:** Pupils make their own graphic dictionary to remember the terms: food web/chain, producer, consumer, predator, prey, herbivore, carnivore.

7C6 Eating and being eaten

National framework/QCA SoW references

National framework – Interdependence: Describe way in which organisms are adapted to their mode of feeding.

QCA SoW

7C Section 3: What is a feeding relationship?
NC links: Sc2 5e.

Lesson starter suggestions

- **How do you get your meal?** Consider ways in which animals get their food.
- **Links:** Pupils discuss and explain the links between triplets of animals or plants, e.g. bramble, rose, thistle; killer whale, great white shark, giant seal; sea eagle, cormorant, osprey; adder, cobra, rattlesnake; humming bird, bumble bee, monarch butterfly.
- **Good, bad, want to know more:** Discuss the statement 'Plants and animals have ways of defeating consumers and predators'.

Learning objectives

- To know that animals have features that are adaptations against predators.
- To know that animals are adapted to their particular food source.
- To collect sufficient data to reduce error and obtain reliable evidence.

Teaching strategy

Key points to highlight

- Predators and prey are adapted to their mode of feeding.
- Predators tend to have forward facing eyes, acute hearing and sense of smell, and limbs and/or teeth suitable for grabbing and tearing their prey.
- Prey animals tend to have all round vision, acute hearing and sense of smell, and sensitivity to movement, and are often camouflaged or poisonous.
- Plants also have strategies for putting off herbivores, i.e. spines, poison, growing pattern.

Difficulties/Misconceptions

With such a wide variety of predators and prey, pupils may be overwhelmed with data. Make sure that they understand the similarities between predators, particularly mammals and birds, i.e. good eyesight and the tools to catch prey. Also they need to know the common defence strategies of prey.

Notes and tips

- This may be the occasion to visit a wildlife sanctuary, or to have a visit from an expert who can bring some animals into school.
- Evolution by natural selection will be studied in a later unit, but pupils may ask about the origin of the adaptations and how predators and prey became suited to their environments.

Extension ideas

More detailed research on a particular predator and its prey.

Activity and technicians' notes

Safety notes:
Pupils should wash their hands before and after touching animals.

Equipment required
Camouflage activity:
- Card, scissors, coloured pens/crayons.

Shape of beak activity:
- Small and large seeds (e.g. millet and sunflower), pointed and blunt tweezers/forceps.

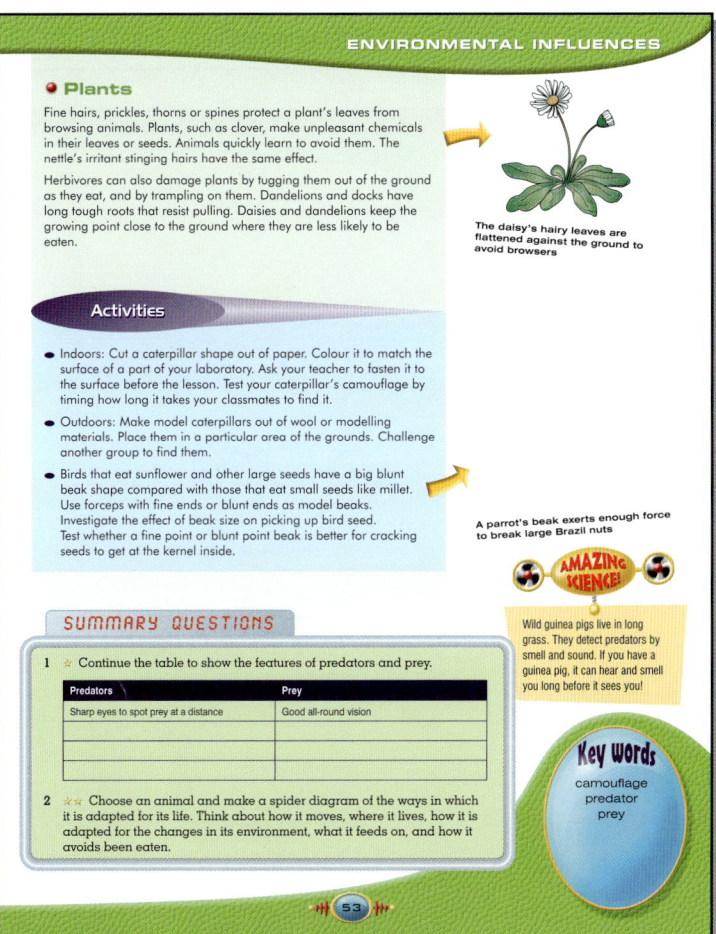

Amazing science
Guinea pigs are native to South America where they were caught as a source of food. They were brought back to Europe in the sixteenth century and have been kept as pets ever since. Guinea pigs were first used as 'guinea pigs' in experiments on diets because, like humans, they need vitamin C in their diets or else they suffer from scurvy.

Learning outcomes
Following this, most pupils will:
▶ Describe ways in which organisms are adapted to their mode of feeding.

Some pupils will also:
▶ Recall the adaptations of particular predators and prey.

Learning styles

Visual
- **Looking at** animals and plants and noting their features.

Auditory
- **Discussing.**

Kinaesthetic
- **Investigating** (camouflage and beak shape).

Interpersonal
- **Collaborating** on activities.

Intrapersonal
- **Reflecting** on humans as highly adapted predators.

Answers to questions

Text
Q1 Suitable examples are: pike, cheetah

Summary
1 for example:

Predators	Prey
acute hearing and/or sense of smell	acute hearing and/or sense of smell
teeth/claws for grasping and tearing	camouflaged
body shape/muscles suitable for stealth or sudden attack	often freeze when they detect movement

2 The diagram should have branches referring to the creature's senses, limbs/body shape, teeth/claws, colouring, food source and eating habits, and environmental conditions in its favoured habitat.

Suggestions for plenaries

- **Build a predator:** Pupils play a game in which they put together a model predator.
- **Man as predator:** Discuss the reasons why humans became a top predator.
- **Fighting for survival:** Summarise the ways that plants and animals are adapted to eat and avoid being eaten.

7C7 Disturbing the web

National framework/QCA SoW references

National framework – Interdependence: Explain why some organisms can live more successfully than others in different habitats.

QCA SoW
7C Section 4: What do food webs tell us?
NC links: Sc2 5d.

Lesson starter suggestions

- **Who's competing?** Pupils look at the food web and decide which animals are competing for particular foods. Then ask them to consider what happens if the number of one of the species increases.
- **Why do numbers increase?** Pupils discuss the reasons for a possible increase in the numbers of a consumer in a food web.
- **Recalling conditions:** Ask pupils questions about how conditions in a habitat may affect food webs.

Learning objectives

▸ To know that factors influencing the number of organisms in one part of a food web have an effect on other parts of the web.
▸ To know that organisms in a habitat compete for resources from the environment.

Teaching strategy

Key points to highlight

- There is competition for food between animals at the same level in a food web.
- Factors, such as availability of food, number of predators, human interference, and pollution, affect the number of a species in a habitat.
- Animals and plants that can adapt to changing conditions are more likely to survive.

Difficulties/Misconceptions

Competition is a difficult concept. Some pupils may find it hard to see that a change in one part of a food web can have a dramatic effect in anther part. For instance, the removal of a crop pest by spraying with a pesticide may ultimately reduce the number of the top predator.

Extension ideas

Further consideration of the roles of organisms in a community.

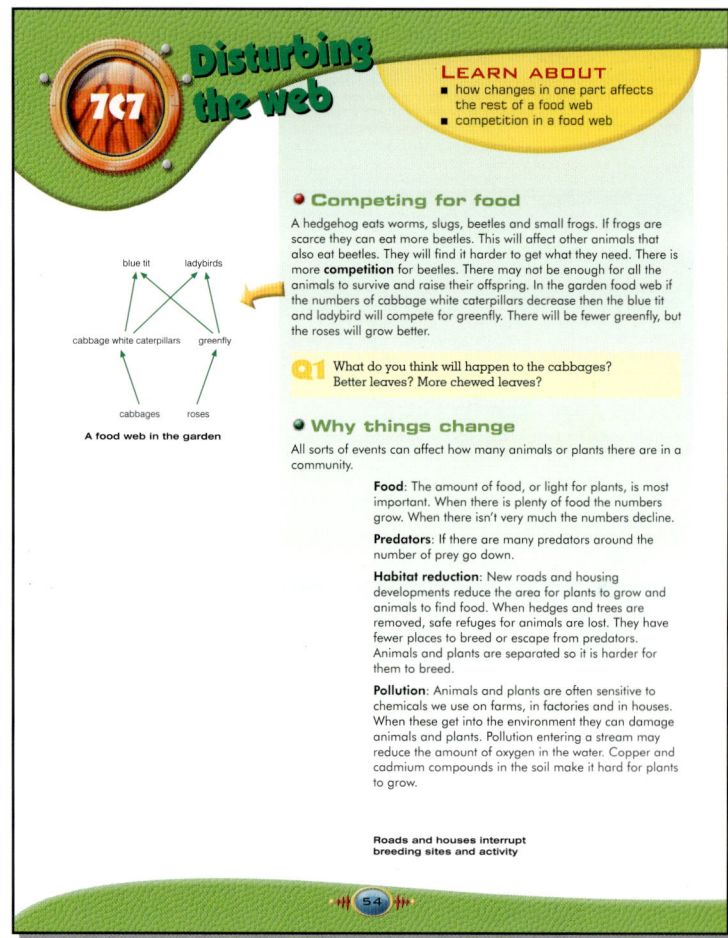

Activity and technicians' notes

Food webs

Equipment required
- Blank paper
- Wool or string
- Drawing pins

Access required to:
- Wildlife magazines
- Photocopier
- Internet

Learning styles

Visual
- **Using pictorial** food webs to illustrate communities and competition.

Auditory
- **Discussing** competition and definitions of terms.

Kinaesthetic
- **Making** a giant food web.

Interpersonal
- **Collaborating** with other pupils in discussion and activities.

Intrapersonal
- **Reflecting** on individual position in a community and responsibility for conservation.

Answers to questions

Text

Q1 better leaves

Summary

a) The crop will grow more since it isn't being eaten, but the number of carabid beetle will diminish because there are fewer flies for them to eat, and robins will diminish because of increased competition for food.

b) There will be fewer nettle and bramble plants. The number of small tortoiseshell and gatekeeper butterfly caterpillars will fall as they have no food. The number of great tits and robins will fall as they compete for a reduced food supply.

Suggestions for plenaries

- **Extinction:** Explain why some animals have become or may become extinct.
- **Changing world:** Pupils discuss the reasons why the number of magpies has increased recently, particularly near roads and populated areas (magpies feed on dead animals).
- **3Cs:** Discuss the meaning of the terms: community, competition, and conservation, as applied to habitats.

Learning outcomes

Following this, most pupils will:
- Predict the effects of altering the numbers of an organism in one part of a food web.
- Recognise that organisms living in a habitat compete with each other for food resources.
- Recognise the importance of plants as the food source at the start of all food chains.

Some pupils will also:
- Describe how different organisms contribute to the community in which they are found.

7C Read all about it!

Teaching strategy

- Read the articles about ecologists and park rangers (perhaps out loud).
- Discuss the following questions:
  1. What skills does an ecologist need?
     For example, good eye sight, the ability to move quietly, knowledge of plants and animals and their habits, the ability to recognise what they see.
  2. Why do ecologists capture animals and put tags on them?
     To find out about their diets, age, distance travelled, diseases carried, injuries sustained.
  3. Why are park rangers needed?
     To manage the countryside (if it was left to itself, parkland would become overgrown and the range of organisms would become smaller); to keep records of the plants and animals that are present; to supervise the people who use it (farmers and members of the public), to lead tours and give talks to visitors about the wildlife in the park.

Difficulties/Misconceptions

- Pupils may think that the people who work as rangers and do jobs such as mending fences and chopping down trees are not well educated and knowledgeable people.
- They may think that examining excreta is a nasty task; it should be emphasised that scientists always take precautions not to contaminate or be contaminated by specimens; e.g. by wearing gloves.

Skills

literacy, discussion

Notes

Rangers do not need to have a degree. They can gain experience of environmental work and be promoted into the position.

Activity: Treasure hunt

- Hide some mini-chocolate eggs or the parts of a message in the room or around the school grounds.
  Pupils search for the 'treasure' without disturbing the surroundings.
- **Safety:** Pupils must be accompanied if they are outside the classroom; make sure they do not go to any unsafe areas.
- **Requirements:** mini-eggs or parts of a message
- **Skills:** observation, coordination to move quietly and carefully

Questions

Ideas and evidence

1. Why is it often difficult to spot animals?
2. What is 'spraint'?
3. What can a hole in a nutshell tell you?
4. What leaves broken shells around a stone?
5. What does an owl pellet contain?
6. What may be found in a shark's stomach?

Scientific people

7. What type of habitat is Sutton Park?
8. What do the letters SSSI stand for?
9. State three activities that take place in the park.
10. What would happen to the heath if the rangers didn't cut back trees?

Answers

1. they are shy
2. the fishy droppings of otters
3. the teeth marks reveal which animal ate it
4. thrushes
5. bones and indigestible bits of animals it has eaten
6. remains of food, plastic containers, tin cans, lifebuoys
7. lowland heath with lakes
8. site of Special Scientific Interest
9. grazing, horse riding, dog walking, cycling, orienteering
10. woods would take over the heath

Danger – common errors

- Animals do not choose where to live because the conditions suit them, but they have adapted to their habitats.
- Trees do not die when they shed their leaves, they are becoming dormant.
- The direction of arrows in a food chain or web should show the direction of energy flow from producer to consumers.
- Organisms on the same level in a food web do not consume each other.

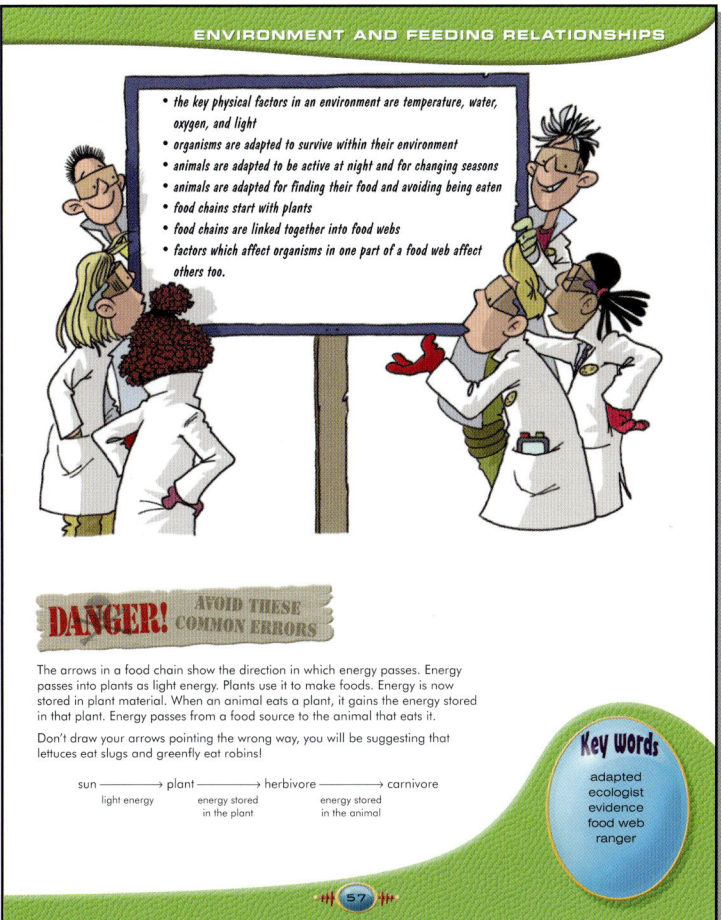

Learning outcomes

Following this unit, most pupils will:
- Identify differences between different habitats and relate these to the organisms found in them.
- Describe ways in which organisms are adapted to daily or seasonal changes in their environment and to their mode of feeding.
- Describe food chains within an environment and combine these into food webs.

Some pupils will also:
- Explain why a variety of habitats is needed in a community.
- Describe how different organisms contribute to the community in which they are found and relate food chains to energy transfer.

In scientific enquiry:

Most pupils will:
- Make a series of measurements of environmental variables appropriate to the task.
- Identify a question to investigate about the activity of an invertebrate, suggesting a suitable approach and sample size.
- Use results to relate animal and plant activity to environmental changes.

Some pupils will also:
- Describe, in terms of approach and sample size, how strongly any patterns or associations identified are supported by the evidence.

7C Unit review

Answers to review questions

1. food web, plants, producers, herbivores, kill, senses, hearing, streamlined, nocturnal
2.
   a. midge larvae and sludge worms
   b. stonefly nymphs, mayfly nymphs, minnow
   c. sludge worms
   d. oxygen
3.
   a. better insulation, reduces loss of heat, to keep warm
   b. to deter herbivores from damaging and eating the plant
   c. The fly mimics the wasp to make it seem that it too can sting/is poisonous and deters predators.
4. temperature (hot/cold)
   moisture (wet/dry)
   wind strength and direction
   The following are not strictly 'physical factors':
   pH of soil
   nutrients (minerals in soil)
   air (oxygen, carbon dioxide content)
   competition from other plants and consumers
5. Answers may include:
   construction of roads, housing estates, factories; rubbish tips; draining land; changing farming practices e.g. larger fields, new crops; introduction of new species of plant and animal that compete with indigenous species; pollution generally

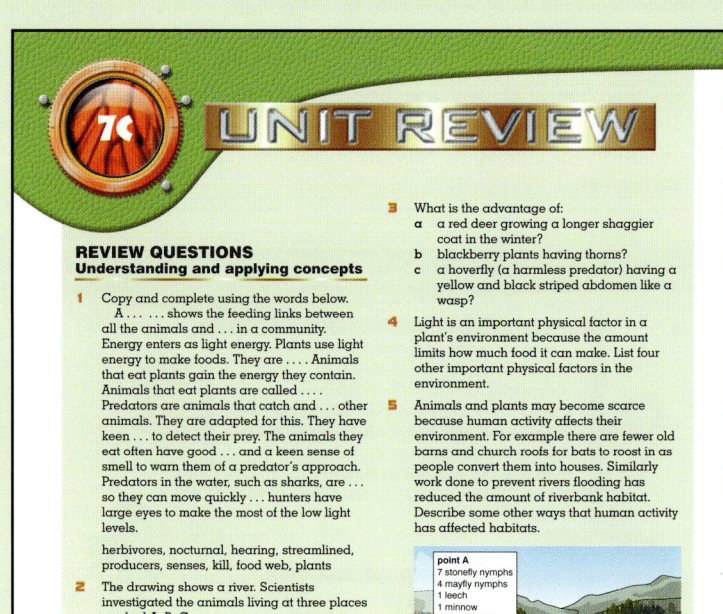

UNIT REVIEW QUESTIONS

SAT-STYLE QUESTIONS

1. The diagram shows a food web that you might find on a farm.
   a. Where does the energy for this food web come from? (1)
   b. One year the population of wood mice almost died out because a small piece of woodland was removed and ploughed. Why would the number of barn owls go down? (1)
   c. What other effects might there be? (2)

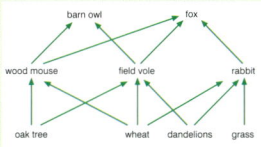

2. Sea turtles are reptiles that spend their lives at sea, eating seaweed and jellyfish. The females come ashore each year to dig a deep hole with their flippers. They lay their eggs in this hole. What features can you see in the diagram that adapt a turtle for this life? (5)

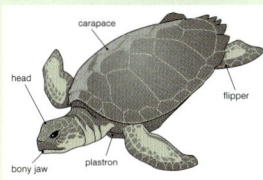

3. Reese and Pip were investigating the animals living in the area behind the Year 7 classrooms. They had found woodlice in a pile of bricks by the fence but there weren't any under a bush nearby. Reese thought that it was not dark enough for woodlice under the branches. They decide to investigate why there were woodlice in one place but not the other. They set up a choice chamber for woodlice as shown below.

They placed 10 woodlice in the centre of the chamber and left them for 5 minutes. After this time they counted how many woodlice were in each part of the chamber.
   a. Write down the question the pupils were investigating. (1)
   b. What would Reese expect to see? This is her prediction. (1)

They found that 7 woodlice were in the dark part and 3 were in the light part.
   c. Draw a bar chart of the data Reese and Pip collected. (3)
   d. Do the data support Reese's prediction? (1)

Pip was not certain that they had found out why there were no woodlice under the bush. It was damp under the bricks but very dry under the bush. She thought woodlice need dampness to live because they are crustacea that breathe with gills.
   e. How could Pip change the apparatus to test her idea? (1)
   f. Write down a new prediction for Pip's investigation. (1)

Key words
Unscramble these:
nummo city
more scun
mint no never
bait hat
to draper

Answers to SAT-style questions

1. a. the Sun (1)
   b. Mice are an important source of food for owls. The owls would be short of food. (1)
   c. Number of field voles may drop, as owls and foxes switch to eating them. Number of rabbits may increase, if the ploughed land was used to grow wheat or grass and more dandelions were allowed to grow. (2)

2. Eyes for finding plants and jellyfish;
   beak suited for eating plants and jellyfish;
   flippers for powerful swimming strokes;
   hard carapace and plastron for protection from predators;
   powerful (front) flippers for digging in sand. (5)

3. a) Do woodlice prefer dark or light? (1)
   b) More woodlice will be found under the dark side of the chamber than in the light. (1)
   c) (Give one mark for suitable scale for number of woodlice and two marks for bars drawn correctly.)

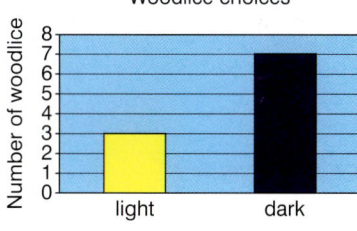

(3)
   d) Yes (1)
   e) Put a piece of damp filter paper on one side of the choice chamber (remove the cover). (1)
   f) More woodlice will be found on the damp side of the chamber than on the dry side. (1)

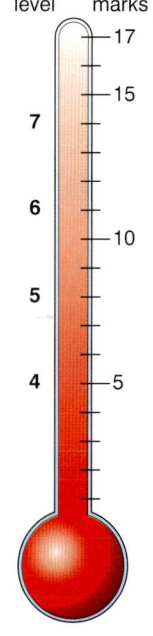

levelometer

7D Variation and classification

National framework/QCA SoW references

National framework – Interdependence:
Explain that organisms can be grouped by their similarities and differences, and that a species is a group of very similar organisms.
Identify some of the main taxonomic groups of animals, describing some common features.

QCA SoW (7D)
How do individuals of the same species differ from each other? (7D4)
What are the causes of variation? (7D4, 7D5, 7D6)
How can we describe living things? (7D1)
How can we sort things into groups? (7D1, 7D2, 7D3)
How do scientists classify living things? (7D1, 7D2, 7D3)
NC links: Sc2 4a, b, c; Sc1 2 a, b, g, i, k, m.

Learning objectives

- To suggest ways that individuals of a species may vary.
- To know some characteristics and to identify inherited characteristics within a family.
- To know that environmental factors can also cause variation.
- To use features to sort organisms into groups.
- To know that scientists classify organisms into groups including plants and animals.
- To know that the animal kingdom is divided into vertebrates and invertebrates and that these groups are sub-divided further.
- To know that each species is given a scientific name that shows which groupings it belongs to.

In scientific enquiry:

- To frame questions to be answered using first-hand or secondary data.
- To make qualitative observations and record these in a variety of ways.
- To draw conclusions from observations and explain these using scientific knowledge.
- To investigate variation between individuals of the same species using an appropriate sample size.

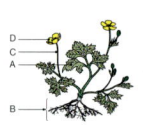

Answers to questions

What do you remember?

1. A leaf
   B root
   C stem
   D flower
2. a stag beetle

Notes to support 'What do you remember?'

In earlier Year 7 units, pupils will have learnt about the features and adaptations of plants and animals in particular habitats and that, in reproduction, characteristics are passed from parents to offspring. In KS2, pupils leant about keys and started to place animals and plants into groups according to their features. Before beginning this unit, it would be useful if pupils can:

- Use and design keys to identify plants and animals.
- Identify the features that help a plant or animal adapt to its habitat.
- Name the parts of a plant.

Teaching strategy

This unit builds on work done in the previous unit 7C 'Environments', and also on work done at KS2 in unit 6A 'Interdependence and adaptation' and 5B 'Lifecycles'. It leads into units 8D 'Ecological relationships' and 9A 'Inheritance and selection'.

Difficulties/Misconceptions

- Pupils may be confused by the distinction between a 'species' and a 'variety', e.g. all dogs are the same species despite their many and varied varieties.
- Pupils may also be unsure about which features can be inherited and which are influenced by the environment, particularly when both may be contributing (e.g. tallness in humans may be inherited and may also be affected by diet and exercise).
- Some pupils may have the Lamarckian idea that acquired characteristics can be passed to offspring, e.g. someone is a smoker because their parents smoke.

Notes and tips

- The order of topics in this unit is different to that in the QCA scheme, and begins with classification before moving on to variation.
- At this stage, pupils are not expected to develop a full knowledge of the classification scheme and Linnaean binomial system; but they should be aware of them.

Learning styles

Visual
- **Using cartoons, pictures and keys** to stimulate recall.

Auditory
- **Discussing** questions.

Interpersonal
- **Collaborating** on activities.

Intrapersonal
- **Recalling** earlier learning.

Launch activity notes – Differences and similarities

This activity introduces the two strands of the unit: using keys and classification systems; looking at variation between individuals. The cartoons and questions can be used to stimulate discussion of these issues.

Answers

a) Discuss the questions: Can dogs with very different appearances breed? What will the offspring look like? Pupils may have their own experiences of mongrels; they may also have knowledge of breeding in other species, such as cats, rabbits, and cows. Note that all varieties of dog are the same species and can theoretically breed. The puppies will display some characteristics of both parents, although one variety may be dominant.

b) 'Bulldozer' is not the right answer, although pupils should see that it fits the clue. Pupils may practise their own versions of 'I spy', describing something that they can see and allowing their partner to guess what it is. Alternatively, groups of pupils can play '20 questions', asking questions with yes/no answers to identify the plant or animal that one of the group has in mind.

c) Pupils may remember that, at fertilisation, factors from both father and mother are joined to form the embryo.

d) Pupils can list the features that they would record in order to find out what the creepy crawly is, such as number of legs, shape of body, colour and pattern. Pupils may be aware of the differences between spiders and insects. Discuss how pupils can find out what an organism is, by using books or websites on classification.

7D1 Animal and plant groups

National framework/QCA SoW references

National framework – Interdependence: Explain that organisms can be grouped by their similarities and differences, and that a species is a group of very similar organisms; identify some of the main taxonomic groups of animals.

QCA SoW

7D Section 3: How can we describe living things?
7D Section 5: How do scientists classify living things?
NC links: Sc2 4b.

Learning objectives

- To compare different styles of writing about observations.
- To spot connections and links between how information is presented in different forms.
- To know how to work safely with living organisms.
- To make and record appropriate observations relevant to a particular piece of work.
- To use observations to make comparisons of living things.
- To sort organisms into groups according to common features.
- To know that there are different ways of classifying living things.
- To know that newly discovered organisms may fit into an already existing system of classification, or extend it.
- To know that scientific classification is important because it is a worldwide labelling system, and provides a means for systematic study of living things.
- To know that two groups of living things are green plants and animals.

Lesson starter suggestions

- **Cave paintings:** How and why can we identify the animals in cave paintings?
- **What are elephants?** Examine a nonsense rhyme to see the importance of accurate descriptions.
- **Characteristics:** What are the characteristics of humans?

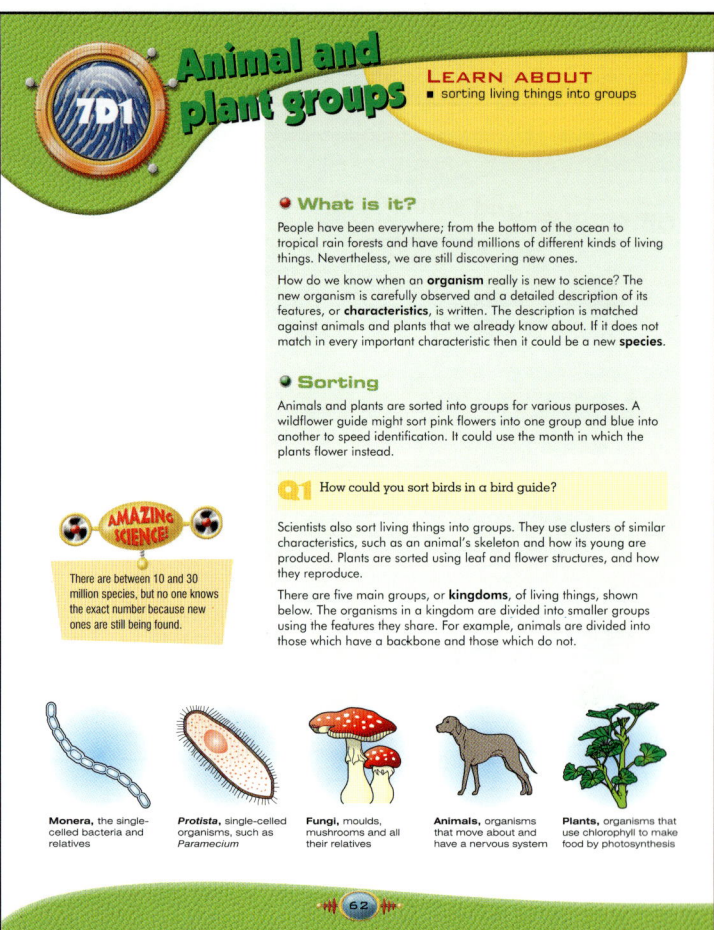

Amazing science

Some scientists believe that we have still only discovered a minority of the species on the planet. Others say that we are losing species faster than we are discovering them. The rate of loss of species at the present time rivals that of some of the great extinctions of the distant past, and could become worse as climate change speeds up.

Teaching strategy

Key points to highlight

- Organisms can be sorted into a small number of groups (five kingdoms).
- Careful descriptions, or drawings, of the characteristics of organisms help them to be sorted.
- Scientists all over the world use the same system of classification of organisms based on the Linnean system.
- A species is a group of animals with the same characteristics that can interbreed.

Difficulties/Misconceptions

- The difference between a 'variety' and a 'species' can give problems, particularly as the definition of a species is open to argument.
- Pupils are not expected to recall the differences between the *Monera* and *Protista*.

Skills

Descriptive writing and diagrammatic drawing.

Activity and technicians' notes

Sorting and describing

Safety notes
- Make sure that animals used in this lesson are not poisonous.
- If pupils are using hamsters, gerbils etc, warn that they may bite if annoyed.
- Pupils should not handle animals more than is necessary.
- You must decide whether or not to allow pupils to bring pets into the lesson.
- Some pupils may have allergies to animal fur, etc.
- Wash hands before and after handling animals.

Equipment required
- Trays to put animals in
- Magnifying lenses or binocular microscopes if very small creatures are being used
- Suitable food for animals
- Animals, e.g. snails, slugs, earthworms, woodlice, spiders, centipedes, hamsters, gerbils, cage birds, fish (in aquaria)

Access required to:
- CD ROMs of pictures of animals
- Field guides to animals

Teaching strategy contd.

Notes and tips
This lesson introduces many ideas that will be revisited in later lessons.

Extension ideas
Writing about animals, i.e. comparing different ways of describing living things.

Learning styles

Visual
- **Using pictures and specimens** to understand the classification groups.

Auditory
- **Listening** to the words used to name the groups of animals.

Kinaesthetic
- **Examining specimens** or moving diagrams around screen.

Interpersonal
- **Discussing** observations and keys.

Intrapersonal
- **Reviewing** the work done.
- **Considering** own place in the classification system.

Answers to questions

Text
Q1 Various possibilities, e.g. habitat, shape of beak/wings/legs, method of feeding (wading, diving, etc.), colour

Summary
1. They are members of the related species of animals, the felines.
2. Any valid answers. Note the monera include all bacteria (e.g. *Salmonella*, *E. coli*), while *Protista* include algae and other single-celled organisms, such as *Paramecium* and *Amoeba*.
4. A species is a group of organisms sharing distinctive features and able to breed successfully with each other.
5. A poodle and a spaniel are varieties of dog, a single species, while chicken and goose are different species.

VARIATION AND CLASSIFICATION

● Classification
Classification is the science of sorting organisms into groups. Carolus Linnaeus (1707–1778) started the classification system. It has been modified as we have learnt more about living things. The Linnean system uses physical features to group organisms together. We now know that organisms in a group are similar because they have evolved from shared ancestors millions of years ago. New organisms are fitted into the system. If they are significantly different to other organisms they may become a new group in the system.

● What is a species?
Members of the same **species** can freely breed together and produce young. They cannot breed with members of another species even if they look very similar. Each species has a unique scientific name used worldwide. Scientific names avoid confusion. Scientists talking about *Calliphora vomitoria* know exactly which fly they are talking about.

Sorting and describing
- Use pictures of ten organisms to find three different ways of sorting them into groups. Each time you sort them, write down the animals in each group.
  a) What was the reason for sorting each group?
  b) Were any of your animals difficult to place?
  c) Why was that?
- Choose one of your sorts to produce a simple identification guide.
- Carefully examine a specimen provided by your teacher. Use a hand lens or binocular microscope to see small features. Write a description of your specimen. Try to make your description objective (i.e. write down what you see).
- Make a diagrammatic drawing of your specimen and label it. (A diagrammatic drawing is like the one on p.8, where you are not trying to make an artistic exact copy but are picking out key features.)
- Compare your specimen with a different one. Make a table of any important differences between the two specimens.
  d) Do you think they belong to the same species?

SUMMARY QUESTIONS
1. ☆ Why do domestic cats and lions have similar features?
2. ☆ Give an example of one member of each of the five kingdoms.
3. ☆ Find out the scientific names of three animals or plants.
4. ☆ Write a definition of 'species'.
5. ☆☆ Why can a poodle and a spaniel breed together, but a chicken cannot breed with a goose?

Key words
characteristic
classification
kingdom
organism
species

Suggestions for plenaries

- **All one species:** Discuss the common features of humans.
- **Comparing observations:** Draw together observations and identification systems from the activities.
- **Aliens!** Pupils to make up some alien creatures, deciding what group they should belong to and thinking up a two-word name for them.

Learning outcomes

Following this, most pupils will:
▸ Explain the importance of classifying living things.

Some pupils will also:
▸ Name some organisms that are not readily classified as plant or animal.

7D2 Sorting animals

National framework/QCA SoW references

National framework – Interdependence: Identify some of the main taxonomic groups of animals describing some common features.

QCA SoW
7D Section 5: How do scientists classify living things?
NC links: Sc2 4b.

Lesson starter suggestions

- **Important differences:** Which are the most important features that separate one group of animals from another? Pupils to look at pairs of pictures of: a lizard and an otter; a housefly and a spider; a bird on a nest and a bat.
- **Alphabet animals:** Pupils to compete to see how quickly they can produce a (nearly) complete alphabet of animals (leave out X).
- **Odd one out:** Pupils to pick out and explain the odd one from the following trios: cow/pig/salmon; adder/blackbird/beetle; butterfly/ladybird/spider.

Learning objectives

▸ To know that animals can be subdivided into vertebrates and invertebrates.
▸ To know that vertebrates include mammals, birds, fish, reptiles and amphibians.
▸ To know that invertebrates can be further subdivided.
▸ To be able to search a CD-ROM database for information.
▸ To be able to transfer written information from one form to another.

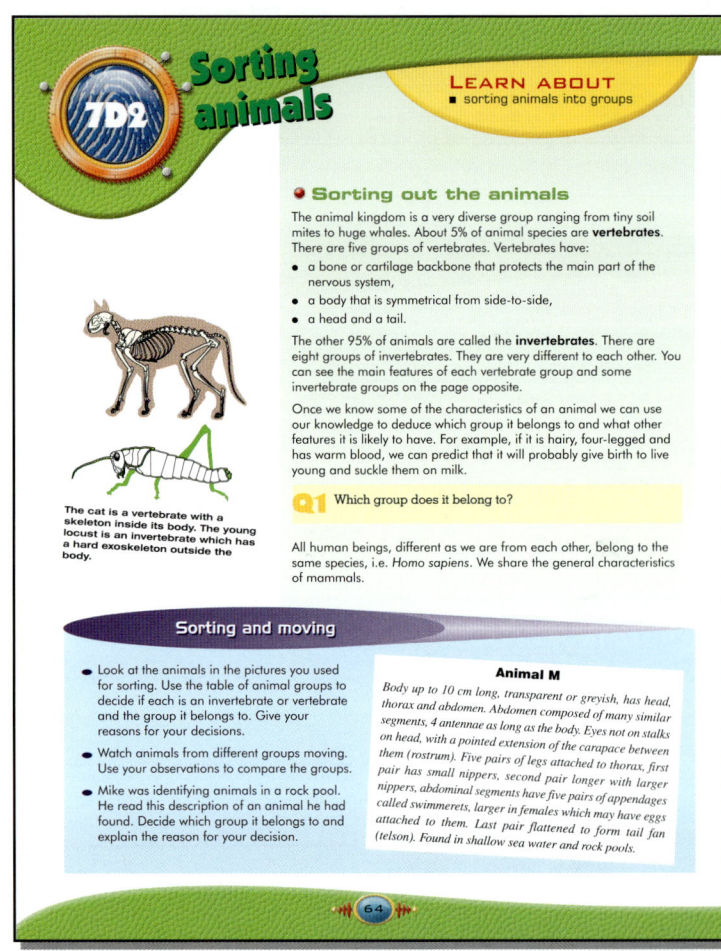

Teaching strategy

Key points to highlight

- Animals are divided into two main groups: vertebrates (with a backbone) and invertebrates (without a backbone).
- Each group is further divided into sub-groups, depending on common features.
- Vertebrates are: fish, amphibians, reptiles, birds, and mammals.
- Invertebrates include arthropods, which are in turn made up of sub-groups such as insects and crustacea.
- Use pictures and descriptions to decide which group and sub-group animals belong to.

Difficulties/Misconceptions

Pupils can be overwhelmed with information and may be confused by the characteristics used to classify different animals (summary question 5 illustrates that each group has some odd members that don't fulfil all the criteria).

Skills

Literacy – reading and writing descriptions
Observation – of features
Using keys

Activity and technicians' notes

Sorting and moving

Access required to:

CD ROMs or websites with information on groups of animals (see ICT challenge).

Teaching strategy contd.

Notes and tips
- Pupils do not need to know the terms phylum, class, order, genus. For non-biologists, there are 23 phyla of animals of which the vertebrates are just one. The five sub-groups of vertebrates are classes. The arthropods are themselves a phylum and divided up into classes, such as insects, crustacea, spiders and myriapods. Each class may be divided into orders. The binomial naming system gives the genus followed by the species. Thus humans (*Homo sapiens*) are in the phylum of vertebrates, the class of mammals, the order of primates and are the sole surviving species of the genus *Homo*.
- Pupils may be confused by the features of some animals that seem to suggest that they should be in different groups, e.g. snakes have backbones but worms do not, bats have wings but are mammals.

Extension ideas
Some pupils may do further research on the invertebrate groups.

Learning styles

Visual
- **Looking at** pictures of animals to recognise common features.
- **Looking at** keys.

Auditory
- **Discussing** in groups.
- **Presenting** results of research, etc.

Kinaesthetic
- **Moving around** in groups to use computers.

Interpersonal
- **Collaborating** in groups.
- **Sharing** information with other groups.

Intrapersonal
- **Reviewing** what has been learnt, e.g. by mind mapping.

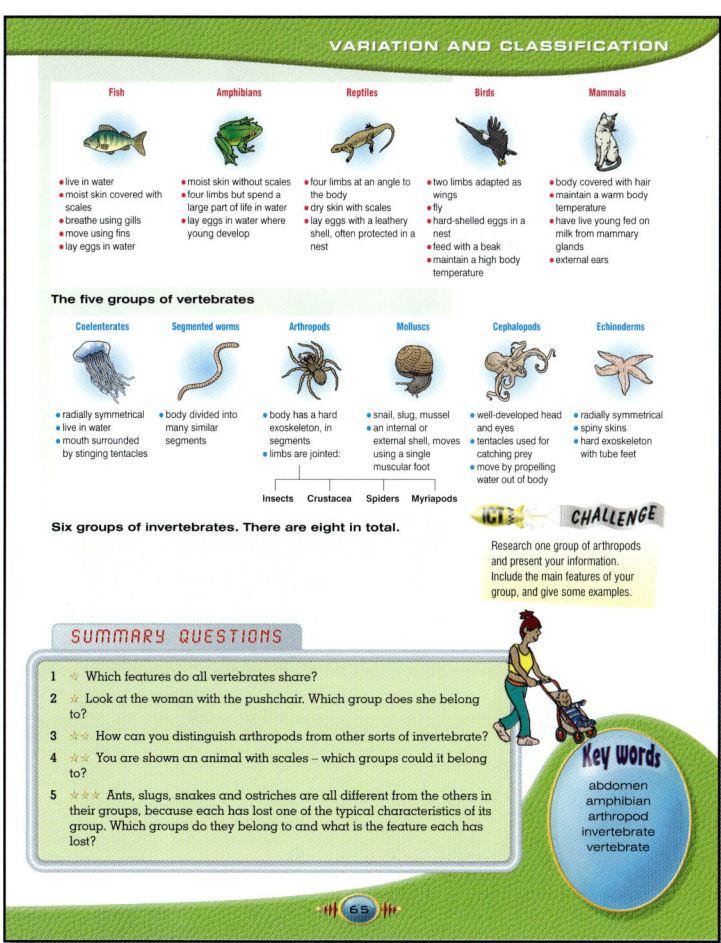

Answers to questions

Text
Q1 mammals

Summary
1. have backbone, lateral symmetry, head (and tail)
2. vertebrates – mammals
3. hard exoskeleton, pairs of jointed legs
4. fish or reptiles
5. ants: invertebrates – arthropods – insects: no wings
   slugs: invertebrates – molluscs: no shell
   snakes: vertebrates – reptiles: no legs
   ostrich: vertebrates – birds: do not fly

Activities
Animal M is an invertebrate (no evidence of backbone or fur) – an arthropod (segmented body with jointed limbs) – a crustacean (limbs on each segment, on the head these are the antennae, five pairs on thorax).

Suggestions for plenaries
- **Animal groups:** A memory exercise linking description to names of groups.
- **Mind mapping:** Pupils to draw up a mind map for the features of the different groups of animals.
- **Presentations:** Presentation of results of research (see ICT challenge).

Learning outcomes

Following this, most pupils will:
▶ Identify some of the main taxonomic groups of animals and describe some features of these.

Some pupils will also:
▶ Be able to assign animals to the groups.

7D3 Sorting plants

National framework/QCA SoW references

National framework – Interdependence: Explain that organisms can be grouped by their similarities and differences.

QCA SoW

7D Section 5: How do scientists classify living things?
NC links: Sc2 4b.

Learning objectives

▶ To know that scientific classification is important because it is a worldwide labelling system, and provides a means for systematic study of living things.

Note that classification of plants is extended in unit 8D 'Ecological relationships'.

Teaching strategy

Key points to highlight

- All plants use sunlight to produce their own food.
- There are five groups of plants.
- The most common group are the flowering plants, which reproduce using flowers (see unit 7A) and form seeds in a fruit.
- Many trees are conifers, which are not flowering plants but produce seeds in cones.

Difficulties/Misconceptions

It is easy to think of all plants as being the same group, because they are all various shades of green. The features that distinguish some of the groups are a little subtle. For instance, the fruits of flowering plants come in many forms from pineapples to dandelion 'clocks'. Most conifer trees are evergreen, but not all trees that are flowering plants lose their leaves in winter.

Skills

Observing and using keys.

Notes and tips

- Make sure that the specimens that pupils examine can be identified by the key provided.
- Lichens can often be seen on rocks or old walls. They look like stains. They are very slow growing so should not be disturbed.

Extension ideas

A closer look at flowering plants and the main divisions into monocotyledons and dicotyledons.

Lesson starter suggestions

- **A woodland scene:** How many different types of plant are there?
- **Recap:** How do plants differ from animals?
- **Useful plants:** What do we use plants for?

Activity and technicians' notes

Sorting and using
Safety notes
Make sure that no poisonous plants are included amongst the specimens.

Equipment required
- Specimens of: grasses, flowering plants (with leaves and flowers or developing seeds), leaves and cones from a conifer, fern fronds, moss
- Magnifiers

Access required to:
Internet for research

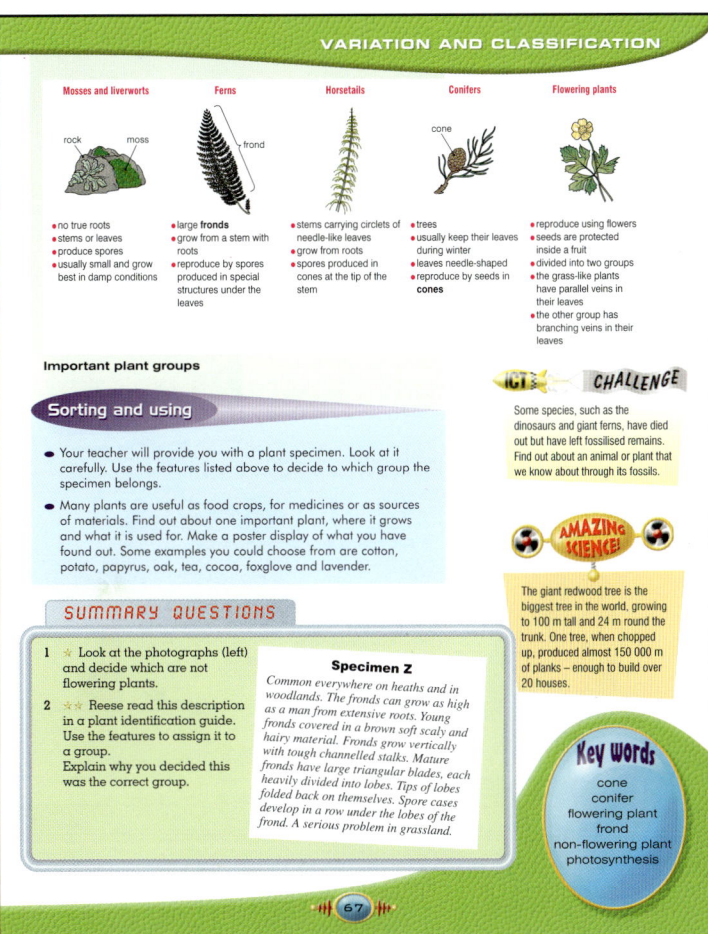

Learning outcomes

Following this, most pupils will:
▶ Identify some of the main taxonomic groups.
Some pupils will also:
▶ Be able to identify which group certain plants belong to.

Learning styles

Visual
- **Examining** pictures and specimens of plants.

Auditory
- **Describing and listening to** descriptions of plants.

Kinaesthetic
- **Examining** specimens.

Interpersonal
- **Sharing** observations.

Intrapersonal
- **Reviewing** what has been learnt about classification.

Answers to questions

Summary
1. Clockwise from the top left.
   a) flowering plant
   b) fern
   c) conifer
   d) flowering plant
   e) flowering plant
   f) flowering plant
2. Specimen Z is bracken, a fern, because it has large fronds and reproduces by forming spores under the leaves.

Amazing science

The Californian Redwood tree is in fact the tallest, one having grown to 112 m, while the related Wellingtonia is the largest, some weighing over 1000 tonnes. Both are evergreen conifers native to California. Both trees have a soft spongy bark that can be 'punched'.

Suggestions for plenaries

- **The kingdoms:** A drag and drop exercise to reinforce the classification system.
- **Sharing observations:** Pupils to describe observations of plants.
- **Summarising:** Pupils to make a summary of what they know about classification.

7D4 Variation

National framework/QCA SoW references

National framework – Interdependence: Explain that organisms can be grouped by their similarities and differences, and that a species is a group of very similar organisms.

QCA SoW

7D Section 1: How do individuals of the same species differ from each other?
7D Section 2: What are the causes of variation?
NC links: Sc2 4a.

Lesson starter suggestions

- **Looking for variations:** A non-verbal reasoning odd-one-out puzzle.
- **Describe yourself:** List your best (and worst) features.
- **How are we different?** Pupils to list the ways in which people's appearances differ.

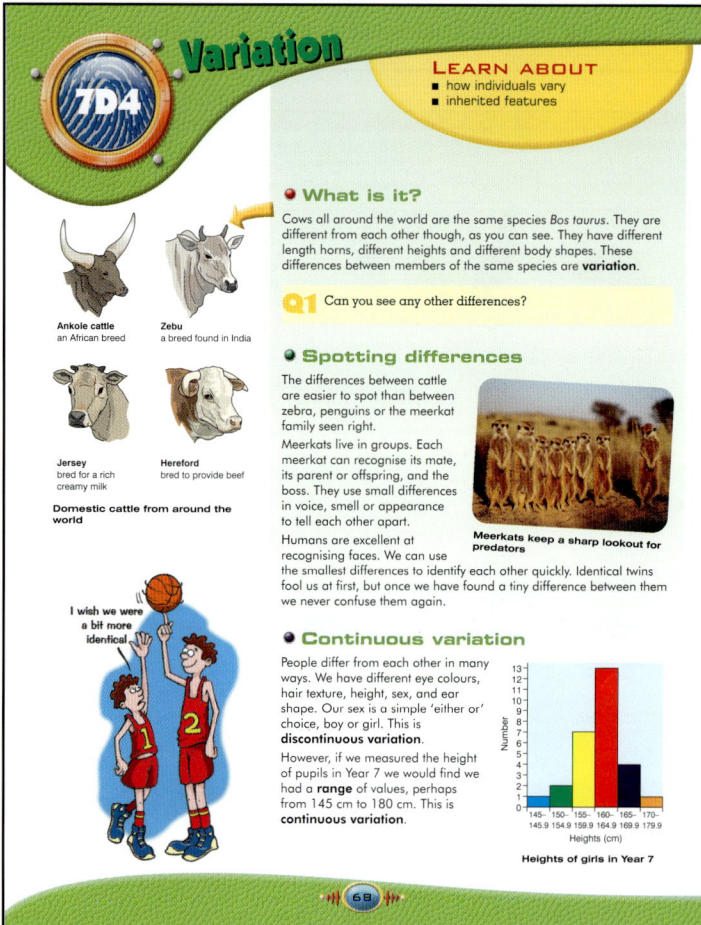

Learning objectives

- To know that individual members of a species may differ in many ways.
- To know that some variations are inherited.
- To know how environmental differences can result in variations in a species.

In scientific enquiry:

- To use a spreadsheet to store data.
- To frame questions that can be investigated.
- To interpret graphs generated by a spreadsheet.
- To decide how confident to be in the evidence.
- To choose a sufficiently large sample size.
- To interpret graphs and draw conclusions from them.
- To interpret secondary data and draw conclusions from it.

Teaching strategy

Key points to highlight

- All members of species show variation.
- Members of communities recognise individuals by their variations in appearance or behaviour.
- Some variations are continuous (e.g. height) while others are discontinuous (e.g. eye colour).
- Some variations are inherited from parents.
- Some variations are a result of environmental differences during the individual's lifetime.
- Data can be collected on variations that can answer questions about the cause of the variation.

Difficulties/Misconceptions

- Pupils may think that the variation in some creatures is sufficient to consider them as separate species, e.g. breeds of cattle or dogs. Remind pupils that members of a species can interbreed, despite their apparent variation. All humans are the same species, despite their variation in colour, size and features.
- Some human variations may be both inherited and environmental, e.g. hair colour is an inherited factor, but people may dye their hair.

Skills

Collecting, analysing and evaluating data.

Notes and tips

- This lesson is concerned with handling data and looking for correlations. Look at simple variation ('Dogs') before considering correlations between two factors ('Holly leaves') or cause and effect ('Plants' in different places in the school grounds).
- Inherited factors may be discontinuous, e.g. blood group, eye colour, or continuous, e.g. height.
- The class itself can form an adequate sample base for investigating the correlation of height, weight, shoe size/length of feet, width of hands, etc. However, some pupils may be sensitive to having their measurements taken. It will be difficult to investigate causal factors amongst a sample of pupils.

Extension ideas

Examining causal links.

Activity and technicians' notes

Investigating variation

Safety notes

Here are some hazards involved in fieldwork:
- Broken glass/pieces of metal on ground
- Wasp/bee stings (note pupils who are allergic to stings)
- Plants that have thorns/stings/or are poisonous (yew, foxgloves, etc.)
- Uneven ground
- Unstable trees/walls.

Equipment required
- Rulers
- Tape measures
- Data logging apparatus (light sensors, thermometers)

Access required to:
- Bathroom scales
- Balances

Learning styles

Visual
- **Looking at** pictures showing variation.
- **Creating** bar charts.

Auditory
- **Discussing** reasons for variation.
- **Planning** investigations.

Kinaesthetic
- **Collecting** samples.
- **Taking** measurements.

Interpersonal
- **Collaborating** on investigations.
- **Presenting** results.

Intrapersonal
- **Reviewing** progress.

Answers to questions

Text

Q1 colour and pattern of colours, length of fur

Summary

1. a) facial expression; clothes
   b) hair colour, nose, mouth, eye-shape
   c) scars, suntan, overweight, underweight, length of hair

2. a)

Features	Cattle breed			
	Ankole	Zebu	Jersey	Hereford
Horn length	long	short	short	long
Horn direction	up	up	up	down
Colour of coat	brown	grey	sandy	brown + white
Patterned or plain	plain	plain	plain	patterned
Fur length	short	short	short	long

   b) all of them
   c) There is enough.

Suggestions for plenaries

- **Same but different:** Pupils to examine a picture of a crowd and point out the variations that allow them to recognise the different people.
- **Causes:** Pupils to list the causes of variation in plants and animals.
- **Reports:** Pupils present their findings of correlations between factors.

Learning outcomes

Following this, most pupils will:
▶ Identify similarities and differences in organisms of the same species and begin to attribute these to environmental or inherited factors.

Some pupils will also:
▶ Recognise that inherited and environmental causes of variation cannot be completely separated.

7D5 Passing on the information

National framework/QCA SoW references

National framework – Interdependence: Explain that organisms can be grouped by their similarities and differences.

QCA SoW
7D Section 2: What are the causes of variation?
NC links: Sc2 4a.

Learning objectives

- To know that some characteristics are inherited.
- To know that, although individuals are like their parents, they are not identical to them.
- To know that offspring from the same parent show considerable variation.

Teaching strategy

Key points to highlight

- The characteristics of an organism are determined by the genes carried on the DNA in the nuclei of every cell.
- The sex cells (sperm and egg) contain half of the parents' genes.
- Offspring inherit genes for every characteristic from their parents; some of the genes that are expressed will have been inherited from the mother and some from the father. (Note that sometimes expression of genes can jump generations, i.e. a gene that the mother inherited from her father may lie dormant but may be expressed in her child.)
- Different offspring of the same parents will inherit different combinations of genes, so they'll have some common features and some that are different.
- Some behaviour patterns, as well as physical characteristics, are inherited.

Difficulties/Misconceptions

- Pupils may think that boys inherit features from fathers, and girls inherit features from mothers. This can be shown to be false by looking at family photographs, or photographs of famous father/daughters, mother/sons (e.g. Princess Diana/Prince William).

Lesson starter suggestions

- **Fertilisation:** What do you remember?
- **Happy families:** Pupils to look at the families in a pack of happy families' cards, what characteristics are common to the members of each family?
- **What did I inherit?** Pupils to find common features of themselves and one parent or a sibling.

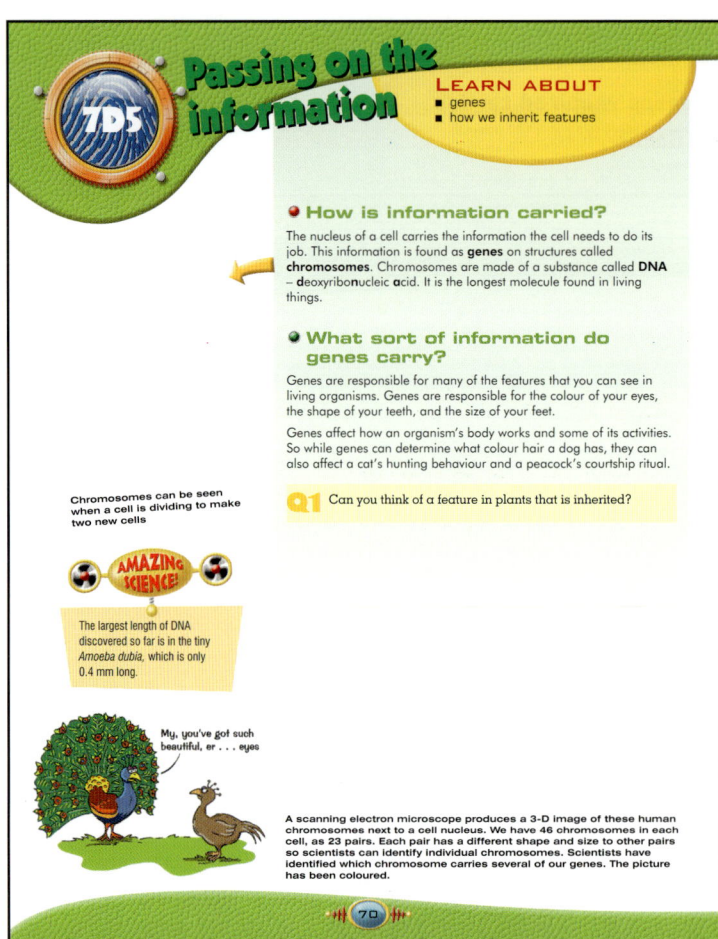

Amazing science

Johann Miescher discovered DNA in pus-soaked bandages in the 1880s. Then he obtained it from fish sperm, particularly from salmon. Later DNA was found in almost every cell, but it wasn't until the 1940s that its role in inheritance was proved.

Teaching strategy contd.

Skills
Observation – noting similarities and differences.

Notes and tips
- Pupils will learn more about genes and inheritance in unit 9A 'Inheritance and selection'.
- It may be inappropriate with some classes to ask pupils to consider their own inheritance, particularly if they are adopted, step-children, in families where there are multiple fathers, or in immigrant families that have lost touch with older members.
- The Royal Family provide a good resource that is in the public domain.
- Magazines such as *Hello!* often have photographs of celebrities with their children, which can be used to look for common features.

Extension ideas
'Behaving like animals': a comparison of inherited and learned behaviour in animals.

Learning styles

Visual
- **Looking at** photos, and at each other, to compare characteristics.

Auditory
- **Discussing** answers to questions.

Kinaesthetic
- **Moving** into groups.

Interpersonal
- **Discussing** each other's features without causing offence.

Intrapersonal
- **Thinking** about features shared with parents and siblings.

Answers to questions

Text
Q1 leaf shape/colour, flower shape and colour, fruit and seed shape, tallness, etc.

Summary
1. sex, build, hair colour/type, eye colour, nose and mouth shape, size and shape of hands/feet, blood type etc.
2. They will have inherited different versions of the gene for eye colour from their parents.
3. Identical twins have the same set of genes, since they started as a single fertilised egg. (Note that some of the differences between identical twins will be environmental, and some due to different genes being activated.)

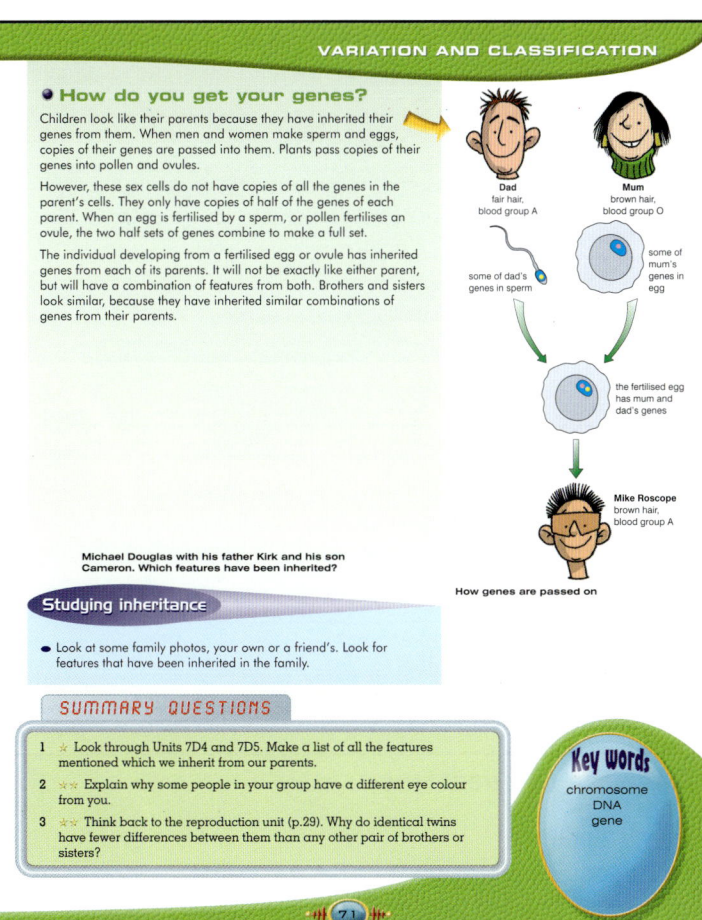

Activity and technicians' notes

Studying inheritance
Access required to:
- Family photos

Suggestions for plenaries

- **Possible children:** Give the pupils characteristics of grandparents and parents, and ask 'What might the children look like?'.
- **Why two parents?** Discuss the advantages and disadvantages of inheriting characteristics from two parents.
- **Unique:** Ask: 'What characteristics do we have that make us unique?'

Learning outcomes

Following this, most pupils will:
▸ Identify similarities and differences in organisms of the same species and begin to attribute these to environmental or inherited factors.

Some pupils will also:
▸ Recognise that inherited and environmental causes of variation cannot be completely separated.

7D6 Making use of genes

National framework/QCA SoW references

National framework – Interdependence: Explain that a species is a group of very similar organisms.

QCA SoW

7D Section 2: What are the causes of variation?
NC links: Sc2 4c.

Learning objectives

▸ To know that some variations are inherited.

Teaching strategy

Key points to highlight

- Variation has occurred in domestic species of animals and plants, because farmers have selected which offspring to keep and breed from and which to discard.
- There are many different varieties of domesticated plants and animals, because in different parts of the world different qualities have been selected for.
- The different varieties of domesticated species retain common features, although some features are wildly different.
- The environment helps to determine which varieties a farmer will select.

Difficulties/Misconceptions

- Pupils may think that all the offspring of a selected pairing will be the same and have the desired features. Only after a number of generations will the offspring 'breed true', and variation will still occur occasionally ('throwbacks').
- Pupils may have difficulty in accepting that large numbers of different varieties of domesticated animals and plants originated from a few wild species. Remind them that the process of selection has been going on for at least 10 000 years (or at least 5000 generations of animals and plants).

Skills

Literacy

Notes and tips

Selective breeding will be covered in more detail in unit 9A 'Inheritance and selection'.

Extension ideas

A closer look at how the environment may influence the features that farmers select for.

Lesson starter suggestions

- **Natural or not?** Is a field of cows natural or man-made?
- **Odd one out:** Ask pupils to find the odd ones out in the following trios: cod/salmon/trout; horse/dog/bear.
- **Good, bad, want to know more:** Pupils to discuss the statement 'most farm animals are bigger than their wild relatives'.

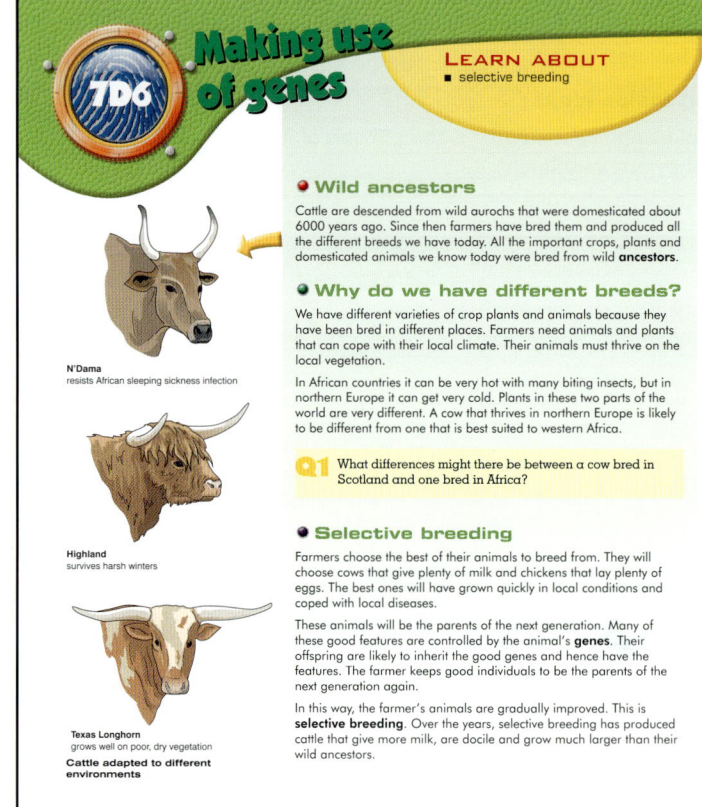

Answers to questions

Text

Q1 The N'Dama cattle of West Africa tolerate the worst effects of sleeping sickness carried by tsetse flies (where other breeds are unable to survive) and are adapted to the tropical climate; whereas Highland cattle have compact shapes with long hair, to cope with cold and wet weather.

Answers to questions contd.

Summary
1. The features are determined by the form of the genes present in the nuclei of every cell. The sex cells (sperm or egg) carry the genes to meet the sex cell of the other parent. (The offspring is built from genes of both parents. However, only half of each parent's genes are passed on to each offspring. The offspring has a pair of each gene, which may be different. So the desired features may not appear in each offspring.)
2. Mature earlier: to get them to market earlier to make money
Last longer: so their sale can be spread over a longer period, to reduce waste, so fruit can travel further to shops
Resistant to pests: less fruit lost to pests, less money spent on pesticides
3. Jersey - small size, produce creamy milk
Hereford – (an old breed), medium size, good beef production
Limousin – medium size, small boned but well muscled, low fat meat

Activity
Breeding programme: Decide on the features required. Find male and female plants that come closest to the required features. Grow these plants. When they flower, take pollen from the male plant and put it on female flowers. Collect seeds and grow the plants. Choose the plants that are closest to the requirements and repeat the procedure.

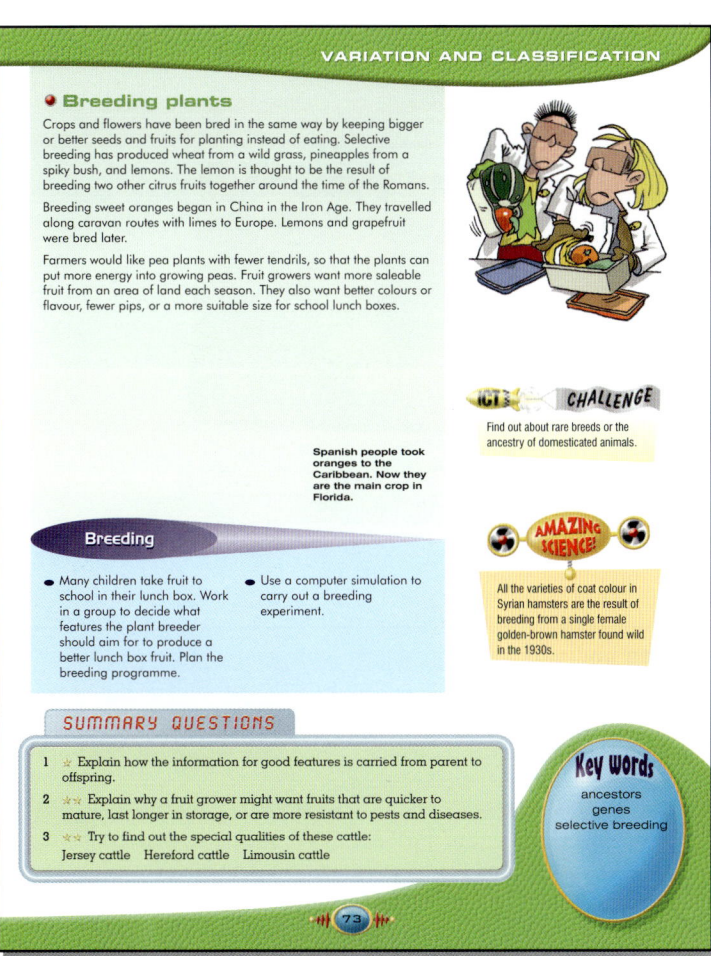

Learning styles

Visual
- **Comparing** pictures of different breeds.

Auditory
- **Debating** issues.

Kinaesthetic
- **Moving** from one group to another.
- **Standing up** to put forward an argument.

Interpersonal
- **Putting forward** an argument in a debate.

Intrapersonal
- **Reviewing** what has been learnt about variation.

Amazing science

The golden-brown or Syrian hamster is one of a number of species of hamster. The Common hamster, found in eastern Europe and west Asia, is a larger, black and brown solitary rodent. It digs extensive burrows and is occasionally treated as a pest, because of the damage it does to crops.

Learning outcomes

Following this, most pupils will:
▶ Identify similarities and differences in organisms of the same species and begin to attribute these to environmental or inherited factors.

Some pupils will also:
▶ Recognise that inherited and environmental causes of variation cannot be completely separated.

Suggestions for plenaries

- **Fit for the job:** Ask 'What have these breeds of dog been bred for? What special features do they have?'
- **Debate:** What are the advantages and disadvantages of selected breeding?
- **Mind map:** Pupils to summarise the last few lessons on variation in species.

7D Read all about it!

Teaching strategy

Conserving genes
- Read the passage (perhaps out loud).
- Discuss these questions:
  1. What are the advantages of new varieties to shops and customers?
     Bigger yields, shorter growing times, longer lasting freshness, better flavour.
  2. What are the dangers of selective breeding?
     Desirable features may be lost; domesticated breeds of animals may not be able to survive without help (e.g. unable to move easily or liable to catch diseases). New diseases could wipe out new varieties.
  3. How can gene banks help?
     Store sperm/eggs/embryos/seeds of many different varieties of plants and animals which can be grown when needed. Rare breeds farms provide similar service.
  4. Why is the conservation of wilderness areas important (e.g. the Amazonian rain forest)?
     Many species and varieties that we do not know about yet may turn out to be useful. They may disappear completely if these areas are misused.
- Individual research.
  Pupils can identify a topic to investigate using books and the Internet.
  Examples of topics are:
  What varieties of apple/tomato/tulip/rice/sheep etc. are there and how do they differ?
  What are/were the wild ancestors of domesticated plants and animals like?
  Where are gene banks and who is looking after them?
  Where are there rare breed parks? What breeds do they have?

Difficulties/Misconceptions
- Pupils may not realise that all domesticated plants (and animals) originated from wild varieties and have been much changed by selective breeding, e.g. the modern tomato barely resembles the wild variety found in Peru.
- Pupils may not realise that banks and libraries of seeds etc. are a short term answer to the problem. All stored material eventually deteriorates, although seeds have a very long shelf-life. The only true resource is the natural world.

Skills
Literacy
Discussion
Research

Notes and tips
- Pupils will be familiar, from trips to the supermarket, that there are many different varieties of fruit and vegetable, e.g. tomatoes, apples and potatoes. They may recognise that from time to time, new varieties appear.
- Pupils should recognise that variation occurs in all species and is a natural process, but that selective breeding speeds up the rate at which variations are established and may encourage variations which would not normally survive.
- Binomial classification names are not required.

Questions
1. Why have older varieties of crops such as apples become rare?
2. Why do some new varieties of crops need more fertile soil?
3. Why do we need to preserve older varieties of crop plants?
4. Why do farmers need crops that can grow on poorer soils?
5. Why are seeds kept at $-20\,°C$?

Answers
1. growers have replaced them with new varieties
2. they produce larger yields/more fruit so have a greater need for nutrients
3. they may have features/genes that we need in the future
4. so that they can grow more food for a growing population
5. to preserve them for a long time/to stop them germinating and growing

Danger – common errors

- A organism cannot belong to more than one classification group, even if it appears to display the features of two groups. Particular features distinguish one group from another.
- Pupils may confuse species and varieties. All the varieties of a species can interbreed.
- Only features controlled by genes can be inherited. Variations caused by environmental differences are not passed on to offspring.
- The distinction between discontinuous and continuous variables: line graphs can only be drawn when two continuous variables are compared.
- Characteristics are not exclusively inherited or environmental; many have an element of both.
- Characteristics are not inherited from a single parent. Every offspring of sexual reproduction has genes from both parents, although the genes of one may be expressed more obviously.

Learning outcomes

Following this unit, most pupils will:
▸ Identify similarities and differences in organisms of the same species, and begin to attribute these to environmental or inherited factors.
▸ Explain the importance of classifying living things.
▸ Identify some of the main taxonomic groups of animals and describe some features of these.

Some pupils will also:
▸ Recognise that inherited and environmental causes of variation cannot be completely separated.
▸ Name some organisms that are not readily classified as plant or animal.

In scientific enquiry:

Most pupils will:
▸ Use observation to identify questions to investigate about variation between individuals.
▸ Suggest data to collect to answer the questions.
▸ Present and analyse the data.
▸ Identify associations or correlations in their data.

Some pupils will also:
▸ Evaluate graphs and tables of data in relation to sample size and describe how strongly any association or correlation is supported.

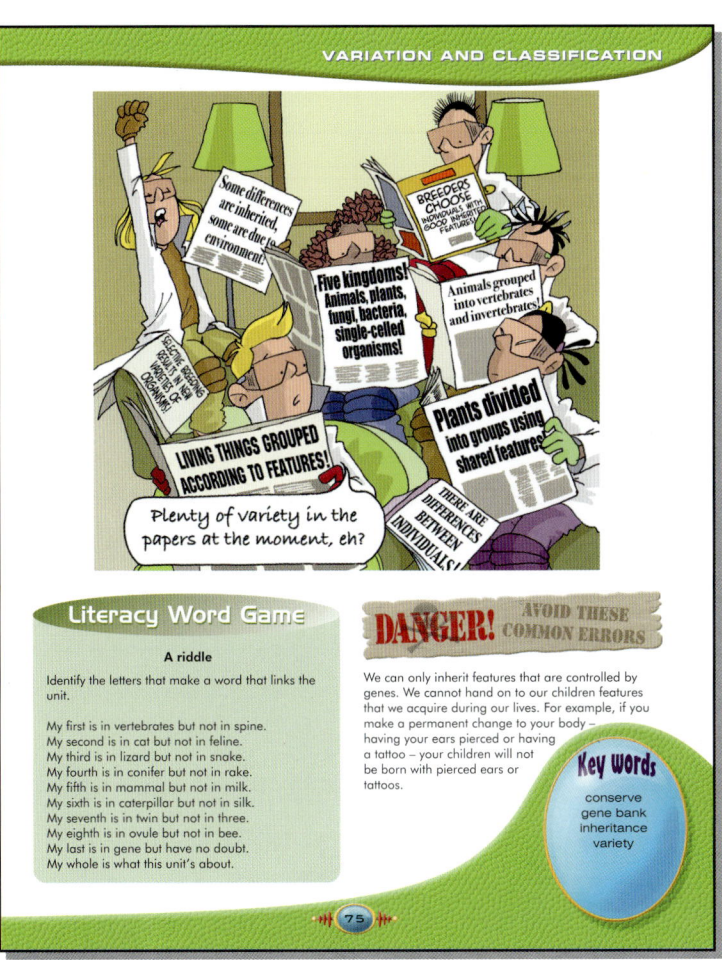

Solution to word game

variation

7D Unit review

Answers to review questions

1. genes, chromosomes, nucleus, father, eggs, parents, selective breeding
2. a amphibians
   b 6 legs, 3 segment body, hard exoskeleton, wings
   c reptiles
   d flowering plants
   e crested newt, sand lizard, common dandelion
3. a all cut from same plant so each cell would have the same genes
   b random environmental differences: more leaves, more photosynthesis, more growth

5.

Insects	Spiders	Myriapods	Crustacea
bee	wolf spider	centipede	woodlouse
ant	scorpion	millipede	lobster
ladybird	garden cross spider		brine shrimps
butterfly	tarantula		crab
daddy-longlegs			
stag beetle			
house fly			
wasp			
dragonfly			
grasshopper			
mosquito			
water skater			

Answers to SAT-style questions

1 (5 marks, −1 for each error)

Characteristic	Animal A	Animal B	Animal C
number of legs	none	10	6
number of wings	none	none	4
body divided into segments	yes	no	yes
eyes	no	yes	yes
antennae	no	no	yes

2 long legs – large stride (1)
large muscles – greater speed (1)
strong heart – supply more blood to muscles (1)
large lungs – supply more oxygen to muscles (1)
strong bones – to withstand the forces of galloping (1)
lively nature – eagerness to race (1)
any valid feature and explanation (1 mark for each, maximum of 6)

3 a data sorted correctly (3 marks, −1 for each error)

Hand span (cm)	Number of people
17.0–17.9	1
18.0–18.9	5
19.0–19.9	6
20.0–20.9	7
21.0–21.9	5
22.0–22.9	3

b good choice of scales (2)
data plotted correctly (3, −1 for each error, allow errors carried forward)

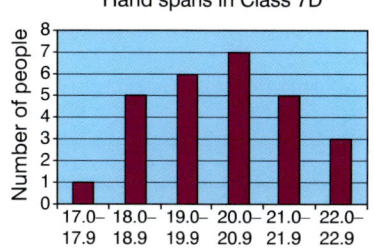

c 20.0–20.9 cm (1)
d 22.8 cm, 17.5 cm (2)
e different genes for hand size, different rates of growth (1)
f continuous (1)

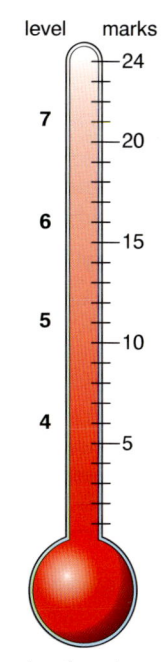

levelometer

UNIT REVIEW QUESTIONS

SAT-STYLE QUESTIONS

1 Use the diagrams and your own knowledge to fill in the table below:

Characteristic	Animal A	Animal B	Animal C
number of legs			
number of wings			
body divided into segments			
eyes			
antennae			

(5)

2 Horse racing is a multi-million pound business. Champion racehorses are worth millions because they could pass on their winner's genes to their offspring. Winning horses are fast, but they need other features too. They have to cope with the challenges of races, travelling long distances round the country, and being ridden by several different jockeys.
Think of six features that a breeder might look for when breeding a future winner and explain why they are important. (6)

A winning racehorse is worth millions for its breeding potential

3 Class 7D measured their hand span. They stretched their fingers out and measured the distance from the tip of the thumb to the tip of the little finger.
There were 27 people in school that day; here is the data they collected:

Hand span (cm)
21.0 20.0 17.5 19.5 20.5
19.3 19.0 20.5 18.2 19.5
21.3 20.2 22.3 21.0 20.8
21.9 18.6 19.2 22.8 18.3
19.5 21.4 18.8 18.8 22.0
20.5 20.4

a Group the data ready to plot a bar chart. Use the class sizes of 17.0–17.9 cm, 18.0–18.9 cm, and so on. (3)
b Plot a bar chart of your data. (5)
c What was the most frequent size of hand span? (1)
d What was the largest hand span? What was the smallest? (2)
e Why do you think there is such a variation? (1)
f Is this continuous or discontinuous variation? (1)

Key words
Unscramble these:
nasic oils facit
frenioc
the rini
escipes
ever trable

7E Acids and alkalis

National framework/QCA SoW references

National framework – Scientific enquiry:
- Use scientific knowledge to decide how ideas and questions can be tested: make predictions of possible outcomes.
- Identify and control the key factors that are relevant to a particular situation.
- Describe and explain what their results show when drawing conclusions: begin to relate conclusions to scientific knowledge and understanding.

QCA SoW (7E)

What are acids and alkalis like and where do we use them? (7E1, 7E2)

How can acids and alkalis be identified and distinguished from each other? (7E3)

Is there a range of acidity and alkalinity? (7E4)

What happens when an acid is added to an alkali? (7E5)

Where is neutralisation important? (7E5)

NC links: Sc1 2a, c, d, g, i; Sc3 3d, f.

Learning objectives

- To learn about acids and alkalis as groups of substances.
- To appreciate that not all acids are hazardous and that some are used in the home.
- To know the common acids and alkalis used in science lessons.
- To recognise common hazard symbols.
- To know how to deal with acids or alkalis if they are spilt or splashed on the skin.
- To know some uses of acids and alkalis in everyday situations.
- To describe the pH scale.
- To appreciate a neutralisation reaction occurs when an acid is added to an alkali.

In scientific enquiry:
- To make indicators and use them to test whether a solution is acidic or alkaline.
- To measure acidity/alkalinity using universal indicator.
- To use data logging to monitor the pH changes during a neutralisation reaction.
- To carry out a fair test to evaluate indigestion remedies.

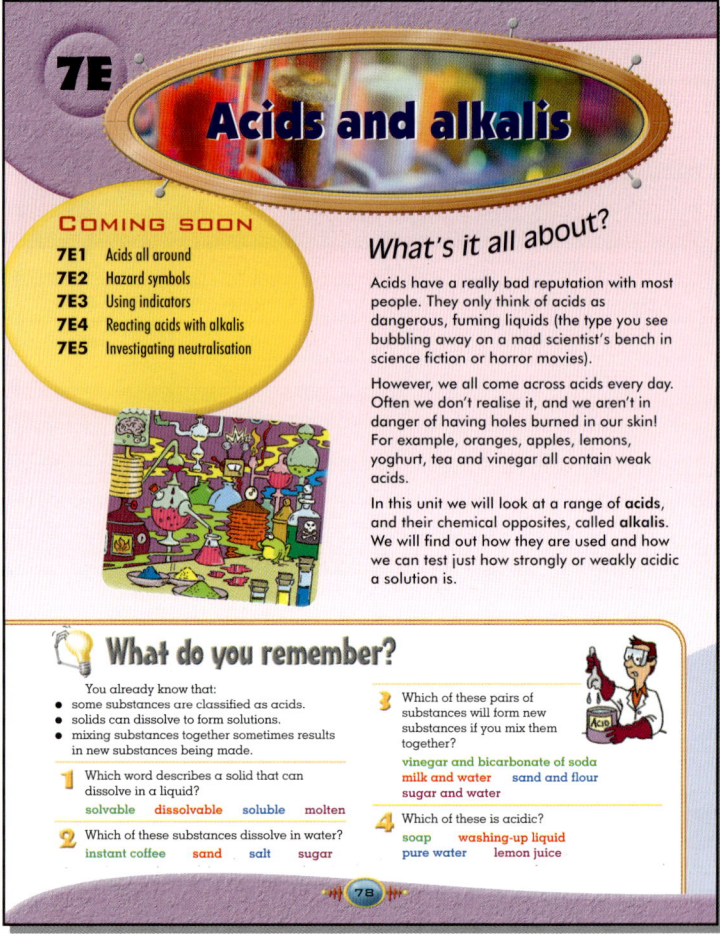

Teaching strategy

This unit builds on the contents of units 6C 'More about dissolving' and 6D 'Reversible and irreversible changes' in the KS2 scheme of work. It also uses ideas that were developed in the KS2 programme of study. The unit provides a foundation for work on carbonate rocks in unit 8G 'Rocks and weathering' and unit 8H 'The rock cycle', and on the reactions of acids in unit 9E 'Reactions of metals and metal compounds'.

Difficulties/Misconceptions

- Many pupils may believe all acids to be unpleasant, corrosive substances; they may take some convincing that many acids are harmless. This would seem to work against the notion that acids are a group of substances that have similar properties. If this becomes an issue, pupils should be told that acids have other properties upon which the grouping is made.
- It is likely that pupils may not be familiar with the term 'alkali'. This should be described as 'the chemical opposite of an acid'; the work on indicators will demonstrate this.
- Many pupils will see the pH scale as an abstract concept. Some pupils may ask questions about why indicators change colour and exactly what it is that a pH probe is measuring. It will not be possible to give meaningful answers to these questions at this stage.

Launch activity notes – Ideas about acids

- This activity will provide an introduction to acids. It is likely that most pupils will be familiar with the term 'acid' from everyday experiences. They could be asked to say in what context they have seen the word, and may respond with things like acid-drop sweets, ant-acid powders, acid rain and acid taste. More observant pupils may have seen the word 'acid' used on the contents lists of food packaging or on the labels of bottles in the laboratory.
- By developing a discussion along these lines, pupils will realise for themselves that not all acids are dangerous. Nevertheless, from the point of view of safety in the laboratory, it is important that pupils appreciate the hazardous nature of mineral acids and the need to wear eye protection when handling them.

Answers

a) Many pupils are likely to think of acids in terms of chemistry laboratories but not in the context of everyday materials, such as food.

b) Pupils would be wise to handle all acids with caution until they learn more about them and are able to differentiate between those that are harmful and those that are not.

c) Chemical names are appropriate in the laboratory, but common names may be more user-friendly for the population as a whole. Pupils will need to get used to dealing with both. Vinegar is only dilute ethanoic acid (5%). Pure ethanoic acid is potentially harmful and should be handled with care.

d) Pupils should realise that products such as shampoos have been thoroughly tested and would not be offered for sale, if they were harmful.

e) Some pupils might already be familiar with indicators. Other suggestions might relate to the different chemical reactions of acids, alkalis and water.

f) Mopping up spillages is largely a matter of common sense. Pupils should make suggestions like using a dry cloth, washing with water, avoiding contact with hands. The unlikely scenario of acid eating through the floor into the room below should help pupils realise that not all acids are highly corrosive liquids.

Learning styles

Visual
- **Observing**.
- **Tabulating** and displaying data.

Kinaesthetic
- **Manipulating** apparatus.
- **Experimenting**.

Intrapersonal
- **Understanding**.
- **Reflecting**.

Answers to questions

What do you remember?

1 Soluble.
2 Instant coffee, salt, sugar.
3 Vinegar and bicarbonate of soda.
4 Lemon juice.

Notes to support 'What do you remember?'

- Pupils will have already been introduced to the idea of grouping substances on the basis of similar properties, so the presentation of acids and alkalis as two groups of substances should not present any problems.
- Pupils should also know that changes sometimes take place when things are mixed, and this can be used to introduce chemical reactions.

7E1 Acids all around

National framework/QCA SoW references

QCA SoW (7E)

7E Section 1: What are acids like and where do we use them?

Learning objectives

▸ To understand that just because a substance is an acid it doesn't necessarily mean that it is harmful.
▸ To appreciate that rainwater is acidic even when it doesn't contain pollutant gases, but is made more acidic by pollutant gases.
▸ To know that we use acids at home and that some of the food we eat contains acids.

Teaching strategy

Key points to highlight

- Acids are a group of substances and have similar properties.
- Rain is always acidic because of the carbon dioxide gas dissolved in it.
- Acid rain is formed when acidic gases dissolve in atmospheric moisture.
- Not all acids are harmful and some are found in the foods we eat.
- Fizzy drinks are acidic because carbon dioxide is used to give them fizz.

Difficulties/Misconceptions

- Many pupils will think that all acids are harmful.
- Pupils will think that pure rainwater is neutral rather than acidic.

Skills

Much of this lesson will be led by the teacher, however pupils could be divided into small groups to discuss their initial ideas about acids. Information panels from various foods would provide pupils with opportunities to use observation skills in identifying the acids found in foods.

Lesson starter suggestions

- **All rain is acidic, why is rain acidic?** When we talk about 'acid rain', we really mean rain that is even more acidic than normal. What are the problems associated with acid rain?
- **Who knows the names of any acids? Where have you seen the names of these acids?** Write down a list.
- **Who knows the names of any acids that are found in the food we eat? Are these acids likely to be harmful?** Draw attention to the potential damage to our teeth caused by eating food containing acids and to the importance of oral hygiene.

Activity and technicians' notes

How much fizz?

The activity requires pupils to gather and measure the volume of carbon dioxide dissolved in two fizzy drinks. They may do this in a variety of ways. A range of apparatus including large beakers, funnels, tubing and some form of calibrated tube, such as a gas syringe or a burette, should be made available.

Answers to questions

Text

Q1 The carbon dioxide is put into the drink under pressure.

Q2 Phosphoric acid, will damage the enamel of your teeth if you drink lots of cola and you don't clean your teeth regularly.

Teaching strategy contd.

Notes and tips

Many pupils will be under the impression that all acids are harmful. This notion should be dispelled, by drawing their attention to the contents' labels on some of the foods that they eat. After a quick read through the contents of a can of coca cola, for example, they will realise that they have been drinking phosphoric acid and citric acid for some years without any ill effect provided they keep their teeth clean.

Extension ideas

Pupils could find how these additives are classified by their E numbers and why they are added to foods: antioxidants, colouring, stabilisers, flavourings, preservatives and sweeteners.

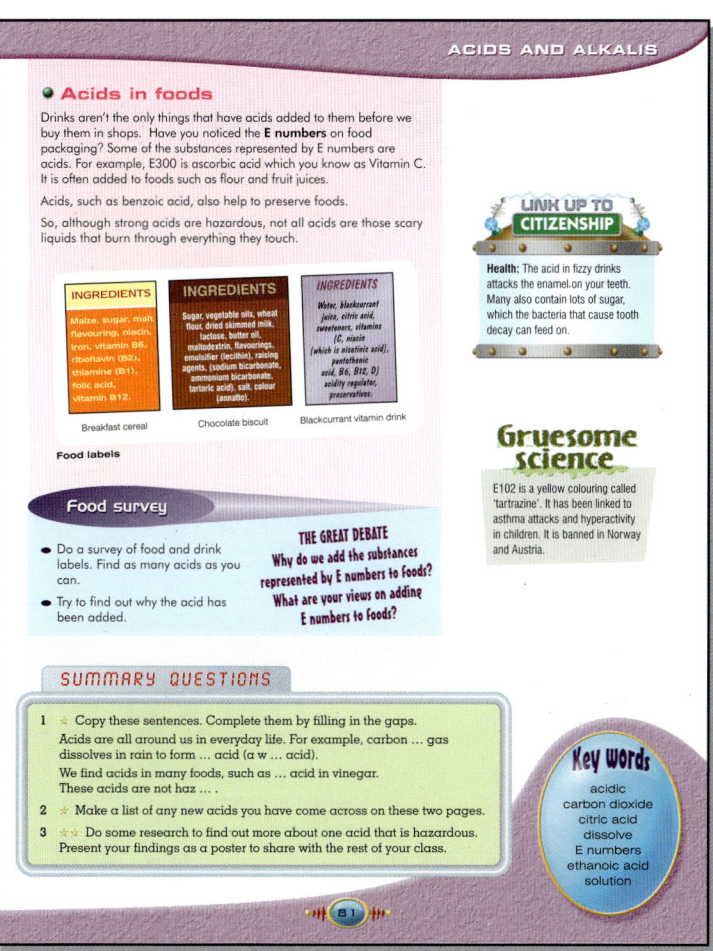

Learning outcomes

Following this, most pupils will:
▶ Be able to name some common acids and describe some of their everyday uses.

Some pupils will also:
▶ Be able to name acids found in common foods.

Suggestions for plenaries

- **Acids are a group of substances**: What other groups of substances are there?
- **Harmful and harmless acids:** Pupils to make lists under these two headings.
- **Are you for or against food additives?** Invite individual comments on food additives and then put it to the vote – should the use of some or all kinds of food additives be banned?

Learning styles

Visual

- **Observing** information on the contents' labels of foods.

Answers to questions contd.

Summary

1 dioxide, carbonic, weak, ethanoic, hazardous
2 carbonic, citric, ethanoic, phosphoric, ascorbic, benzoic, folic, tartaric, nicotinic, pantothenic

Gruesome science

The role of some food additives as causal agents of conditions such as asthma and hyperactivity in young people has attracted considerable attention in recent years. There is a wealth of information available, both in published form and from the Internet. If time allows, this could be used as the basis of an investigation in which groups research different aspects of the problem and report back to the whole class.

The great debate

- Many pupils may not be aware that the food they eat contains additives of different kinds. Initially, they should discuss the reasons why they are added.
- Food additives may be conveniently divided into the following groups: antioxidants, colouring, stabilisers, flavourings, preservatives and sweeteners.
- Pupils might feel that some food additives are more justified than others. For example, an additive that delays the decay of food might be perceived as being more justified than one that improves the colour of a product.
- Pupils should be made aware of the allergic reactions caused to some people by some food additives. They could discuss whether providing information on food packaging, as is the current practice, or an outright ban would serve the greater good.
- The list of acceptable food additives is not the same from country to country. For example, Yellow 7G is allowed in the UK but not in Australia and the USA. Pupils should discuss the reasons for this in terms of different testing procedures, and express an opinion on whether the UK procedures are too lax or those of other countries are too stringent.

7E2 Hazard symbols

National framework/QCA SoW references

QCA SoW (7E)
What are acids and alkalis like and where do we use them?

Learning objectives

▶ To be able to recognise some common hazard symbols.
▶ To know what to do in an emergency, if acids or alkalis are spilt or splashed on the skin.

Teaching strategy

Key points to highlight

- Containers of hazardous substances carry warning or hazard symbols.
- The hazard symbol indicates the nature of the substance.
- Road tankers carry information about the substance they are transporting, so that any spillage can be dealt with effectively.
- Acids and alkalis are made less harmful by diluting them with water.

Difficulties/Misconceptions

- Many pupils will have seen hazard symbols but may not be aware of their meaning.
- Pupils may not appreciate that hazardous substances are dealt with in different ways, depending on the nature of the hazard.

Skills

The lesson content is given over to observation skills, however the content can be made more interactive by getting pupils to make deductions about the nature of the substances from their hazard symbols. Pupils could also be given information about substances and asked to predict what hazard symbol should appear on their containers.

Lesson starter suggestions

- **Describe any warning labels you have seen on containers at home:** What sorts of substances were in the containers? In what ways were the substances dangerous?
- **What substances are transported in road tankers?** Why might it be dangerous to transport materials by road? Discuss the transport of fuels and chemicals.
- **Suppose there was a rail or road accident involving the spillage of a hazardous material:** What information would the emergency services need to have do deal with the accident? How would the hazardous material be contained or made harmless?

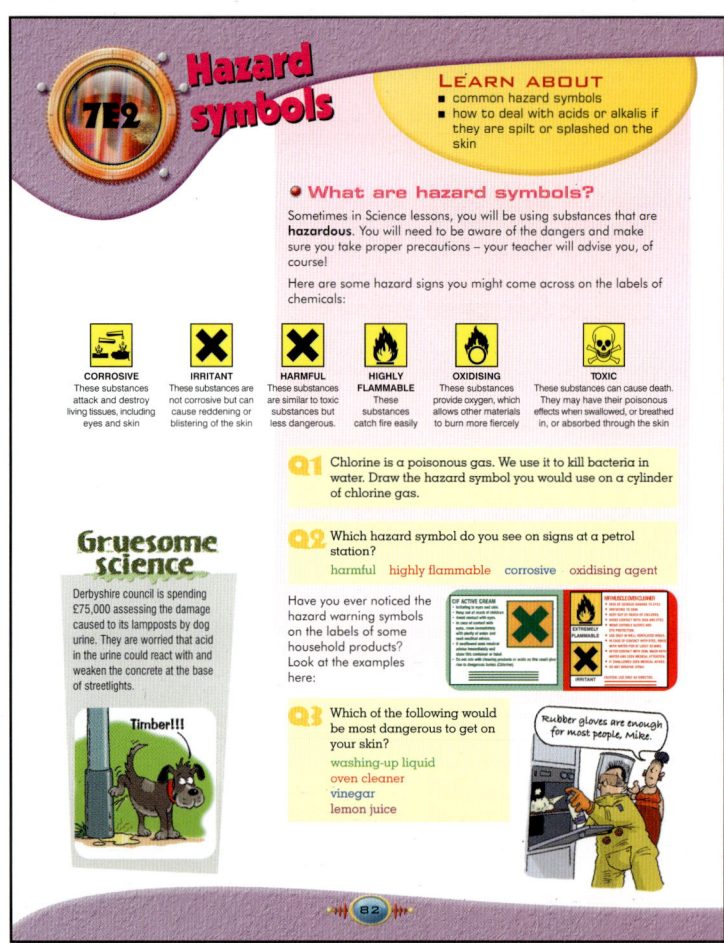

Gruesome science

- The problems anticipated by the Derbyshire council illustrate the corrosive nature of acids and the problems caused by acid rain, and acids from other sources.

Activity and technicians' notes

Concentrated and dilute acids

Safety notes

Safety glasses must be worn when handling both dilute and concentrated sulphuric acid. The laboratory should be well ventilated or the experiment carried out in a fume cupboard.

Equipment required
- Dilute sulphuric acid
- Concentrated sulphuric acid
- Sugar
- Two medium-sized beakers

PUPIL SHEET

Gruesome science

- Pupils should be made aware that the handling concentrated alkali is just as hazardous as the handling of concentrated acid. Oven cleaners provide a good means of illustrating this point. Pupils can be asked to look at the instructions on aerosol oven cleaners they may find at home. These cans carry warnings about not getting cleaner on the skin or near the eyes, and the need to treat splashes immediately with copious amounts of water.

Teaching strategy contd.

Notes and tips

- Many household materials carry hazard symbols. This lesson provides an ideal opportunity to link science into the everyday lives and experiences of pupils, so they can see that what they learn inside the classroom is also relevant outside.
- Pupils could find out more about safety signs and colour codes by looking in publications such as *Signs, Symbols and Systematics* (ASE).

Extension ideas

Pupils could discuss the properties of magnesium, identifying potential hazards, and then design warning signs for a road tanker carrying magnesium powder.

Learning styles

Visual
- **Observing** hazard symbols and their significance.
- **Observing** the reaction between sugar and concentrated sulphuric acid.

Answers to questions

Text

Q1 hazard sign for 'Toxic'
Q2 highly flammable
Q3 bleach or oven cleaner
Q4 Wear protective clothing. Dilute spillage with fine spray of water and evacuate area.

Summary

1 Tell your teacher. Dilute with water and mop up with cloth.
2 Wearing breathing apparatus and fire-fighting kit, dilute spillage with fine spray of water.

Learning outcomes

Following this, most pupils will:
- Be able to identify some common hazard symbols.
- Know how to deal with acids and alkalis if they are spilt or splashed on the skin.

Some pupils will also:
- Be able to predict the hazard symbol(s) that should be placed on the containers of different hazardous materials.

Suggestions for plenaries

- **Many acids and alkalis are harmful substances:** What should you do if you spill or splash acid or alkali on the bench; on your clothes; on you skin?
- **Who can describe/draw hazards symbols for these?** Corrosive, harmful, highly inflammable, irritant, oxidising agent, toxic.
- **Why do road tankers carry hazard signs?** What information do they give?

7E3 Using indicators

National framework/QCA SoW references

QCA SoW (7E)
How can acids and alkalis be identified and distinguished from each other?
Is there a range of acidity and alkalinity?
NC links: Sc1 2i; Sc3 3d.

Learning objectives

- To learn the names of some of the acids and alkalis used during lessons.
- To understand that an indicator is a substance that is a different colour in acids and alkalis.
- To make an indicator by extracting the coloured pigment from plants.
- To test solutions with an indicator to determine if they are acidic or alkaline.
- To learn that acidity–alkalinity is measured on the pH scale and that this runs from 0–14.
- To understand that universal indicator has a range of colours so it indicates how acidic or how alkaline a substance is.

Teaching strategy

Key points to highlight

- Indicators are substances that are different colours in acidic and in alkaline solutions.
- Indicators can be extracted from vegetable and plant material by crushing, adding water and filtering.
- Universal indicator has a range of colours and shows whether a solution is strongly or weakly acidic or alkaline.
- Acidity and alkalinity are measured on the pH scale.
- Universal indicator can give the approximate pH of a solution.
- Common laboratory acids are hydrochloric acid, nitric acid and sulphuric acid.
- Common laboratory alkalis are calcium hydroxide solution, ammonia solution, potassium hydroxide solution and sodium hydroxide solution.
- The common laboratory acids and alkalis are used in many chemical reactions.

Lesson starter suggestions

- **How can a flower's dye show whether a solution is acidic or alkaline?** Introduce the idea that the colour of a dye may be different in acidic and alkaline solutions. The result is a substance that 'indicates' whether something is an acid or an alkali.
- **How can the dye be extracted from a flower?** What processes are needed to break up the plant material and extract the colour from it?
- **Who can name the acids and alkalis used in science lessons?** Write down the names of the common acids: hydrochloric acid, nitric acid and sulphuric acid; and the common alkalis: sodium hydroxide solution, potassium hydroxide solution, calcium hydroxide solution and ammonia solution (ammonium hydroxide).

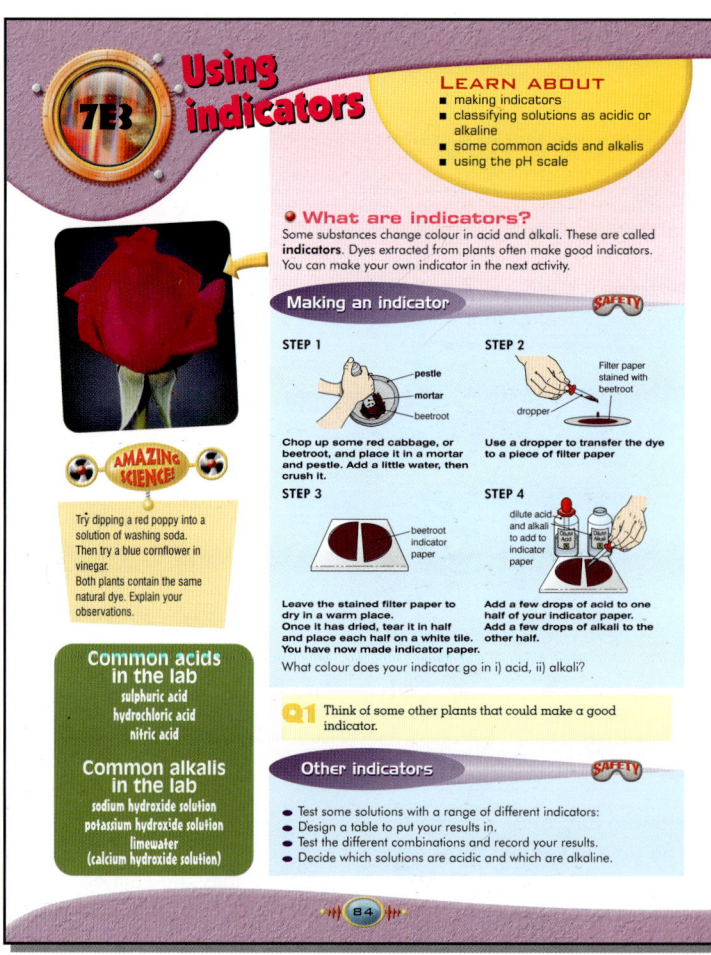

Activity and technicians' notes

Making an indicator
Safety notes

- Many plant dyes will permanently stain clothing. Pupils should be warned not to splash the dye solution onto their cloths when crushing the plant material.
- Safety glasses must be worn when handling both dilute acids and dilute alkalis.
- The acid and alkali should be very dilute.

Equipment and materials required

- Supply of plant material, such as beetroot or red cabbage
- Any dilute acid, any dilute alkali

For each group, one of each of these:
- Mortar and pestle, dropper pipette, filter paper, white tile

Activity and technicians' notes contd.

Other indicators

Safety notes

Safety glasses must be worn when handling both dilute acids and dilute alkalis.

Equipment and materials required

- A range of common indicators, e.g. litmus, methyl orange, phenolphthalein, in bottles fitted with dropper pipettes
- Common acids and bases, e.g. dilute sulphuric acid, dilute nitric acid, dilute hydrochloric acid, dilute sodium or potassium hydroxide solution, dilute ammonia solution, calcium hydroxide solution (limewater)

For each group:
- A set of test tubes
- Dropper pipettes to transfer the acid and alkali

Testing the pH of solutions

Safety notes

Safety glasses must be worn when handling both dilute acids and dilute alkalis.

Equipment and materials required

- Universal indicator solution in a bottle fitted with a dropper pipette
- Common acids and bases, e.g. dilute hydrochloric acid, dilute nitric acid, dilute sulphuric acid, calcium hydroxide solution, dilute potassium solution, dilute sodium hydroxide solution, ammonia solution

For each group:
- A set of test tubes
- Dropper pipettes to transfer acid and alkali

Teaching strategy contd.

Difficulties/Misconceptions

- Pupils frequently get the idea that the pH scale runs from 1 to 14 rather than 0–14. Emphasise that the scale runs from 0.
- Pupils should understand that ph, Ph and PH are not acceptable alternatives for pH.

Learning styles

Kinaesthetic
- **Preparing** the vegetable/plant extract
- **Carrying out** tests on unknown solutions

Visual
- **Observing** colour changes
- **Making** deductions

Answers to questions

Text

Q1 any coloured plant material

Q2 pH 1

Summary

1. The dye must be different colours in acidic and alkaline solutions.
2. a) It has a range of colours that show how acidic or how alkaline a solution is.
   b) 12, 8, 7, 6, 3, 2
   c) i) 7 ii) green

Learning outcomes

Following this, most pupils will:
▸ Name some common acids and alkalis.
▸ Classify solutions as acidic, alkaline or neutral, using indicators and pH values.

Some pupils will also:
▸ Relate the pH value of an acid or alkali to its hazards and how corrosive it is.

Suggestions for plenaries

- **What are indicators and how are they used?** How is universal indicator different to other indicators?
- **What is the pH scale?** What is the range of the scale? What value is neutral and which direction is increasing acidity; increasing alkalinity?
- **What are the names of the common laboratory acids and alkalis?** Make two lists under the headings 'acid' and 'alkali'.

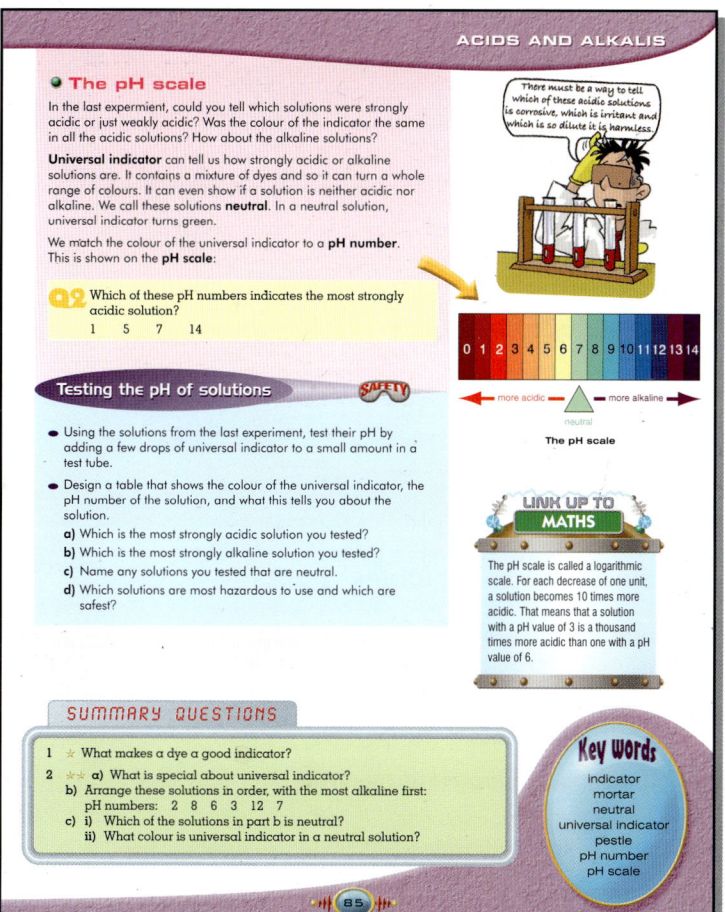

7E4 Reacting acids with alkalis

National framework/QCA SoW references

QCA SoW (7E)
What happens when an acid is added to an alkali?
NC links: Sc3 3f.

Learning objectives

▸ To find out how acids and alkalis are used in some everyday situations.
▸ To follow how the pH of an alkaline solution changes when acid is added to it.
▸ To appreciate that when an acid is added to alkali a chemical reaction takes place.

Teaching strategy

Key points to highlight

- In some ways, acids and alkalis are chemical opposites.
- An acid and an alkali react together in a neutralisation reaction.
- The progress of a neutralisation reaction is followed by measuring the pH of the reaction mixture.
- Neutralisation is complete when the pH of the reaction mixture is 7.
- During neutralisation, heat is given out.
- The production of heat is evidence that a chemical reaction is taking place.
- Acids and alkalis are found in many everyday contexts.

Difficulties/Misconceptions

Following the progress of a neutralisation reaction using universal indicator is not very accurate, because the colour changes only give an approximate pH value.

Skills

This lesson is mainly given over to practical activities:
- Initially, pupils will need to use their manipulative and observational skills in following the progress of a neutralisation reaction using universal indicator.
- The experiment on temperature change also requires them to use their observational skills to record temperatures accurately using a thermometer.

Lesson starter suggestions

- **Acids and alkalis are chemical opposites:** Discuss acids and alkalis in terms of their different effects on indicators, and their reactions together.
- **Using pH to follow the reaction of an acid and an alkali:** The pH of an alkali decreases as an acid is added to it. At some point, when there are equivalent amounts of acid and alkali, the pH equals 7 and the reaction mixture is neutral. How can the changing pH of the reaction mixture be monitored?
- **What evidence is there that a chemical reaction takes place when an acid is mixed with an alkali?** Heat is given out when an acid and an alkali react together. How can this be used to monitor the progress of the reaction and determine when there are equivalent amounts of acid and alkali?

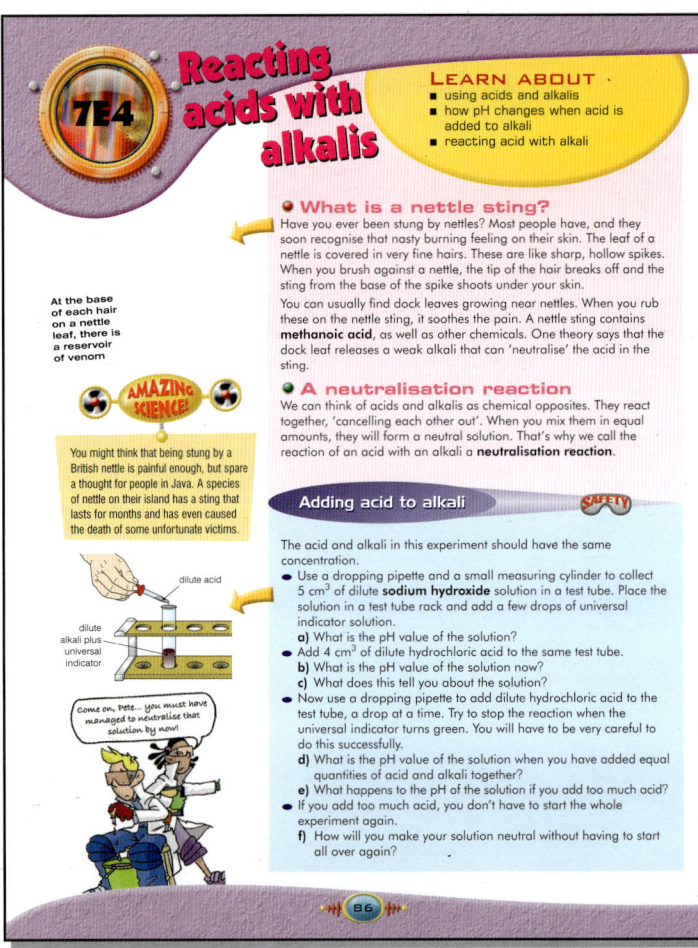

Activity and technicians' notes

Adding acid to alkali

Safety notes

- Safety glasses must be worn when handling both dilute acids and dilute alkalis.

Equipment and materials required

- Universal indicator solution
- Dilute sodium hydroxide (0.4 M) and dilute hydrochloric acid (0.4 M)

For each group:
- Dropping pipette
- Small measuring cylinder
- Test tube
- Test tube rack

Activity and technicians' notes contd.

Temperature changes
Safety notes
- Safety glasses must be worn when handling both dilute acids and dilute alkalis.
- Pupils should be warned to handle thermometers with care. Any mercury spillage resulting from a broken thermometer should be dealt with immediately.

Equipment and materials required
- Dilute sodium hydroxide (0.4 M) and dilute hydrochloric acid (0.4 M)

For each group:
- A small measuring cylinder
- A thermometer
- Two test tubes
- Test tube rack

Useful acids and alkalis
Access required to:
Some of these:
- Books
- CD ROMs
- Videos
- Internet
- Leaflets

Teaching strategy contd.
- Pupils are also required to carry out research using whatever reference materials are available.

Extension ideas
Pupils could investigate traditional and modern treatments for bee stings. Bee stings are said to be acidic; the pain is said to be relieved by neutralising with a weak alkali such as bicarbonate of soda (sodium hydrogencarbonate). Ask the pupils if this is true or simply an 'old wives' tale'?

Learning styles
Kinaesthetic
- **Carrying out** a neutralisation reaction.

Visual
- **Observing** the change in pH of a neutralisation reaction using universal indicator.

Gruesome science
Gruesome science
The old names for methanoic acid (formic acid) and ethanoic acid (acetic acid) are still widely used. The common names of these acids are used because they are easier to say and remember than their systematic names, e.g. citric acid is rather more user-friendly than 2-hydroxypropane-1,2,3-tricarboxylic acid!

Answers to questions
Summary
1. lower, react, neutral, 7, neutralisation
2. a) It was dissolved in the water.
   b) Evaporate the water from the solution.

Suggestions for plenaries
- **What happens in a neutralisation reaction?** How can the progress of a neutralisation reaction be followed?
- **How does the pH of a mixture change when an acid is slowly added to an alkali?** What is the range of the pH scale? At which end of the scale are acids and at which end of the scale are alkalis?
- **Who can give an example where acids and/or alkalis are found in everyday life?** Where are they to be found?

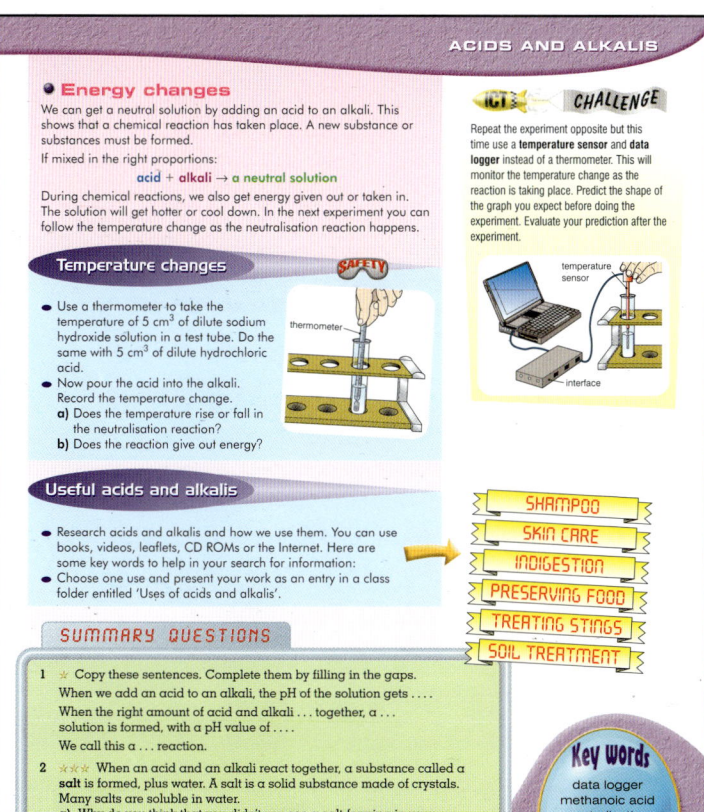

Learning outcomes

Following this, most pupils will:
- Describe some everyday uses of acids, alkalis and neutralisation.
- Describe what happens to the pH of a solution when it is neutralised.
- Appreciate that a chemical reaction occurs during neutralisation.

Some pupils will also:
- Explain how a neutral solution can be obtained.

7E5 Investigating neutralisation

National framework/QCA SoW references

National framework – Scientific enquiry:
- Use scientific knowledge to decide how ideas and questions can be tested: make predictions of possible outcomes.
- Identify and control the key factors that are relevant to a particular situation.
- Describe and explain what their results show when drawing conclusions: begin to relate conclusions to scientific knowledge and understanding.

QCA SoW (7E)
Where is neutralisation important?
NC links: Sc1 2a, c, d, g, i, k; Sc3 3f.

Learning objectives
- To learn how to use data logging equipment in order to monitor a chemical reaction.
- To compare indigestion remedies by carrying out a 'fair test'.

Teaching strategy

Key points to highlight
- Neutralisation has many important applications, several of which are given in the pupils' book.
- A pH probe gives a much more accurate value for the pH of a solution than universal indicator.
- Indigestion results when too much acid is produced in the stomach, often as a result of eating too much. Indigestion remedies contain mild alkalis that neutralise the excess acid. This is why they are called 'antacids' (anti-acids).

Difficulties/Misconceptions
When pupils carry out the neutralisation reaction themselves using a pH sensor, they must ensure that the acid and alkali are properly mixed by stirring; otherwise the pH value they record will not be accurate.

Skills
This lesson is mainly given over to practical activities:
- Initially, pupils will need to use their manipulative and observational skills in following the progress of a neutralisation reaction using a pH sensor.
- The investigation of indigestion remedies requires them to apply the knowledge they have gained in this unit in order to plan and carry out an investigation.

Lesson starter suggestions

- **How does the pH value change during a neutralisation reaction?** How does the pH change when we add an acid to an alkali? How does the pH change when we add an alkali to an acid?
- **How can we monitor a neutralisation reaction?** What changes take place during a neutralisation reaction? How can we monitor pH changes?
- **How do remedies for indigestion work?** What causes our stomachs to make too much acid?

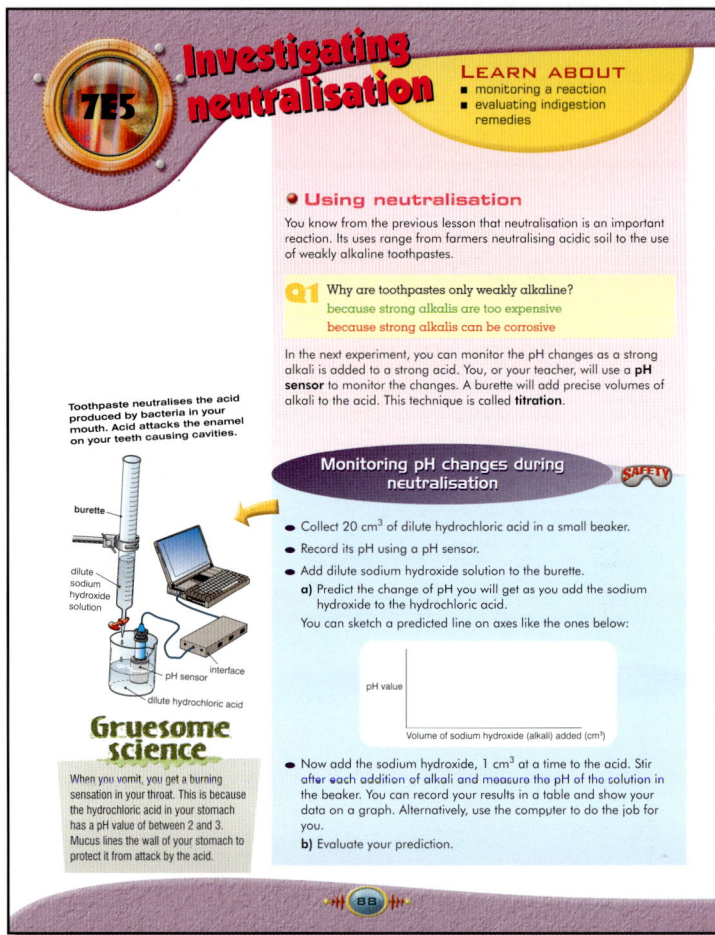

Activity and technicians' notes

Monitoring pH changes during neutralisation

Safety notes
Safety glasses must be worn when handling both dilute acids and dilute alkalis.

Equipment and materials required
- Dilute hydrochloric acid and dilute sodium hydroxide solution

For each group (or as a demonstration):
- A small beaker
- A burette
- A 50 cm³ measuring cylinder
- A stirring rod
- A pH sensor

Teaching strategy contd.

- Subsequently, they will devise a suitable means of recording and displaying results and evaluating their worth.
- Finally, they will need to make valid comparisons between their results and those of others.

Notes and tips
Pupils should look at the contents' labels on indigestion remedies and identify the weak alkali it contains. Some help may be needed where old names are used, for example, magnesia instead of magnesium hydroxide.

Extension ideas
Pupils could predict the shapes of pH curves for neutralisation reactions in which the conditions differ from above. For example, using more of less concentrated solutions, or adding acid to alkali rather than alkali to acid.

Activity and technicians' notes contd.

Investigating indigestion remedies
Equipment and materials required
- Several indigestion remedies
- The activity requires pupils to devise methods of comparing some different indigestion remedies. They may do this in a variety of ways. Universal indicator solution and/or pH sensors should be available, together with a range of common apparatus including a balance, test tubes and small measuring cylinders.

Learning styles

Kinaesthetic
- **Carrying out** a neutralisation reaction.
- **Carrying out** an investigation of indigestion remedies.

Visual
- **Monitoring** pH during a neutralisation reaction.
- **Observing** the actions of different indigestion remedies.

Intrapersonal
- **Rationalising** observations.
- **Making conclusions** about the investigation into indigestion remedies.

Answers to questions

Text
Q1 because strong alkalis can be corrosive

Summary
1. a) hydrochloric acid
   b) by taking a remedy that neutralises excess acid

Suggestions for plenaries

- **How does the pH of a mixture change during a neutralisation reaction?** Who can draw a graph of pH against volume of sodium hydroxide solution added? Explain the shape of the graph.
- **Which is better for following the pH of a neutralisation reaction?** A pH probe or universal indicator? Why is it better?
- **How do indigestion remedies work?** What causes indigestion? What sort of substance does an indigestion remedy contain?

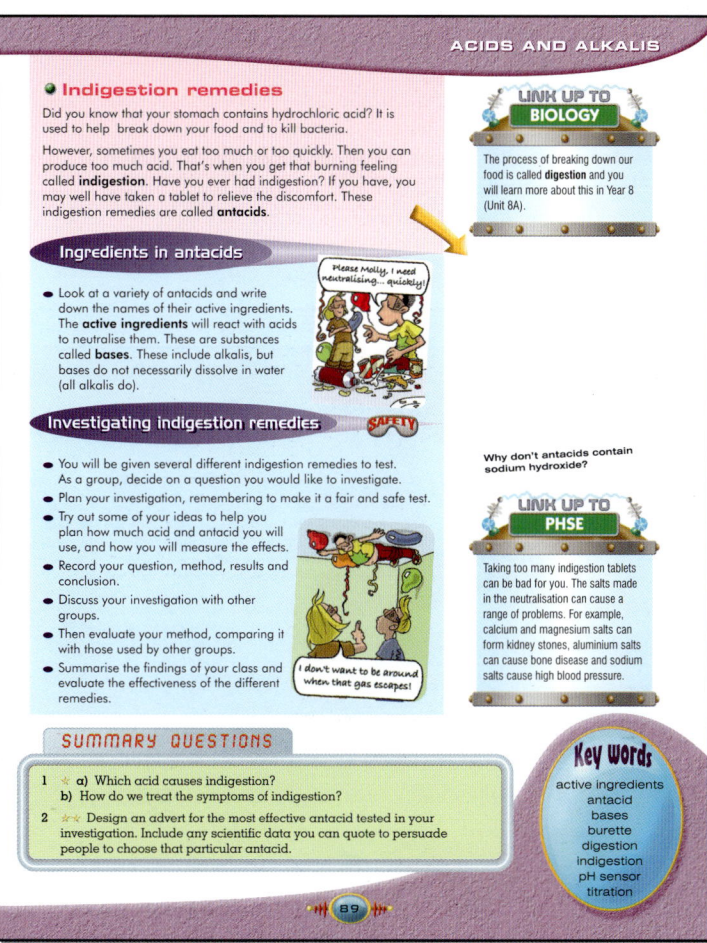

Learning outcomes

Following this, most pupils will:
- Describe what happens to the pH of a solution when it is neutralised.
- Obtain and present quantitative results in a way that helps to show patterns.
- Suggest how to investigate a question about antacids, planning and making a fair comparison.

Some pupils will also:
- Explain how a neutral solution can be obtained.
- Explain how the conclusions of an investigation match the evidence obtained.
- Suggest ways in which the data collected could be improved.

7E Read all about it!

Teaching strategy

Acids for health/The story of aspirin
Difficulties and misconceptions

- It is likely that all pupils will have heard of aspirin. Unfortunately, 'aspirin' is sometimes assumed to be a generic name for all painkillers. The situation is further complicated by the fact that aspirin may be an active ingredient in other analgesics.
- Pupils should be advised that aspirin is one of a range of painkillers that includes other examples with which they may be familiar, such as paracetamol and ibuprofen. In some products, a mixture of painkillers is used.

Notes and tips

- Aspirin is one of a group of analgesics that may be familiar to pupils. The group also contains ibuprofen, and paracetamol. Analgesics are taken to relieve a wide range of aches and pains, including headaches, toothache, period pain, and rheumatic pain.
- Aspirin works by interfering with synthesis of prostaglandins, which are associated with inflammation and fever. It does this by blocking the enzyme cyclo-oxygenase, which stimulates prostaglandin release.
- Unfortunately, aspirin has a tendency to cause gastric irritation, so paracetamol is often taken instead. It is thought to work in the same way as aspirin but without the same adverse effects, and is particularly effective against common headaches.
- There is some controversy over the role of vitamin and mineral supplements. Marketing companies are keen to extol the virtues of their products and the benefits to health. However if a person is eating a balanced diet consisting of a broad range of foods, then why would they need additional vitamins and minerals? The role of vitamin and mineral supplements in our society is a fertile area for research and class debate.

Skills

This unit has opportunities for pupils to develop the Sc1 skills of manipulating apparatus, making observations, recording and displaying results and evaluating their worth. The content encourages pupils to become more aware of information all around them by drawing attention to hazard signs and the contents labels of foods and medicines. There are also opportunities for group discussion and research into a variety of issues.

Extension ideas

Pupils could work together to research in more detail one of the following topics:

- What painkillers can you buy at the local shops and what are the active ingredients in each of them?
- How do painkillers work?
- What vitamin and mineral supplements can you buy at the local shops and what do they contain?
- How does the body use vitamins and minerals?
- Are vitamin and mineral supplements really necessary?

Questions

1. What is ascorbic acid commonly known as?
2. Name two other acids often found in vitamin supplements.
3. What type of acids make up all proteins?
4. Why do you think that 9 of acids that make up proteins are described as 'essential'?
5. What were the problems associated with the use of salicylic acid and its salt as a painkiller?
6. Besides pain relief, name some other benefits of aspirin.
7. What did the ancient Greeks use to relieve pain? Why do you think it took so long to discover the active ingredient?
8. Carry out some research to find some brief details about the development of paracetamol and ibuprofen.

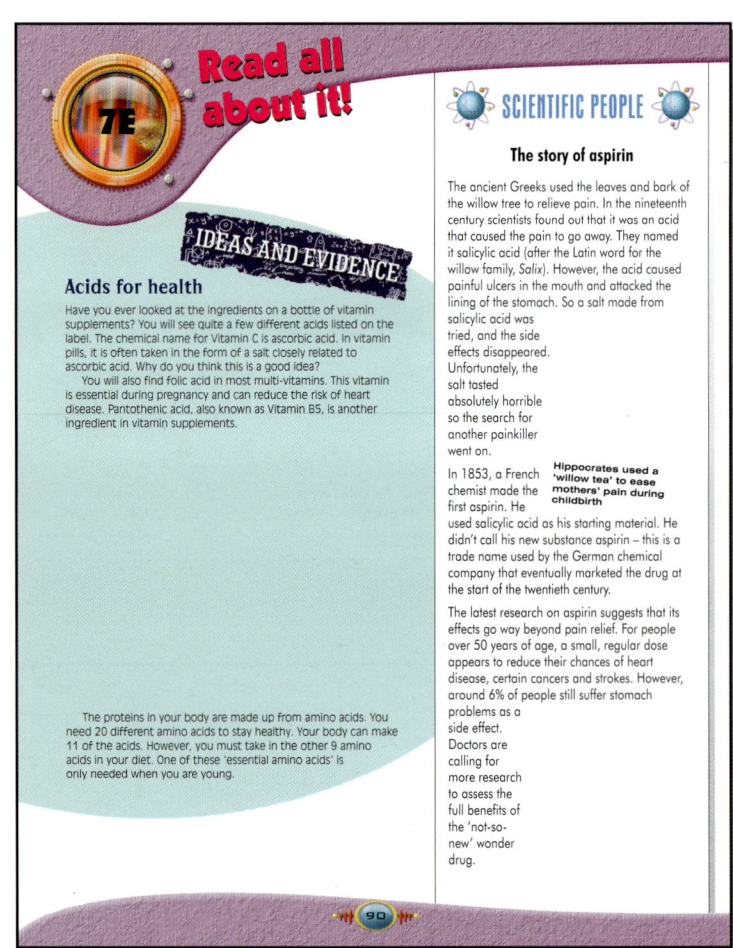

Danger – common errors

The pH scale is a cause of confusion for some pupils. They incorrectly associate increasing pH value with increasing acidity. They may find it easier to think of the pH scale as two scales which both starting at pH 7; one scale, measuring acidity, falling to 0 and the other scale, measuring alkalinity, rising to 14.

Some pupils may believe that any mixture of acid and alkali produces a neutral solution. In order to avoid this misconception, neutralisation could be described in terms of adding alkali to an acid until the pH has risen to 7; or adding an acid to an alkali until the pH has fallen to 7. In either case, emphasise that the solution is only neutral when the pH equals 7, and that at this point there are 'equal' amounts of acid and alkali present.

Learning styles

Visual
- **Observing** information on pain killers and vitamin and mineral supplements.

Interpersonal
- **Discussing** the role of vitamin and mineral supplements.

Intrapersonal
- **Understanding** the way in which pain killers work and the role of vitamins and minerals in the body.

Learning outcomes

Following this unit, most pupils will:
- Name some common acids and describe some of their everyday uses.
- Identify some common hazard symbols and know how to deal with acids and alkalis if they are spilt or splash on their skin.
- Describe some everyday applications of neutralisation.

Some pupils will also:
- Name acids found in common foods.
- Predict the hazard symbol(s) that should be placed on the containers of different hazardous materials.
- Relate the pH value of an acid or alkali to its hazards and corrosiveness.
- Explain how a neutral solution can be obtained.

In scientific enquiry:

Following this unit, most pupils will:
- Classify solutions as acidic, alkaline or neutral, using indicators and pH values.
- Describe what happens to the pH of a solution when it is neutralised, obtaining and presenting quantitative results in a way which helps to show patterns.
- Suggest how to investigate a question about antacids; planning and making a fair comparison.

Some pupils will also:
- Explain how the conclusions of an investigation matches the evidence obtained and suggest ways in which the data collected could be improved.

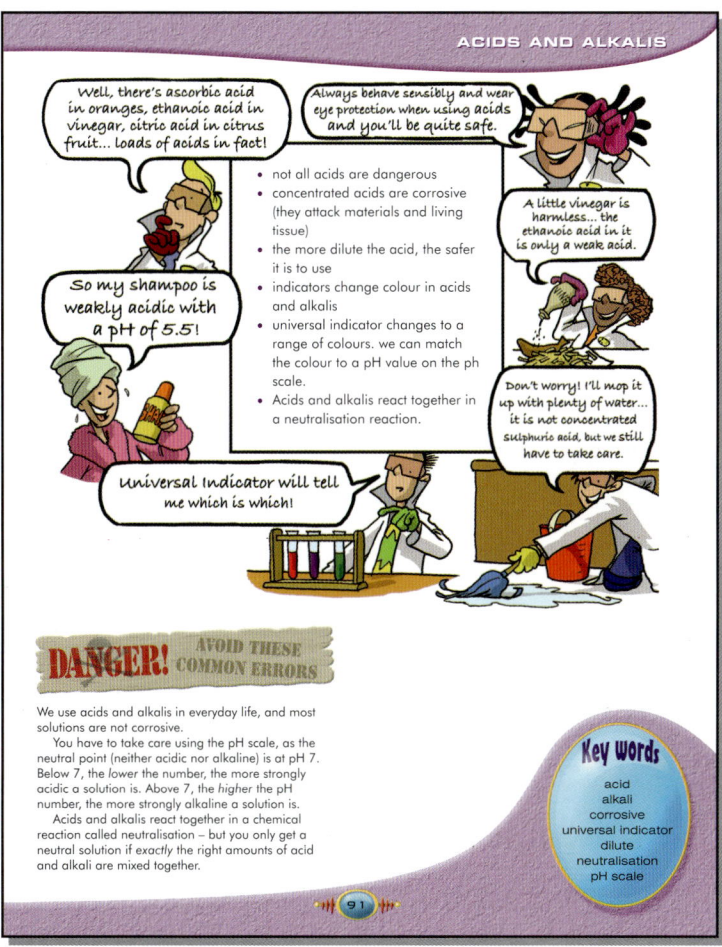

Teaching strategy contd.

Answers

1. Vitamin C
2. folic acid, pantothenic acid
3. Amino acids
4. Essential amino acids cannot be made by the body, so they must form part of the diet
5. salicylic acid caused mouth ulcers; the salt of salicylic acid had a very unpleasant taste
6. it appears to reduce the risk of heart disease, strokes and certain types of cancer in people over 50
7. they used the bark and leaves of the willow tree. It was only when modern methods of analysis and separation were developed that it was possible to isolate the active ingredient.

7E Unit review

Answers to review questions

1. **a** i) B ii) D iii) A iv) C v) E
   b

Solution	pH value within the range
A	7
B	0–4
C	8–10
D	5–6
E	11–14

   c C
   d E
   e D
   f i) C and E
      ii) neutralisation
   g corrosive

5. **a** 0.4 g
   b 20 g

6. 10 000 times

7. **a**

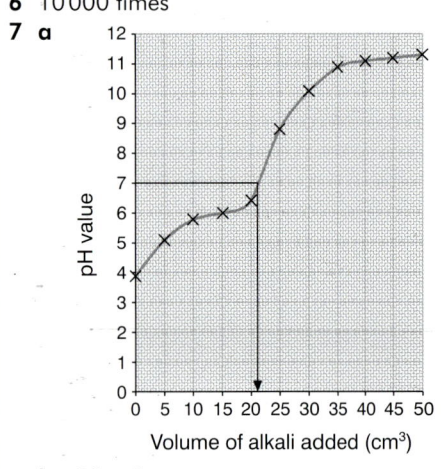

   b 22 cm^3

8. Strong acids: hydrochloric acid, nitric acid, and sulphuric acid
   Weak acids found in food: ascorbic acid, carbonic acid, citric acid, ethanoic acid, folic acid, methanoic acid, nicotinic acid, pantothenic acid, phosphoric acid and tartaric acid

c (diagram labelled: burette, clamp, alkali, interface or pH meter, pH probe/sensor, acid)

UNIT REVIEW QUESTIONS

a Draw a graph of their results.
b Use your graph to find out how much alkali was needed to produce a neutral solution.
c Draw the apparatus the group could have used to gather their data.

Extension questions

8 Make a list of all the acids you have come across in this unit.
   Organise them into groups and explain the reasoning behind your decisions.
9 Do some research to find out the products that we can manufacture from sulphuric acid. Present your findings on a poster to display to the rest of your class.

SAT-STYLE QUESTIONS

1 You can try to make your own indicator using some coloured flower petals.
   The first step involves crushing and grinding the petals to release the colour into a little water.
   a What apparatus would you use to grind up the petals with water? (2)

A group of students made some indicator solution from three different colours of petal. They added their indicator to an acid and an alkali. Here are their results:

Colour of petals and indicator solution in water (pH 7)	Colour in a solution of pH 1	Colour in a solution of pH 14
yellow	yellow	yellow
red	red	green
purple	pink	blue

b What colour would the red petal indicator be in a solution of sulphuric acid? (1)
c What colour would the purple petal indicator be in a solution of salt, which is neutral? (1)
d What colour would the purple petal indicator be in a solution of sodium hydroxide? (1)
e Explain which colour of flower petal would make the best indicator for both acids and alkalis. (2)

2 The table shows the pH values of five solutions whose labels have been lost.

Solution	pH value
A	6.0
B	7.5
C	7.0
D	4.5
E	8.0

a Which solutions are acidic? (2)
b Soap solution is weakly alkaline. Which of the solutions could be soap solution? (1)
c i) Give two solutions that would react together. (1)
   ii) What do we call this type of reaction? (1)
   iii) Two new substances are made in the reaction. One is a salt, what is the other substance formed? (1)

3 A group of students monitored the pH of the solution formed as they added acid and alkali together.
   Look at their experiment and the graph produced below:

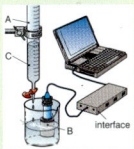 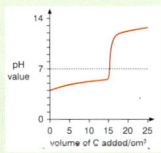

a What do we call the apparatus labelled A? What is its function? (2)
b What do we call the apparatus labelled B? (1)
c Did the students start with an acid or an alkali in the beaker? How can you tell? (1)
d What was the pH of the solution in the beaker when 10 cm³ of solution C had been added? (1)
e How much of solution C was added to get a neutral solution? (1)
f Name a substance that could be solution C. (1)

Key words

Unscramble these:
icad
rcoorvies
lrnonatiue sait

Answers to SAT-style questions

1 a mortar and pestle (2)
   b red (1)
   c purple (1)
   d blue (1)
   e purple, because it changes colour in both acid and alkali (2)
2 a A and D (2)
   b B (1)
   c i)   A or D with B or E (1)
      ii) neutralisation (1)
      iii) water (1)
3 a burette – adds precise volumes of solution (2)
   b pH sensor/probe (1)
   c acid – its pH is 4/below 7 (1)
   d 5–6 (1)
   e 15 cm³ (1)
   f sodium hydroxide/potassium hydroxide solution (1)

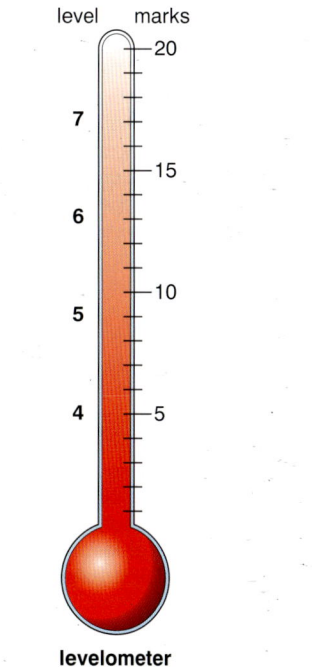

levelometer

7F Simple chemical reactions

National framework/QCA SoW references

National framework – Scientific enquiry:
- Use scientific knowledge to decide how ideas and questions can be tested: make predictions of possible outcomes.
- Use repeat measurements to reduce error and check reliability.
- Present and interpret experimental results through the routine use of tables, bar charts and simple graphs, including line graphs.
- Describe and explain what their results show when drawing conclusions; begin to relate conclusions to scientific knowledge and understanding.
- Evaluate the strength of evidence: e.g. in bar charts and graphs: indicate whether increasing the sample would have strengthened the conclusions.

QCA SoW (7F)
What is a chemical reaction? (7F1)
How do acids react with metals? (7F2)
How do acids react with carbonates? (7F3)
What new substances are made when materials burn in air or oxygen? (7F4)
What is produced when fuels burn? (7F5)
What is needed for things to burn? (7F4, 7F5)
NC links: Sci 1 2a, e, f, g, h, i, k, l, p; Sc3 3a, e, g.

Learning objectives

- To appreciate the sorts of changes that may take place during an experiment and how best to observe them.
- To recognise observations that indicate a chemical reaction is taking place.
- To be aware that in a chemical reaction 'reactants' are changed into 'products'.
- To appreciate that when acids react with metals and with carbonates new substances are formed.
- To discuss some uses of carbon dioxide.
- To appreciate what is needed for things to burn and recall the fire triangle.
- To learn how to extinguish fires.
- To appreciate that when things burn new substances called 'oxides' are formed.
- To understand that substances burn more vigorously in pure oxygen than in air.
- To be aware that coal, oil and natural gas are all examples of fossil fuels.
- To identify the new substances formed when fuels burn.
- To write word equations for all of the reactions in this unit.

In scientific enquiry:
- To record observations carefully and accurately and know how to interpret them.
- To observe the changes that take place when acids react with metals and with carbonates.
- To carry out a simple test for hydrogen gas.
- To carry out a simple test for carbon dioxide gas using limewater.
- To carry out combustion reactions carefully and safely.
- To carry out a simple test for presence of water using blue cobalt chloride paper or anhydrous copper sulphate.
- To carry out an investigation on the combustion of a night-light.
- To spot patterns in data that are collected or given.
- To be aware that some results may not fit into a pattern.

Notes to support 'What do you remember?'

- Pupils will already be familiar with the idea that some changes are easily reversible whilst others are not, and that irreversible changes involve making new substances.
- They should also be aware of the existence of different gases and may be able to name some common examples.
- The pH scale should be fresh in the mind as it was introduced in the previous unit.

Answers to questions

What do you remember?
1 hydrogen
2 paper burning
3 11

Teaching strategy

This unit builds on ideas in units 5C 'Gases around us' and 6D 'Reversible and irreversible changes' in the KS2 scheme of work. It is closely related to the previous unit 7E 'Acids and alkalis', and follows on well from it.

Further work on the reactions of acids and on burning as a chemical change is to be found in units 9E 'Reactions of metals and metal compounds' and unit 9F 'Patterns of reactivity'. Unit 9H 'Using chemistry' includes work on burning in the context of conservation of mass in chemical reactions.

Difficulties/Misconceptions

- Many pupils think of a chemical reaction as something that only takes place in a test tube in the laboratory. They may never have considered chemical reactions in the context of everyday processes, such as cooking and the burning of fuels.
- Pupils may think the term 'observation' is limited to those changes that can be seen. They should appreciate that in chemistry it is used in a more general sense about any change that can be detected, such as a change in temperature, or the evolution of a pungent gas. Pupils should be encouraged to make their observations as detailed as possible. For example, they will have little trouble in observing the evolution of gas when a metal reacts with dilute acids however, they may not appreciate that this reaction can be more or less vigorous depending on the reactivity of the metal.
- Pupils will not appreciate just how reactive oxygen is because they are used to seeing combustion taking place in air, which contains only about $\frac{1}{5}$ of oxygen.
- Pupils should appreciate that the tests they carry out for hydrogen, carbon dioxide, oxygen and water indicate a presence but give no indication of purity.

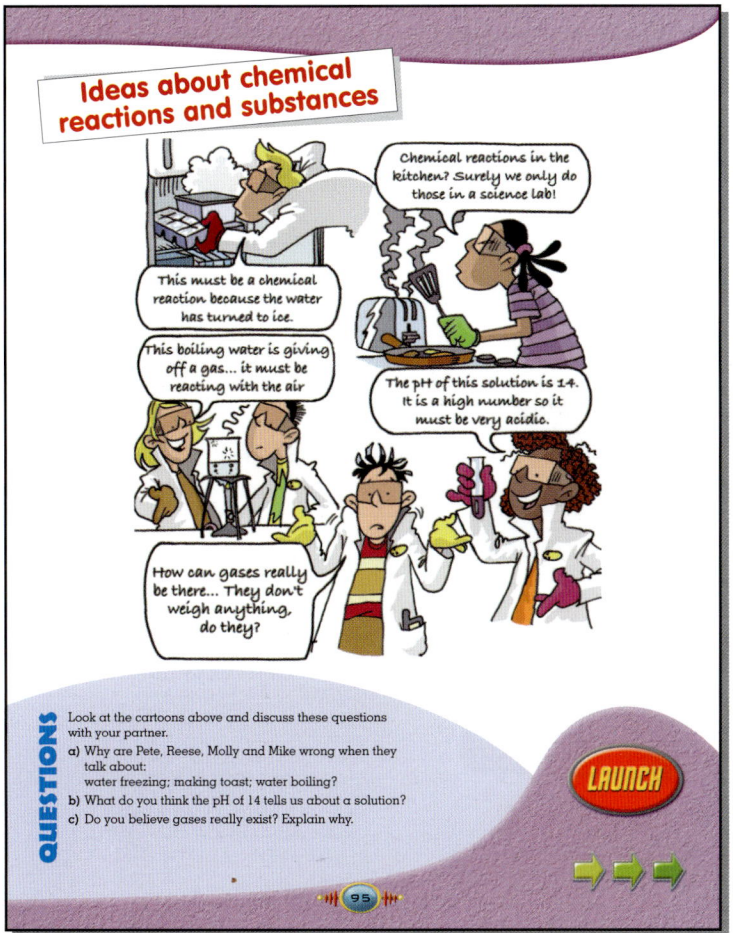

Launch activity notes – Ideas about chemical reactions and substances

- This activity presents pupils with a series of common misconceptions. They can be discussed either in small groups, who subsequently report back their finding to the class, or within the class as a whole.
- Pupils should already have some knowledge of physical and chemical change and be able to identify suitable examples. This activity should get them thinking in more general terms about chemical reactions in everyday contexts as well as in the laboratory.
- It will also provide an introduction to gases and provide a suitable starting point for naming some common gases and discussing their properties and how they might be detected using simple tests.

Learning styles

Visual
- **Observing.**
- **Tabulating**.
- **Displaying** data.

Kinaesthetic
- **Manipulating** apparatus.
- **Carrying out** experiments.

Intrapersonal
- **Understanding.**
- **Reflecting.**

Answers

a) Water freezing is not a chemical reaction. Chemical reactions occur around us and inside us, not just in the laboratory. When water boils, it doesn't react with air but simply becomes a gas.
b) Pupils will already be familiar with the pH scale from work in 7E. Some pupils may think that acidity increases with pH so that a pH of 14 is a strong acid rather than a strong alkali.
c) Pupils are unlikely to have seen a coloured gas so their perceptions are built around substances they cannot in fact see. The discussion should be directed towards indirect evidence such as something burning in air showing that a gas is present that supports combustion. Also moving a paper windmill through the air causes it to go around as the invisible air pushes on it.

7F1 Observing reactions

National framework/QCA SoW references

QCA SoW (7F)
What is a chemical reaction?

Learning objectives

▸ To appreciate the sorts of changes that may take place during an experiment and how best to observe them.
▸ To know how to interpret observations.
▸ To recognise the observations that indicate a chemical reaction is taking place.
▸ To be aware that in a chemical reaction 'reactants' are changed into 'products'.

Teaching strategy

Key points to highlight

- Chemical reactions are going on all around us and inside us.
- Commonly observed reactions include the combustion of fuels and the cooking of food.
- During a chemical reaction, substances undergo change.
- In a chemical reaction, the starting materials are called the 'reactants' and they are converted into 'products'.
- Evidence that a chemical reaction is taking place includes change in physical appearance, colour change, the evolution of a gas and the evolution of heat.

Amazing science

The colours we see in a firework display are the result of using particular metals and metal compounds. The elements used include: barium (green), calcium (red/orange), copper (blue), sodium (yellow) and strontium (red).

Lesson starter suggestions

- **What chemical reactions might we observe in our everyday lives?** Discuss processes such as cooking, the burning of fuels, the use of soaps and detergents, fireworks and any other examples suggested by pupils.
- **What evidence is there that a chemical reaction is taking place?** What changes do we observe?
- **What chemical reactions did we see in the previous unit?** Remind pupils of the neutralisation reaction between acids and alkalis.

Activity and technicians' notes

Heating iron wool

Safety notes

Iron is a good conductor of heat. Iron wool must be held using tongs when it is heated.

Equipment and materials required

- Iron wool

For each group:
- A Bunsen burner
- A heatproof mat
- A pair of tongs

Teaching strategy contd.

Difficulties/Misconceptions
- Many pupils perceive a chemical reaction to be something that only takes place in a test tube in the laboratory. They may have never considered chemical reactions in the context of everyday processes, such as cooking and the burning of fuels.
- Pupils may think the term 'observation' is limited to those changes that can be seen. They should appreciate that in chemistry it is used in a more general sense about any change that can be detected, such as a change in temperature, or the evolution of a pungent gas.

Skills
This lesson is mainly given over to practical activities:
- Initially, pupils will need to use their manipulative and observational skills when heating iron wool.
- The experiment on observing reactions also requires them to use their observational skills.

Extension ideas
Pupils could find out more about the rockets used at firework displays. For example, how they are propelled into the air and how the various colours and patterns are achieved.

Learning styles

Kinaesthetic
- **Carrying out** chemical reactions.

Visual
- **Observing** the changes that take place during chemical reactions.

Activity and technicians' notes contd.

Observing reactions
Safety notes
Safety glasses must be worn when handling the chemicals in these reactions, if they come into contact with the eyes some will cause irritation.

Equipment and materials required
- Plaster of Paris
- Lemon juice
- Bicarbonate of soda (sodium hydrogen-carbonate)
- Baking power

For each group:
- A yoghurt pot and wooden stirrer
- Two small beakers

Answers to questions

Text
Q1 Heat the sparkler.
Q2 No, because the substance on the sparkler has changed into a different substance.

Summary
1. substances are formed
2. a) reactants
   b) products
3. e.g. burning, cooling
4. A chemical reaction has taken place.

Suggestions for plenaries

- **What evidence might there be that a chemical reaction is taking place?** What might we observe?
- **How do substances change during a chemical reaction?** What changes can we see? How else might we detect changes?
- **How can we represent a chemical reaction?** Reactions changing to products, introducing the idea of a word equation.

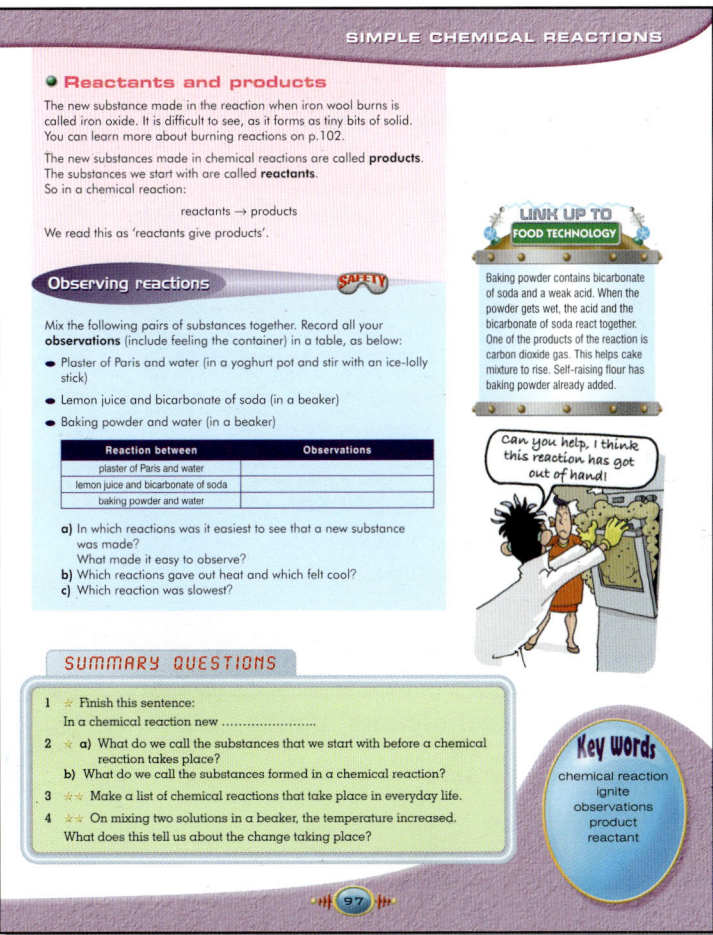

Learning outcomes

Following this, most pupils will:
▸ Identify that some new materials are formed during a chemical reaction.
▸ Understand that during a chemical reaction, reactants are turned into products.

Some pupils will also:
▸ Predict the products of a reaction.

7F2 Acids and metals

National framework/QCA SoW references

National framework – Scientific enquiry:
▶ Present and interpret experimental results through the routine use of tables, bar charts and simple graphs, including line graphs.

QCA SoW (7F)
How do acids react with metals?
NC links: Sc1 2i; Sc3 3a, e, g.

Learning objectives

▶ To appreciate that acids corrode metals.
▶ To observe the changes that take place when some metals react with acids.
▶ To record observations carefully and accurately.
▶ To spot patterns in data that are collected or given.
▶ To be aware that some results may not fit into a pattern.
▶ To appreciate that when metals react with acids, new substances are formed.
▶ To carry out a simple test for hydrogen gas.

Teaching strategy

Key points to highlight

- Most, but not all, metals react with common laboratory acids.
- When a metal reacts with an acid, one of the products is hydrogen gas.
- Hydrogen is less dense than air.
- When mixed with air and ignited, hydrogen explodes with a loud 'pop'.

Difficulties/Misconceptions

- Pupils will have little trouble in observing the evolution of gas when they react metals with dilute acids, however, they may not appreciate that this reaction can be more or less vigorous depending on the metal used. They should make their observations as detailed as possible.
- Pupils may think that acid–metal reactions occur with all metals. They should be made aware that it works for most, but not all metals.

Skills

This lesson is mainly given over to practical activities involving reacting metals with dilute acids. Pupils are also required to carry out a test for hydrogen gas.

Lesson starter suggestions

- **Are all metals the same?** Do they have the same physical properties? How can we find out if they have the same chemical properties?
- **What might we observe if a chemical reaction takes place between a metal and an acid?** Evolution of gas, increase in temperature.
- **What is hydrogen?** How can we make it? What are its properties? How can we test for it?

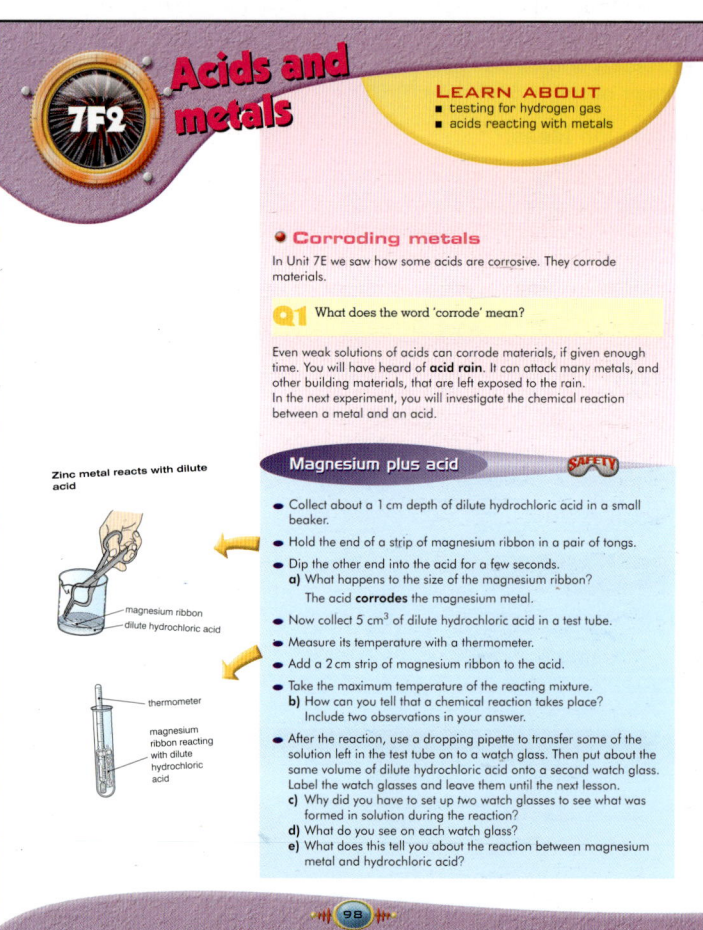

Activity and technicians' notes

Magnesium plus acid

Safety notes

- Safety glasses must be worn when handling dilute hydrochloric acid.
- Pupils should be warned to handle thermometers with care, and not to use them to mix the dilute hydrochloric acid and magnesium. Any mercury spillage resulting from a broken thermometer should be dealt with immediately.

Equipment and materials required

- Dilute hydrochloric acid (0.4 M)
- Magnesium ribbon

For each group:
- A small beaker
- A test tube
- A pair of tongs
- A thermometer
- A dropper pipette
- Two watch glasses
- A boiling tube
- A wooden splint

Activity and technicians' notes contd.

Testing for hydrogen gas
Safety notes
- Safety glasses must be worn when handling dilute hydrochloric acid.
- Remind pupils not to put the lighted splint inside the boiling tube.

Equipment and materials required
- Dilute hydrochloric acid
- Magnesium ribbon

For each group:
- Boiling tube
- Test tube
- A wooden splint

Investigating metals and acids
Safety notes
- Safety glasses must be worn when handling dilute acids.
- Pupils should not touch calcium metal.

Equipment and materials required
- Dilute hydrochloric acid (0.4 M)
- Dilute nitric acid (0.4 M)
- Dilute sulphuric acid (0.4 M)
- Zinc foil
- Copper filings
- Iron filings
- Calcium pellets

For each group:
- A set of test tubes

Teaching strategy contd.

Notes and tips
The relative vigour of the acid–metal reactions will reflect the position of each metal in the reactivity series. The difference in reactivity could be pointed out, but any reference to the reactivity series should not be made until later in the course.

Extension ideas
Pupils could carry out simple calculations on proportionality, in order to predict the volume of hydrogen gas produced by a given mass of magnesium.

Learning styles

Kinaesthetic
- **Carrying out** chemical reactions between acids and metals.

Visual
- **Observing** the reaction of acids on metals.

Answers to questions

Text
- **Q1** eaten away by a chemical reaction
- **Q2** magnesium, dilute hydrochloric acid
- **Q3** hydrogen, magnesium chloride

Summary
1. acids, hydrogen, burning, pop

Amazing science
Galaxies are thought to form from huge clouds of hydrogen gas that condense and are then compressed by gravity. This initiates nuclear reactions leading to the formation of other elements. The average density of the Universe is thought to be in the region of one atom of hydrogen per cubic kilometre.

Suggestions for plenaries
- **What happens when a metal is added to an acid?** What might we observe? Name the products.
- **How do we make and test for hydrogen?** Using an acid–metal reaction.
- **Do all metals react in exactly the same way with acids?** Which are most reactive? Which don't react at all? Does it matter which acid we use?

Learning outcomes

Following this, most pupils will:
▶ Generalise that hydrogen is formed when acids react with metals and describe a test for hydrogen.
▶ Work safely with acids.

Some pupils will also:
▶ Name the products of acid–metal reactions.

7F3 Acids and carbonates

National framework/QCA SoW references

National framework – Scientific enquiry:
▸ Describe and explain what their results show when drawing conclusions: begin to relate conclusions to scientific knowledge and understanding.

QCA SoW (7F)
How do acids react with carbonates?
NC links: Sc1 2e, g, i, k; Sc3 3e.

Learning objectives

▸ To observe the changes that take place when metal carbonates react with acids.
▸ To record observations carefully and accurately.
▸ To spot patterns in observations and make generalisations from them.
▸ To appreciate that when metal carbonates react with acids new substances are formed.
▸ To carry out a simple test for carbon dioxide gas using limewater.
▸ To discuss some uses of carbon dioxide.

Teaching strategy

Key points to highlight
- Carbonates are compounds of metals.
- All carbonates react with common laboratory acids.
- When a carbonate reacts with an acid, one of the products is carbon dioxide gas.
- Carbon dioxide is more dense than air.
- Carbon dioxide turns limewater milky.
- Carbon dioxide is used to make drinks fizzy.
- Carbon dioxide is used in fire extinguishers because it is denser than air and doesn't support combustion.

Difficulties/Misconceptions
The reaction between calcium carbonate and dilute sulphuric acid subsides quite quickly, due to the formation of a layer of insoluble calcium sulphate on the calcium carbonate.

Skills
- This lesson is mainly given over to practical activities involving reacting metal carbonates with dilute acids.
- Pupils are also required to carry out a test for carbon dioxide gas.

Lesson starter suggestions

- **What are metal carbonates?** How can we find out about their chemical properties?
- **What might we observe if a chemical reaction takes place between a metal carbonate and an acid?** Evolution of gas, increase in temperature.
- **What is carbon dioxide?** How can we make it? What are its properties? How can we test for it?

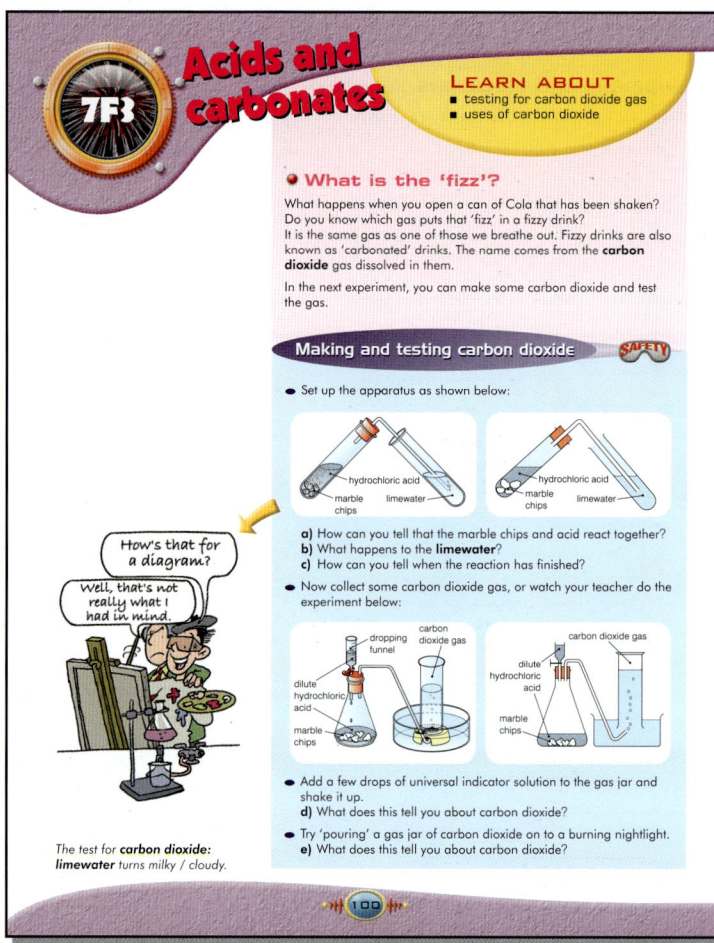

Activity and technicians' notes

Making and testing carbon dioxide

Safety notes
Safety glasses must be worn when handling dilute hydrochloric acid.

Equipment and materials required
- Dilute hydrochloric acid (2 M)
- Marble chips
- Limewater
- Universal indicator

For each group:
- Two test tubes
- A conical flask
- A dropping funnel
- A delivery tube
- A gas jar

Teaching strategy contd.

Notes and tips

If pupils are careful, they will be able to pour carbon dioxide from one test tube into another. Since carbon dioxide is colourless they will not be able to observe anything until a small amount of limewater is placed in the second tube, which will turn milky on shaking.

Extension ideas

Pupils could investigate why limewater goes cloudy and what happens if you continue to bubble carbon dioxide through limewater after it has gone cloudy.

Activity and technicians' notes contd.

Investigating carbonates and acid
Safety notes

Safety glasses must be worn when handling dilute hydrochloric acid.

Equipment and materials required

- Dilute hydrochloric acid, dilute nitric acid, dilute sulphuric acid
- Calcium carbonate, copper carbonate, iron carbonate, lead carbonate, sodium carbonate, zinc carbonate
- Limewater
  Other acids and carbonates may be added or substituted depending on availability and the time available.

For each group:
- Two test tubes
- A delivery tube

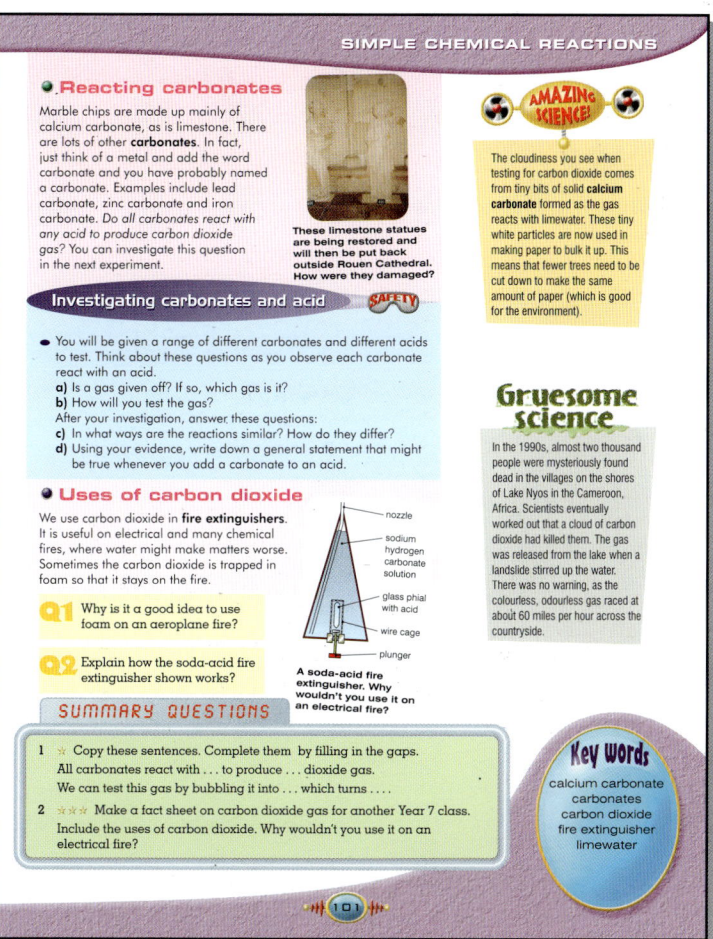

Learning styles

Kinaesthetic
- **Carrying out** chemical reactions between acids and carbonates
- **Carrying out** a test for carbon dioxide

Visual
- **Observing** the reaction of acids on carbonates
- **Observing** the test for carbon dioxide

Answers to questions

Text

Q1 The foam contains bubbles of carbon dioxide that stay on the fire and prevent oxygen from getting to it.

Q2 When they are mixed, the soda and the acid react together giving off carbon dioxide gas. This creates a pressure which forces the solution out of the nozzle.

Summary

1. acids, carbon, limewater, milky
2. Carbon dioxide is a colourless, odourless gas. Carbon dioxide is a chemical compound formed from carbon and oxygen. The chemical formula for carbon dioxide is CO_2. Carbon dioxide is more dense than air, so a balloon filled with carbon dioxide will fall to the ground. Carbon dioxide does not burn. Carbon dioxide is used in fire extinguishers to put out fires. Carbon dioxide is produced by burning fossil fuels.

Learning outcomes

Following this, most pupils will:
▶ Generalise that carbon dioxide is released when acids react with carbonates.
▶ Describe a test for carbon dioxide.
▶ Work safely with acids.

Some pupils will also:
▶ Name the products of acid–carbonate reactions.

Suggestions for plenaries

- **What happens when a metal carbonate is added to an acid?** What might we observe? Name the products.
- **How do we make and test for carbon dioxide? What is carbon dioxide used for?** Using an acid–carbonate reaction. Testing for carbon dioxide.
- **Do all metal carbonates react in exactly the same way with acids?** Do they all react? Are any more reactive than others? Does it matter which acid we use?

7F4 About combustion

National framework/QCA SoW references

QCA SoW (7F)
What new substances are made when materials burn in air or oxygen?
What is needed for things to burn?
NC links: Sc3 3a.

Lesson starter suggestions

- **What are the three parts of the fire triangle?** What happens in combustion reactions? How can fires be put out?
- **What are the main gases found in air?** Does air behave like dilute oxygen? How would things burn in pure oxygen?
- **What kind of substances form during combustion reactions?** Reaction with oxygen to form oxides.

Learning objectives

▶ To appreciate what is needed for things to burn and recall the fire triangle.
▶ To learn how to extinguish fires.
▶ To carry out combustion reactions carefully and safely.
▶ To appreciate that when things burn, new substances called 'oxides' are formed.
▶ To understand that substances burn more vigorously in pure oxygen than in air.
▶ To write word equations for combustion reactions.

Teaching strategy

Key points to highlight

- Oxygen is needed for things to burn.
- Starving a fire of oxygen will extinguish it.
- Burning is a chemical reaction in which a large amount of heat is given out.
- When substances burn, they combine with oxygen to produce oxides.
- The air around us is composed of approximately $\frac{1}{5}$ oxygen and $\frac{4}{5}$ nitrogen.
- Air behaves like dilute oxygen.
- Magnesium burns in air with a bright flame to form magnesium oxide.
- A word equation has the form:
  reactants → products
- magnesium + oxygen → magnesium oxide
- Combustion reactions are much more vigorous in pure oxygen than in air.
- Non-metal oxides dissolve in water to form acids.
- Metal oxides dissolve in water to form alkalis.
- Oxygen relights a glowing wooden splint.

Difficulties/Misconceptions

Pupils will not appreciate just how reactive oxygen is, because they are used to seeing combustion taking place in air, which contains only about $\frac{1}{5}$ of oxygen.

Skills

This lesson involves some practical activity on burning magnesium, however the majority of the time will be taken up with discussion of combustion and a demonstration of combustion in pure oxygen. Pupils will need to use organisational skills to tabulate their observations, and to compile word equations.

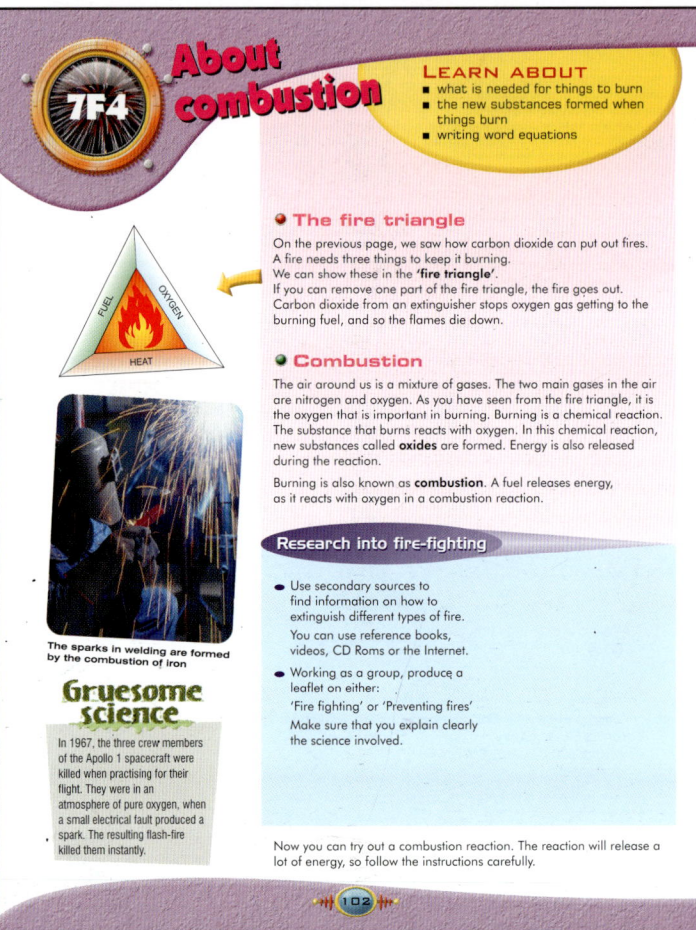

Gruesome science

Pure oxygen is a very reactive substance, however liquid oxygen is even more reactive. Liquid oxygen is a pale blue liquid that boils at −183 °C. It is used in rocket fuels and in liquid oxygen explosives (LOX), where is combines with fine particles of carbon.

Teaching strategy contd.

Notes and tips
- Burning substances in pure oxygen is even more spectacular if carried out in a darkened room.
- Part of the demonstration could be burning magnesium in air alongside burning it in pure oxygen, so pupils can see the contrast.
- As part of the demonstration, the oxides formed by non-metals and metals are tested with universal indicator. Pupils may see the pattern of non-metals giving acids and metals giving alkalis, but this should not be taken any further at present.

Extension ideas
Pupils could use the knowledge they have gained about burning to carry out research into the best ways of dealing with fires involving a variety of materials.

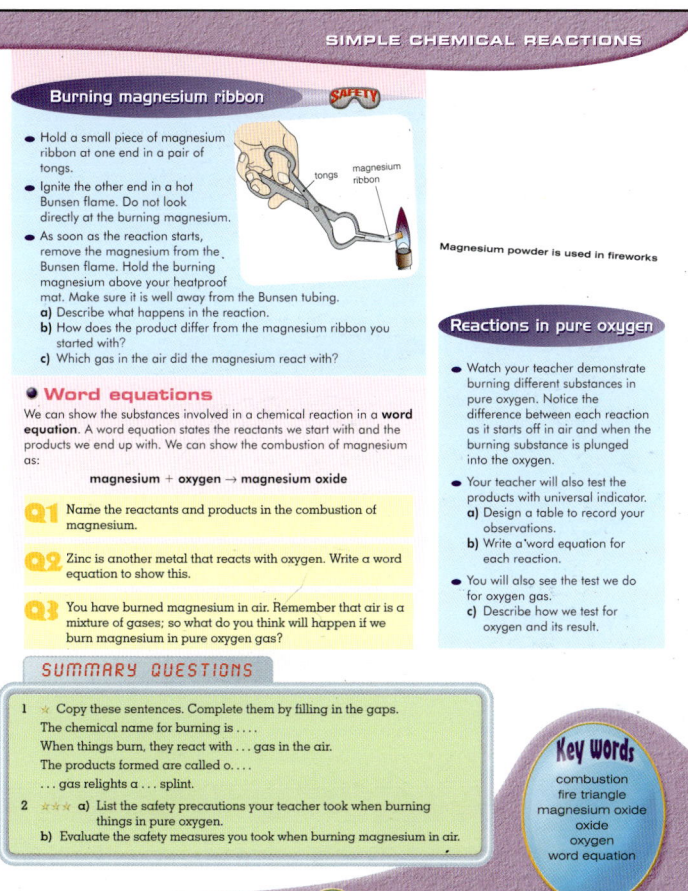

Activity and technicians' notes

Burning magnesium ribbon
Safety notes
- Magnesium gets very hot when burning in air. Magnesium ribbon must be held using tongs when it is heated.
- Burning magnesium produces a very bright light and should not be looked at directly.

Equipment and materials required
- Magnesium ribbon

For each group:
- A pair of tongs
- A Bunsen burner

Reactions in pure oxygen
Equipment and materials required
- Magnesium cut into 2 cm lengths
- Sulphur
- Carbon
- Iron wool
- Wooden spill
- Deflagrating spoon
- Gas jars of oxygen
- Universal indicator

Learning styles

Kinaesthetic
- **Burning magnesium** in air

Visual
- **Observing** the burning of magnesium in air.
- **Observing** the combustion of other substances in pure oxygen.
- **Observing** the test for oxygen.

Learning outcomes

Following this, most pupils will:
- Generalise that oxides are formed when materials burn.
- Describe burning as a reaction with oxygen.
- Work safely when burning materials.

Some pupils will also:
- Be able to use word equations to represent reactions in which materials burn.

Answers to questions

Text
- **Q1** Magnesium and oxygen are the reactants; magnesium oxide is the product.
- **Q2** zinc + oxygen → zinc oxide
- **Q3** It will burn more vigorously.

Summary
1. combustion, oxygen, oxides, oxygen, glowing

Suggestions for plenaries
- **What gas is needed for things to burn?** Oxygen from the air.
- **Why do substances burn more vigorously in pure oxygen than they do in the air?** Air acts like dilute oxygen.
- **How can we present combustion reactions using word equations?**
  fuel + oxygen → products

7F5 Investigating burning

National framework/QCA SoW references

National framework – Scientific enquiry:
- Use scientific knowledge to decide how ideas and questions can be tested: make predictions of possible outcomes.
- Use repeat measurements to reduce error and check availability.
- Present and interpret experimental results through the routine use of tables, bar charts and simple graphs, including line graphs.
- Describe and explain what their results show when drawing conclusions: begin to relate conclusions to scientific knowledge and understanding.
- Evaluate the strength of evidence, e.g. in bar charts and graphs: indicate whether increasing the sample would have strengthened the conclusions.

QCA SoW (7F)
What is produced when fuels burn?
What is needed for things to burn?
NC links: Sc1 2a, f, g, h, i, k, l, p.

Learning objectives

- To be aware that coal, oil and natural gas are all examples of fossil fuels.
- To identify the new substances formed when things burn.
- To write a word equation for the combustion of methane and other fuels.
- To carry out a simple test for presence of water using blue cobalt chloride paper or anhydrous copper sulphate.
- To carry out an investigation on the combustion of a night-light.

Teaching strategy

Key points to highlight

- Fuels are substances that burn and give out energy as heat and light.
- Burning fuels uses up oxygen.
- Fossil fuels were formed by the decay of dead plants and animals.
- Coal, oil and natural gas are fossil fuels.
- Fossil fuels contain carbon, which becomes carbon dioxide when they are burnt.
- Methane is the main gas found in natural gas.
- When methane burns, carbon dioxide and water are formed.
- Water turns blue cobalt chloride paper pink.
- Water turns white anhydrous copper sulphate blue.

Difficulties/Misconceptions

Pupils should appreciate that testing for water using blue cobalt chloride paper, or anhydrous copper sulphate, shows the presence of water but gives no indication of its purity.

Lesson starter suggestions

- **What is a fuel?** In particular, what is a fossil fuel? Who can give an example?
- **What are the products of burning a fuel?** If the fuel is just carbon, like coal? If the fuel contains carbon and hydrogen, like methane?
- **What determines how long something will burn?** Will a candle in a jar burn forever?

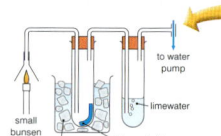

Activity and technicians' notes

Burning methane
Equipment and materials required

- A glass funnel with delivery tube at right angles
- Two boiling tubes fitted with glass tubing to act as traps
- A beaker of crushed ice
- Limewater
- Blue cobalt chloride paper
- Anhydrous copper sulphate
- A micro-burner
- Stands and clamps

Gruesome science

Fires at oil wells are often extinguished by carrying out explosions next to the wellhead. This produces a blast of air and removes the oxygen from around the fire. It is a bit like blowing a candle out on a birthday cake.

Teaching strategy contd.

Skills

This lesson begins with a demonstration of the combustion of methane and a test for water. However, the main part of the lesson is a practical activity:
- Initially, pupils will need to use their manipulative and observational skills in discovering how long a night-light candle burns under different-sized beakers.
- Subsequently, they will need to devise a suitable means of recording and displaying results.
- Finally, they will need to evaluate their method and suggest ways in which it could be improved.

Notes and tips

If pupils decide to carry out the experimental procedure on each beaker more than once, as they should, they need to be certain that the air in the beaker is refreshed each time; otherwise subsequent readings will be inaccurate.

Extension ideas

Pupils could guess which candle would be extinguished first when candles of different heights are placed under a large beaker.

Learning styles

Visual
- **Observing** the combustion of methane.
- **Observing** the test for water.

Kinaesthetic
- **Carrying out** an investigation of a night-light using apparatus.

Intrapersonal
- **Reflecting on and evaluating** an investigation.

Answers to questions

Text

Q1 three of: coal, coke, fuel oil, natural gas, wood

Q2 methane and oxygen

Summary

1. carbon, oxygen, energy, combustion, carbon dioxide, water
2. butane + oxygen → carbon dioxide + water
3. propane, natural gas, petrol, diesel, kerosene, fuel oil

Activity and technicians' notes contd.

Investigating burning

Equipment and materials required

For each group:
- A night-light candle
- A variety of different-sized beakers
- A stopwatch, or sight of a clock reading to the nearest second

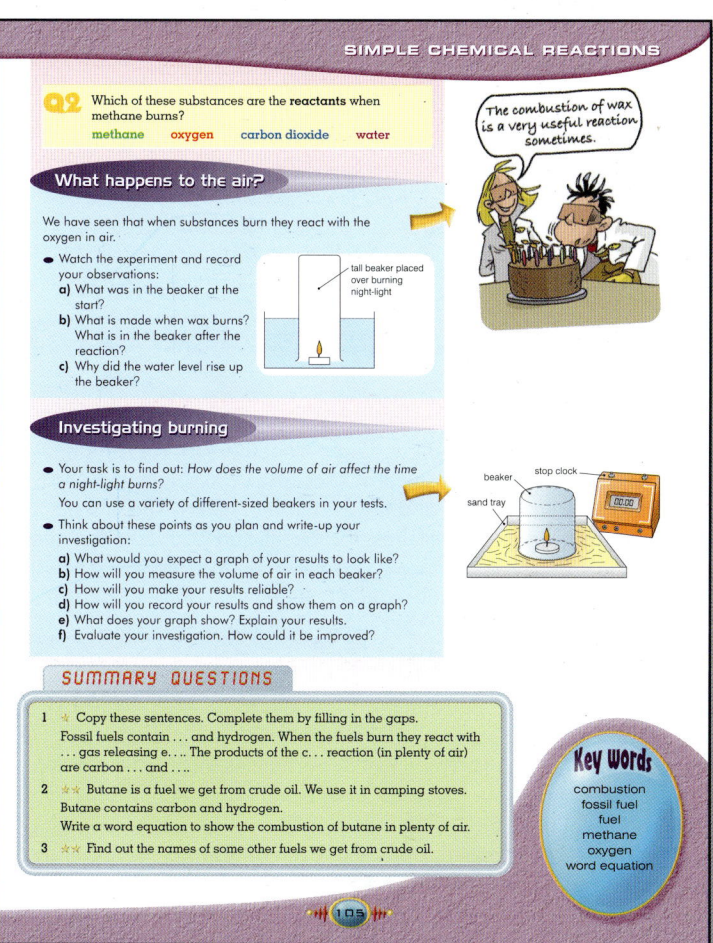

Learning outcomes

Following this, most pupils will:
- Describe burning as a reaction with oxygen.
- Describe a test for the presence of water.
- Obtain and present quantitative results.
- Identify patterns in these quantitative results.
- Work safely when burning materials.
- Suggest how to test an idea about burning.
- Obtain results that can be represented as a line graph.

Some pupils will also:
- Predict that carbon dioxide and water will be made when a hydrocarbon burns.
- Use word equations to represent reactions in which materials burn.
- Evaluate how well ideas about burning match the data.

Suggestions for plenaries

- **What happens when a fuel burns?** In what forms is energy released? What are the products of burning?
- **What are fossil fuels?** From what are they formed?
- **What is the relationship between the volume of available air and the time something will burn?** Do things burn longer when there is more air?

7F Read all about it!

Teaching strategy

The Hindenburg disaster
Difficulties and misconceptions

- Some pupils may assume that hydrogen and helium have similar properties because they can both be used in balloons. This is true of their physical properties to some extent but pupils should appreciate that their chemical properties are very different.
- Pupils researching into the uses of hydrogen may come across the term 'hydrogen bomb'. They should not be misled into thinking that such a device is simply a large canister of hydrogen that explodes when ignited. The source of energy in a hydrogen bomb is a nuclear fusion reaction.

Notes and tips

- In recent years hydrogen has emerged as an important fuel for use in fuel cells. These are devices in which the energy from a chemical reaction is converted directly into electricity; but, unlike an electric cell or battery, a fuel cell does not run down or need recharging. It continues to operate as long as it is supplied with fuel and an oxidizer.
- A fuel cell consists of:
  - an anode which fuel such as hydrogen is supplied
  - a cathode to which an oxidant such as oxygen is supplied

  The two electrodes of a fuel cell are separated by an ionic conductor electrolyte. The overall reaction in the cell is:

  Hydrogen + Oxygen → Water

  The fuel cell voltage in this case is about 1.2 V but decreases as the load is increased. The water produced at the anode has to be removed continuously in order to avoid flooding the cell.
- The potential use of hydrogen as the fuel of the future provides an opportunity to discuss the problems of global warming associated with carbon-based fuels.
- Pupils will not be familiar with electrolysis. This is a good opportunity to introduce the process as a means of bringing about chemical change using an electric current.
- The use of helium in modern balloons could be used to introduce the noble gases as a group of very unreactive gases. Pupils may have heard of some of the other noble gases such as argon and neon because of their use in lighting.
- The discovery of helium is an interesting story since it was discovered on the Sun before it was discovered on Earth. Helium was first identified by Sir Joseph Lockyer in

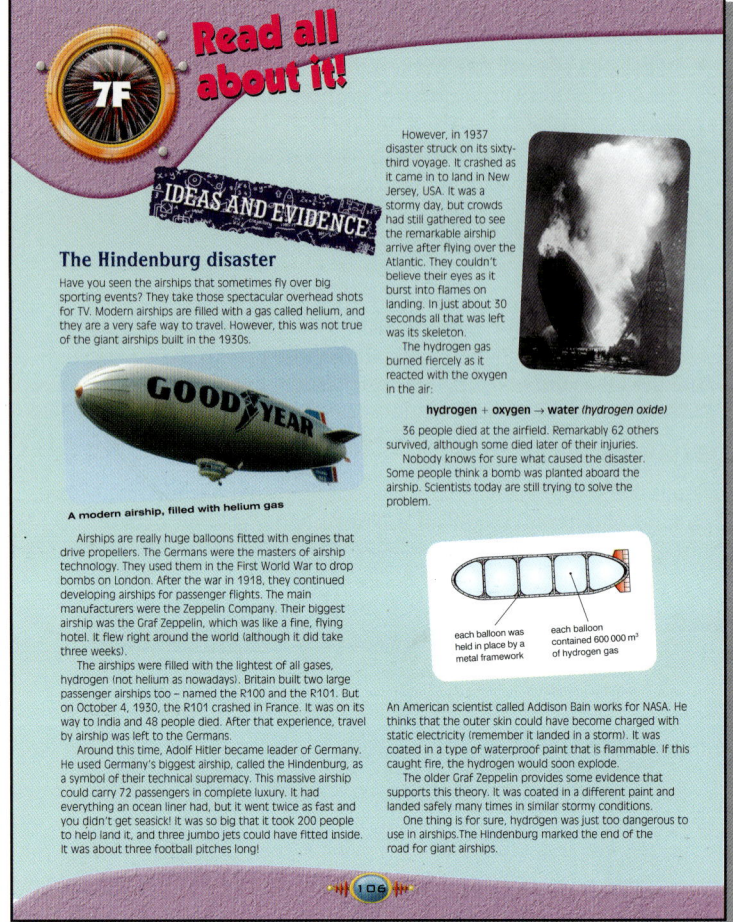

1868 as a result of analysing light from the Sun. He realised that a particular set of spectral lines was from an 'as then' undiscovered element and it was named helium. The first part of the name comes from Helios, the Greek sun god, and the –ium ending was given because it was assumed that the element was a metal. Helium was subsequently found on Earth in 1895 by Sir William Ramsay and turned out to be a gas!

Skills

This unit has many opportunities for pupils to develop the Sc1 skills of manipulating apparatus, making observations and recording and displaying results. There are opportunities for pupils to make generalisations about the reaction of acids with metals and with carbonates. Pupils may work individually or in groups researching some of the issues raised, and as a result of research, and perhaps debate, to evaluate and draw conclusions.

Extension ideas

Pupils could work together to research in more detail one of the following topics:
- How is hydrogen obtained from water?
- Could hydrogen take the place of fossil fuels in the future?
- Do hot-air balloons work on the same principle as hydrogen-filled and helium-filled balloons?
- How different are the chemical properties of hydrogen and helium?
- Is there a role for huge airships like the Hindenburg in modern society?
- How is hydrogen used to power space rockets?

Danger – common errors

The differences between physical changes and chemical changes are sometimes a source of confusion. In reaching a decision, pupils should focus on two questions:
1 has any new substance (as opposed to the same substance in a different state) been formed?
2 is the change easy to reverse?

Learning styles

Visual
- **Researching** information on hydrogen and balloons.

Interpersonal
- **Discussing** the possible role of airships in modern society.

Intrapersonal
- **Understanding** how hydrogen is obtained from water.
- **Evaluating** the potential of hydrogen as a fuel.

Learning outcomes

Following this unit, most pupils will be able to:
▸ State that some new materials are formed during a chemical reaction.
▸ Recall that during a chemical reaction, reactants are turned into products.
▸ Generalise that hydrogen is formed when acids react with metals.
▸ Generalise that carbon dioxide is formed when acids react with carbonates.
▸ Generalise that oxides are formed when materials burn.
▸ Describe burning as a reaction with oxygen.

Some pupils will also:
▸ Name the products of a given chemical reaction.
▸ Use word equations to represent given reactions.

In scientific enquiry:

Following this unit, most pupils will:
▸ Describe a test for hydrogen.
▸ Describe a test for carbon dioxide.
▸ Describe a test for the presence of water.
▸ Obtain and present quantitative results, display data and identifying patterns.
▸ Carry out practical activities safely.

Some pupils will also:
▸ Predict that carbon dioxide and water will be made when a hydrocarbon burns.
▸ Evaluate how well predictions match collected data.

Teaching strategy contd.

Questions
1 What gas is used in modern airships? Why is this better than using hydrogen?
2 How were airships used in the First World War?
3 Give an advantage and a disadvantage of flying by airship in the 1920s compared with flying by an airliner today.
4 What evidence is there to support Addison Bain's idea about the disaster of the Hindenburg?
5 Design a poster for a company trying to start up passenger flights by airship again.

Answers
1 Helium; it is better than hydrogen because it is non-flammable
2 to drop bombs on London
3 they were very luxurious to travel in, but very slow
4 the older Graf Zeppelin was coated in a different paint and landed safely many times in stormy conditions.

7F Unit review

Answers to review questions

1. **a** combustion
   b flour + oxygen → carbon dioxide + water
   c Flour dust reacts very quickly releasing a lot of energy, carbon dioxide and water vapour.
   d highly flammable (or explosive)
   e The match would probably go out, because there wouldn't be enough oxygen in the flour for it to burn.

2. **a** oxygen
   b sulphur, oxygen
   c sodium oxide
   d carbon dioxide, water

6. **a**

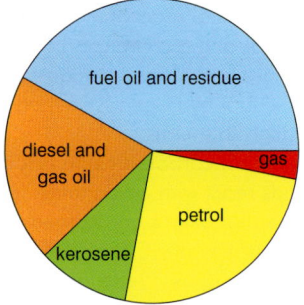

b **i)**

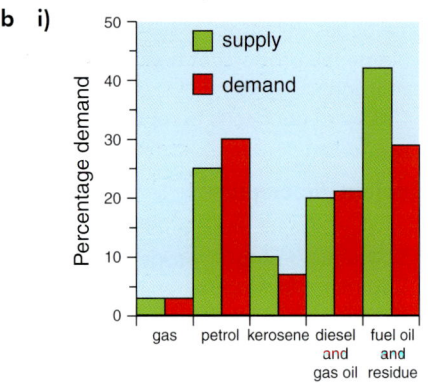

ii) The demand for petrol, diesel and gas oil is greater than the supply, while the demand for kerosene, fuel oil and residue is less than the supply.

7. **a**

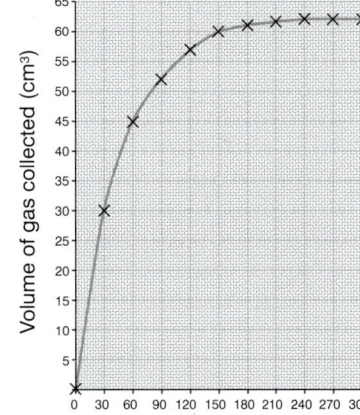

UNIT REVIEW

REVIEW QUESTIONS
Understanding and applying concepts

1. When flour burns in a good supply of oxygen, we get carbon dioxide and water (in the form of gas) produced. Inside flour mills there are strict safety measures in place. Any naked flame or spark could cause an explosion because of flour 'dust' in the air.
   a What is the scientific word for 'burning'?
   b Write a word equation to show flour burning.
   c Explain fully how flour 'dust' in a confined space could cause an explosion.
   d What hazard signs would you display inside a flour mill?
   e What do you think would happen if you dropped a lighted match into a bowl of flour? Explain your answer.

2. Copy and complete the following word equations:
   a copper + ... → copper oxide
   b ... + ... → sulphur dioxide
   c sodium + oxygen → ...
   d methane + oxygen → ... + ...

3. Draw a concept map linking these terms together. Don't forget to label your links, explaining what the connection is.
   burning   product   oxides   acid
   carbonate   hydrogen   reactant   oxygen

Ways with words

4. Write a short poem about burning. Make sure you use your knowledge of the fire triangle within your poem.

5. Write your own definitions of the following words:
   a fuel **b** oxygen **c** limewater **d** ignite

Making more of maths

6. Crude oil is a source of many different fuels.
   a Draw a pie chart that shows the different fuels we get from this sample of crude oil from the North Sea:

Fuel	Approximate percentage
gas	3
petrol	25
kerosene	10
diesel and gas oil	20
fuel oil and residue	42

   b The supply of fuels from crude oil does not always match the demand for them. Look at this data:

Fuel	Demand (%)
gas	3
petrol	30
kerosene	7
diesel and gas oil	21
fuel oil and residue	29

   i) Think of a way to display visually the data in both tables above, highlighting the differences between the supply and demand of fuels in North Sea oil.
   ii) Comment on the differences between the supply and demand of different fuels in North Sea oil.

7. Pete and Pip looked at the reaction between magnesium and dilute sulphuric acid. They measured how much hydrogen gas was given off every 30 seconds. Here are their results:

b It starts very rapidly and gradually slows down until it stops.

c

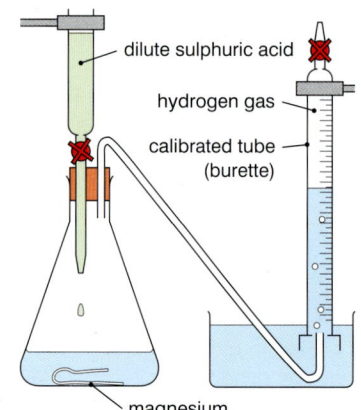

d concentration of sulphuric acid, volume of sulphuric acid used, temperature of sulphuric acid, mass of magnesium used

UNIT REVIEW QUESTIONS

Time (s)	Volume of gas collected (cm³)
0	0
30	30
60	45
90	52
120	57
150	60
180	61
210	61.5
240	62
270	62
300	62

a Draw a line graph, using a line (curve) of best fit, to show their results.
b What does your graph tell you about the reaction between magnesium and dilute acid?
c Draw the apparatus that Pete and Pip could have used to carry out their experiment.
d They wanted to find out if cutting the magnesium up into smaller pieces made any difference to the reaction. What variables would they have to control (keep the same) to make sure it was a fair test?

Extension question

3 Find out about the chemical reaction used by brewers to make alcohol, and by bakers to make bread rise. Write an information sheet on the reaction to help shoppers understand that chemical reactions play an important part in everyday life.

SAT-STYLE QUESTIONS

1 Here is a key you can use to identify gases:

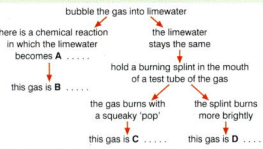

a What are the missing words, A to D? (4)

b i) What would you see if you burned a piece of magnesium ribbon in gas D? (1)
ii) What safety precautions would you (or your teacher) take when doing this experiment? (2)
iii) Write a word equation for the reaction in part b i). (2)

2 A candle is burned under a gas jar in a sand tray:
a When the candle burns a chemical reaction takes place.
Name two of the products of this reaction. (2)

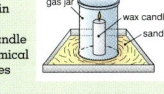

b As the candle burns, wax gets used up. Name two other ways in which we can tell that a chemical reaction takes place. (2)
c Explain what you see happen in the experiment above. (2)
d What safety precaution has been taken in the experiment? Why? (2)

3 A group of students were looking at the reaction between calcium carbonate and dilute hydrochloric acid. They decided to take the mass every minute.
a Which gas is given off in the reaction? (1)
b The students took readings from the balance for 10 minutes, by which time the reaction had been finished for 2 minutes. Use axes like the ones shown to sketch the shape of the graph you would expect. (2)

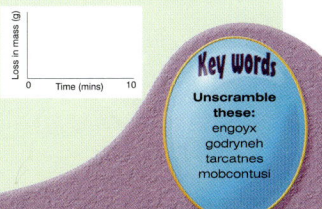

Key words
Unscramble these:
engoyx
godryneh
tarcatnes
mobcontusi

Answers to SAT-style questions

1 a A = milky (1)
B = carbon dioxide (1)
C = hydrogen (1)
D = oxygen (1)
b i) bright white light/white solid forms (1)
ii) wear eye protection and don't look directly at the flame (2)
iii) magnesium + oxygen → magnesium oxide (2)

2 a water and carbon dioxide (2)
b Heat and light are given off. (2)
c The candle burns while there is oxygen in the gas jar, but once the oxygen has been used up the candle goes out. (2)
d A sand tray catches the hot wax as it drips off the candle. (2)

3 a carbon dioxide (1)
b

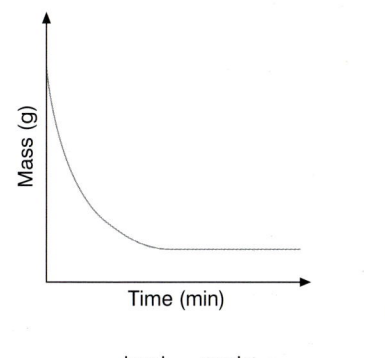

(2)

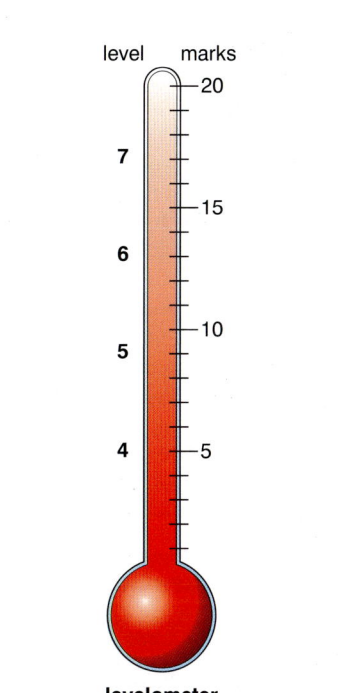

levelometer

7G The particle model

National framework/QCA SoW references

National framework – Scientific enquiry:
▸ Consider early scientific ideas, including how experimental evidence and creative thinking have been combined to provide scientific explanations.

National framework – Particles:
▸ Describe a simple particle model for matter, recognising:
 – the size, arrangement, proximity, attractions and motion of particles in solids, liquids and gases;
 – the relationship between heating and movement of the particles.
▸ Use the simple particle model to explain:
 – why solids and liquids are much less compressible than gases;
 – why heating causes expansion in solids, liquids and gases;
 – why diffusion occurs in liquids and gases;
 – why air exerts a pressure;
 – why changes of state occur;
 – why mass is conserved when substances dissolve to form solutions;
 – why temperature increases are likely to result in substances dissoving more quickly;
 – formation of a saturated solution.

QCA SoW (7G)

How can we explain evidence from experiments? (7G1)
How are theories created? (7G2)
What are the differences between solids, liquids and gases? (7G3)
How can the particle model explain the differences between solids, liquids and gases? (7G3)
How can the particle model explain other phenomena? (7G4)
NC links: Sc1 1c; Sc3 1b.

Learning objectives

▸ To be able to classify materials as solids, liquids and gases; but appreciating that some materials show a mixture of properties and are difficult to classify.
▸ To interpret and explain the results of experiments and appreciate how useful it is to discuss ideas and results with others.
▸ To understand how theories can be based on results obtained from experiments.
▸ To evaluate how well a theory fits the evidence. and appreciate that sometimes theories and models need to be modified in the light of new evidence.
▸ To use a model that has been based up things we cannot see directly, but we have inferred from the evidence collected.
▸ To appreciate that solids, liquids and gases are made up of particles, and to be able to explain their properties in terms of the behaviour of particles.
▸ To use the particle theory to explain phenomena, such as diffusion and pressure, in terms of the motion of particles.

Answers to questions

What do you remember?
1 freezing
2 condensing
3 It fills the shape of the container.

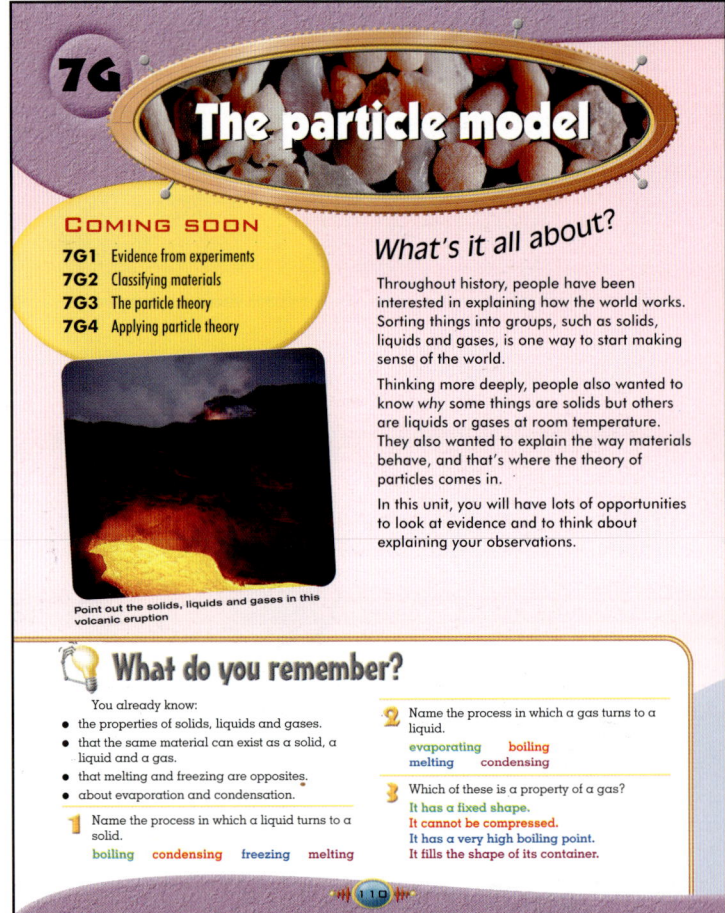

Notes to support 'What do you remember?'

- Pupils have already studied solids, liquids and gases in KS2 and they should be familiar with common properties of materials from their everyday experience.
- Reversible and irreversible changes also feature in the KS2 programme of study. The various reversible and chemical (irreversible) changes covered in units 7E and 7F should provide pupils with a sufficient reminder of their characteristics.

Teaching strategy

This unit builds on the contents of units 4D 'Solids, liquids and how they can be separated', 5C 'Gases around us', 5D 'Changing state' and 6C 'More about dissolving' in the KS2 scheme of work. It also uses ideas that were developed in the KS2 programme of study.

The unit provides a foundation for work in later units that lend themselves to explanation in terms of particle theory: for example; 7H 'Solutions' (dissolving), 8A 'Food and digestion' (digestion), 8H 'The rock cycle' (crystal size related to rate of cooling), 8I 'Heating and cooling' (changes of state) and 9L 'Pressure and moments' (behaviour of gases).

Difficulties/Misconceptions

Problems in this unit are likely to result from the abstract nature of the particle theory. It is difficult for pupils to accept a model based on inference rather than direct observation. Pupils should be encouraged to visualise the motion of particles in different states and consider how this can account for the differing properties.

Learning styles

Interpersonal
- Debating.
- Discussing.
- Evaluating.

Intrapersonal
- Considering.
- Understanding.
- Reflecting.

Kinaesthetic
- **Carrying out** experiments.

Visual
- Imagining.
- Observing.

Launch activity notes – Ideas about solids, liquids and gases

This unit is about the particle theory and how it is used to explain the properties and behaviour of substances. The launch activity is aimed at getting pupils thinking in terms of particles and trying to explain familiar behaviour in terms of particle motion.

Answers

a) Pupils will be familiar with the idea that some materials are more dense or less dense than others, although this is likely to be expressed as 'lighter' or 'heavier'. Explanations could be related to floating and sinking in water.

b) The particle theory is conceptually difficult and initially pupils have to accept much on trust. Explanations about elasticity may centre on the idea of pulling particles away from each other – but not completely away or the elastic would snap.

c) Although they won't think of it in terms of physical change, pupils will be familiar with how easy it is to change water into ice or steam and that the processes are readily reversible. From this they may deduce that it is not the particles themselves that change but the way in which they are arranged.

d) This idea may initially have some appeal. However, it does not explain changes of state. Pupils could be asked to extend this idea to changes of state, thereby realising that it cannot explain what one observes.

7G1 Evidence from experiments

National framework/QCA SoW references

National framework – Particles:
- Use the simple particle model to explain:
  - why solids and liquids are much less compressible than gases.
  - why heating causes expansions in solids liquids and gases.

QCA SoW (7G)

How can we explain evidence from experiments?

Learning objectives

- To be able to classify materials as solids, liquids or gases from their properties.
- To interpret and explain the results of experiments.
- To appreciate how useful it is to discuss ideas and results with others.
- To understand how theories can be based on results obtained from experiments.

Teaching strategy

Key points to highlight

- Substances may be grouped together as solids, liquids or gases.
- Each group of substances has characteristic properties.
- Solids, liquids and gases expand when they are heated.
- In forming a solution, solute particles gradually mix with solvent particles.
- Objects of the same size may not have the same mass – mention density.
- Some solids, like elastic, deform when a force is applied to them

Difficulties/Misconceptions

Some pupils may think that the reason it is easier to move through air than water is because you can't see air, there is nothing there. They should be made aware that as speed increases, air offers more resistance to motion. This could be illustrated by discussing such things as: why cars are built in streamlined shapes to minimise air resistance; why racing cyclists wear tight-fitting clothing and streamlined helmets; why the bottom of a space shuttle is covered in ceramic tiles.

Lesson starter suggestions

- **What are the characteristic properties of solids, liquids and gases?** Shape, ease of compression and ease of flow.
- **Why do solids, liquids and gases behave in different ways?** Arrangement and movement of particles.
- **What happens when you try to run through water?** Resistance.

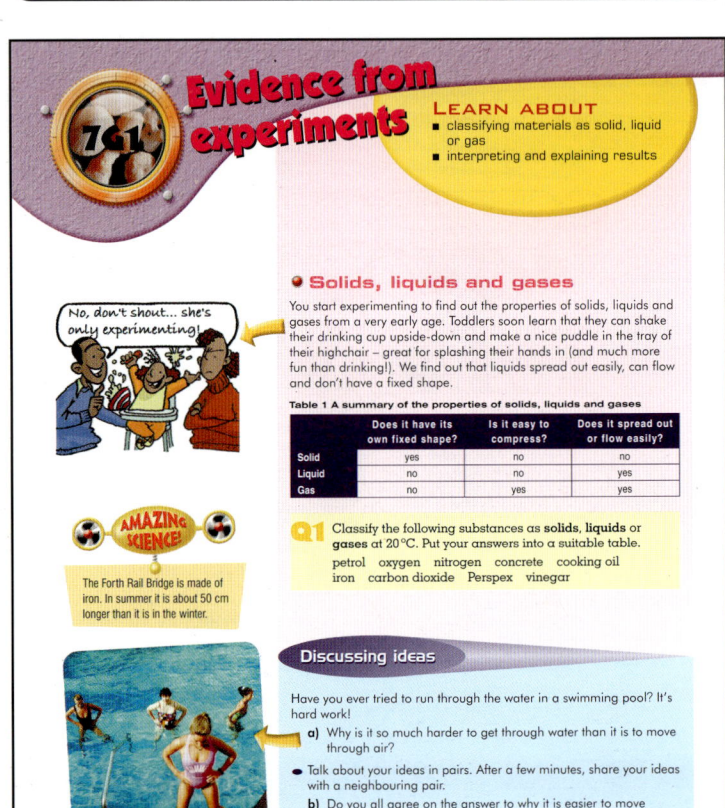

Activity and technicians' notes

Discussing ideas

- Many pupils will be aware that it is almost impossible to run through water without falling forward, from their experiences at swimming pools.
- They should be aware that the particles in water are held closely together, while those in air are far apart. Use this fact to explain why it is easier to move through air.
- Pupils may not be aware that when an object is in water a force, upthrust, acts on the object in the opposite direction to gravity.
- Exercising in water allows the movement of joints and muscles, without the full weight of the body acting down.
- Exercising in water is not restricted to people; it is widely used for race horses, allowing exercise without the whole weight of the horse having to be taken by injured legs.
- The use of water to treat injuries and ailments is called 'hydrotherapy'.

Teaching strategy contd.

Skills
- Some of this lesson will be given over to discussing the properties of solids, liquids and gases.
- The remainder will require practical skills and observation in carrying out a series of experiments and noting the results.

Notes and tips
Pupils should not spend too much time debating movement through water, as there are several different experiments within the investigation of solids, liquids and gases. It may be necessary to demonstrate some of these experiments, such as the 'bar and gauge' or the 'ball and ring', if there is insufficient apparatus and/or time for all groups to carry them out.

Extension ideas
- Pupils could compare the advantages and disadvantages of mercury and alcohol thermometers.
- Alternatively, they could carry out some research into mercury, the only metal that is a liquid at room temperatures.

Learning styles

Interpersonal
- **Debating** the relative ease of movement through air and water.
- **Discussing and evaluating** the results of investigations into solids, liquids and gases.

Kinaesthetic
- **Carrying out experiments** on solids, liquids and gases.

Intrapersonal
- **Considering** some of the characteristics of solids, liquids and gases.
- **Understanding and reflecting** on the particle model.

Visual
- **Imagining** what particles in materials would look like.

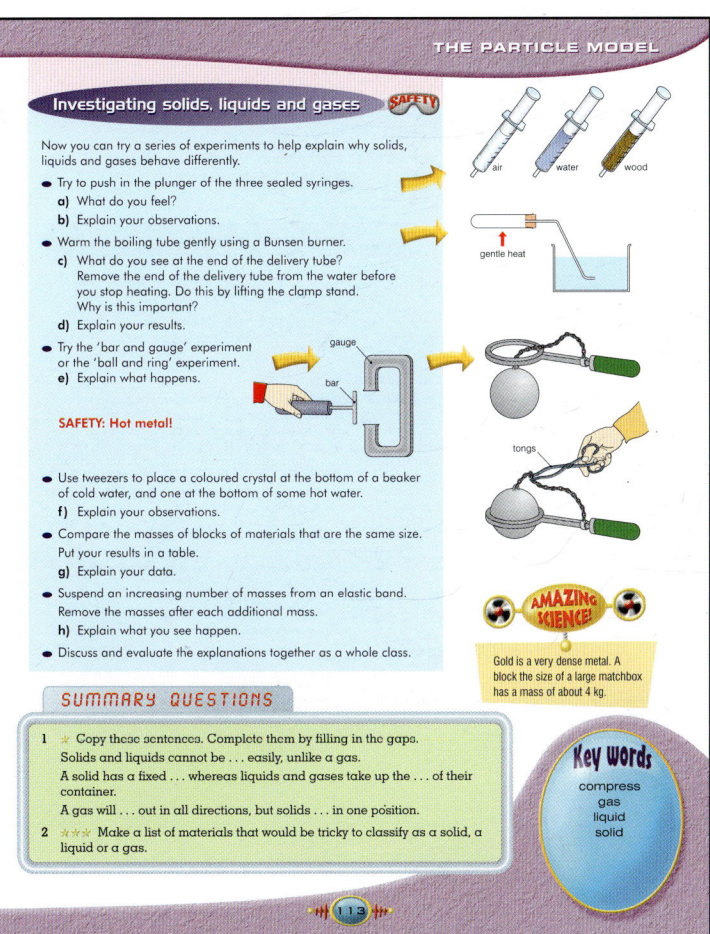

Activity and technicians' notes contd.

Investigating solids, liquids and gases

Equipment and materials required
- Potassium manganate(VII) crystals

For each group:
- Three sealed syringes
- A boiling tube fitted with bung and delivery tube
- A test tube and a Bunsen burner
- 'Bar and gauge' or 'ball and ring' apparatus
- Two medium-sized beakers
- A set of blocks – with the same dimensions, but different densities
- An elastic band
- A set of small masses

Answers to questions

Text

Q1

Solids	Liquids	Gases
concrete	petrol	oxygen
iron	cooking oil	nitrogen
Perspex	vinegar	carbon dioxide

Summary

1 compressed, shape, shape, spread, stay

Suggestions for plenaries

- **The properties of solids, liquids and gases can be explained in terms of their particles:** How are the arrangement and motion of the particles different?
- **Why is moving through water more difficult than moving through air?** Differing densities of water and air.
- **What advantages are there to discussing observations and results with others?** Sharing ideas and opinions.

Learning outcomes

Following this, most pupils will:
- Appreciate the different properties of solids, liquids and gases.
- Describe and explain observations.

Some pupils will also:
- Predict the properties of materials.

7G2 Classifying materials

National framework/QCA SoW references

National framework – Scientific enquiry:
▸ Consider early scientific ideas, including how experimental evidence and creative thinking has been combined to provide scientific explanations.

QCA SoW (7G)
How are theories created?
What are the differences between solids, liquids and gases?
NC links: Sc1 1c.

Learning objectives

▸ To be able to classify materials as solids, liquids and gases, but appreciating that some materials show a mixture of properties and are difficult to classify.
▸ To appreciate that theories can be based on evidence obtained from experiments.
▸ To evaluate how well a theory fits the evidence.
▸ To appreciate that sometimes theories and models need to be modified in the light of new evidence.

Teaching strategy

Key points to highlight

- Solids, liquids and gases are three states of matter.
- Some materials are difficult to classify because they show a mixture of properties.
- In science, a model is sometimes used to help explain observations.
- A model may be modified in the light of new evidence from experiments.
- We use a particle model to explain the different properties of solids, liquids and gases.

Difficulties/Misconceptions

There may be disagreement over the classification of some of the tricky materials. For example, flour is composed of particles of solid but it has no definite shape and flows like a liquid. This should be used to illustrate that, in science, it isn't always possible to provide a definitive answer. In this case, it shows the limitations of having only three categories of substance.

Skills

Much of this lesson will be led by you, however pupils could be divided into small groups to discuss their initial ideas about what particles in different materials would look like.

Lesson starter suggestions

- **What is a model in science?** An aid to understanding and explaining observations.
- **Some materials have a mixture of properties:** For example gels, etc.
- **What would the particles in materials look like?** Use of imagination.

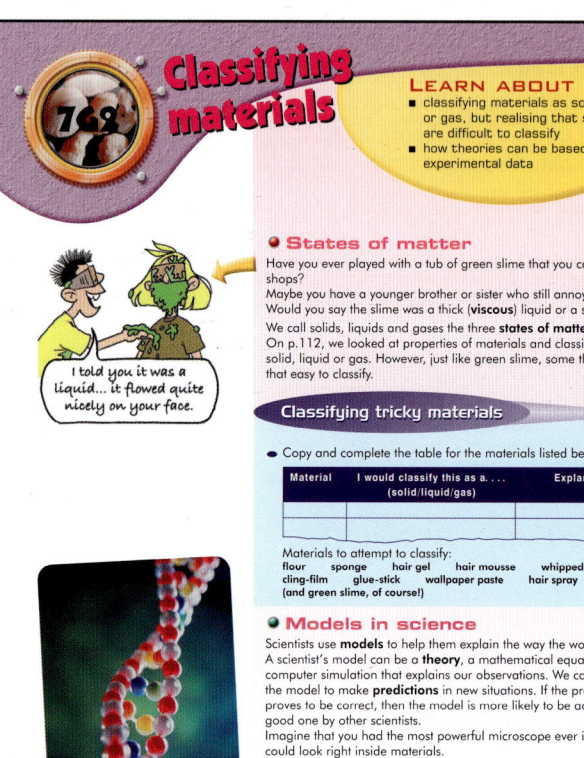

Teaching strategy contd.

Notes and tips
The Greeks were great thinkers but not great practical scientists.

Extension ideas
Pupils could suggest a model that would explain the differences between solids, liquids and gases, perhaps using some fundamental substance.

Learning styles

Intrapersonal
- **Understanding and reflecting** on the particle model.

Visual
- **Imagining** what particles in materials would look like.

Answers to questions

Summary
1 models, particles

Suggestions for plenaries

- **What use is a model?** Does it help you to understand what is going on?
- **How does the particle model explain the properties of solids, liquids and gases?** How are the particles arranged in solids, liquids and gases?
- **Can all substances easily be classified as solid, liquid or gas?** Can you name some substances that are difficult to classify?

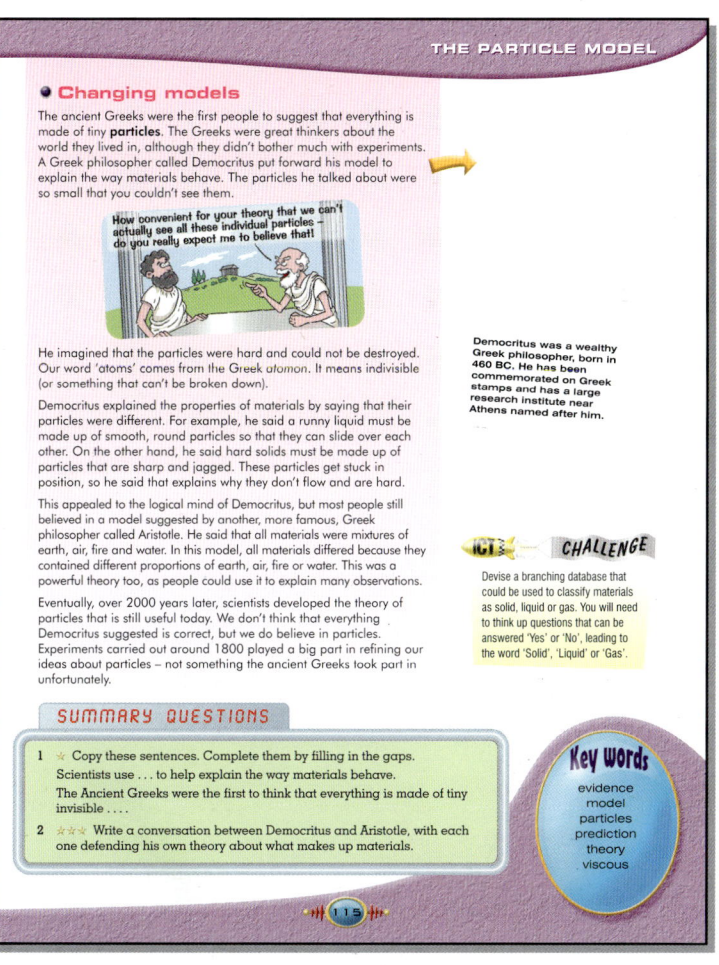

Learning outcomes

Following this, most pupils will:
- Classify materials as solid, liquid or gas.
- Explain their classification of some 'difficult' materials.
- Describe materials as being made of particles.

Some pupils will also:
- Use the particle model to explain some of the properties of solids, liquids and gases.

7G3 The particle theory

National framework/QCA SoW references

National framework – Particles:

▶ Describe a simple particle model for matter recognising:
- the size, arrangement, proximity, attractions and motion of particles in solids, liquids and gases;
- the relationship between heating and movement of the particles.

QCA SoW (7G)

7G Section 3: How can the particle model explain the differences between solids, liquids and gases?
NC links: Sc3 1b.

Learning objectives

▶ To use a model that has been based upon things we cannot see directly, but we have inferred from the evidence collected.
▶ To appreciate that solids, liquids and gases are made up of particles.
▶ To be able to explain the properties of solids, liquids and gases in terms of the behaviour of particles.

Teaching strategy

Key points to highlight

- John Dalton explained chemical reactions in terms or particles.
- Using a microscope, Robert Brown observed the effects of air particles bumping into pollen grains and this is called Brownian motion.
- Air particles are far too small to see, but Brown could see the much larger pollen grains jiggle about as invisible air particles bumped into them.
- Smoke is a suspension of tiny particles in air. When examined under a microscope, smoke particles exhibit Brownian motion.
- In solids, particles are close together. They are in fixed positions but are able to vibrate.
- In liquids, particles are still close together but can move over each other.
- In gases, particles are far apart from each other and free to move.
- Movement of particles in liquids and gases is entirely random.

Lesson starter suggestions

- **What evidence is there to support particle theory?** Observation of properties.
- **How can we see the effects of particles but not the particles themselves?** Discussion of size.
- **How are the particles arranged in a solid, a liquid and a gas?** Are they in a fixed position, or can the particles move about?

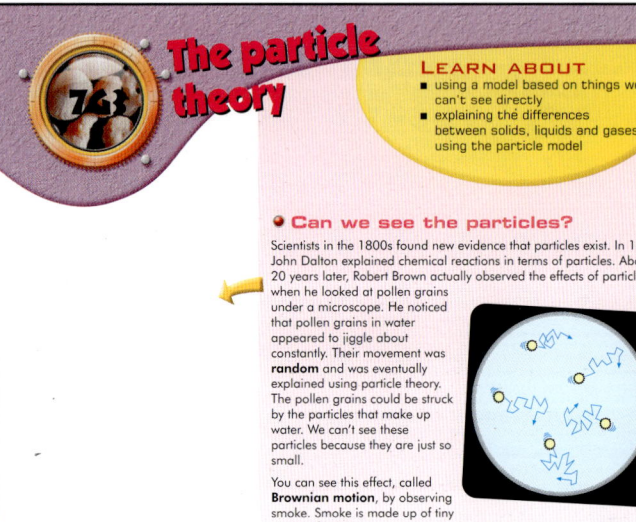

Activity and technicians' notes

The smoke cell experiment

Equipment and materials required

For each group (or as a demonstration):
- A microscope
- A smoke cell
- A smouldering wax straw or similar smoke source

Amazing science

Atoms are so small that they cannot be revealed by any optical microscope, no matter how powerful. However, they can be seen using an instrument called a 'scanning tunnelling microscope' (STM). The STM is able to map the profile of a surface on the atomic scale, by detecting an electric current flowing from the surface to the point of a fine metal probe.

Teaching strategy contd.

Difficulties/Misconceptions
- Smoke cells can be temperamental. A good source of smoke can be made by rubbing a wax candle on a piece of paper and then rolling it up like a straw. When the top is lit, plenty of smoke comes out at the bottom. If several groups of pupils are going to view a single demonstration microscope, it will be necessary to refresh the smoke cells at regular intervals.
- An alternative to the smoke cell is to view 'Aquadag': a suspension of fine particles of carbon in water that can be diluted to give optimum results.

Skills
Much of this lesson will be led by the teacher, however pupils could be divided into small groups to discuss Brownian motion and explain what is going on in terms of moving particles.

Notes and tips
Brownian motion gives indirect evidence of the presence of particles in air. These particles cannot be seen, however their presence can be inferred by the effect that they have on the pollen grains.

Extension ideas
Pupils could research and report on the work of the famous chemist John Dalton of the nineteenth century, and comment on how his working methods might compare to those of a modern chemist.

Learning styles

Visual
- **Observing** Brownian motion.

Interpersonal
- **Debating** Brownian motion and how it can be explained in terms of the movement of particles.

Intrapersonal
- **Considering** how the characteristics of solids, liquids and gases can be accounted for in terms of particles.

Answers to questions

Text
- **Q1** vibrate more (vigorously)
- **Q2** The particles are able to slide over and around each other, so the liquid can change shape continuously as it is poured.
- **Q3** There is lots of space between the particles.

Summary
1. particles, close, fixed, vibrate, close, slide, move, particles, walls

Learning outcomes

Following this, most pupils will:
- Appreciate that materials are made of particles.
- Describe the movement and arrangement of particles.
- Describe some properties in terms of the particle model.

Some pupils will also:
- Describe and explain all of their observations using the particle model of matter.

Suggestions for plenaries

- **What are the characteristics of solids, liquids and gases?** Do they have a fixed shape? Can they be poured? Can they be compressed?
- **How does particle theory account for the properties of solids, liquids and gases?** How are the particles arranged and how do they move?
- **What is Brownian motion?** Indirect evidence of particle motion.

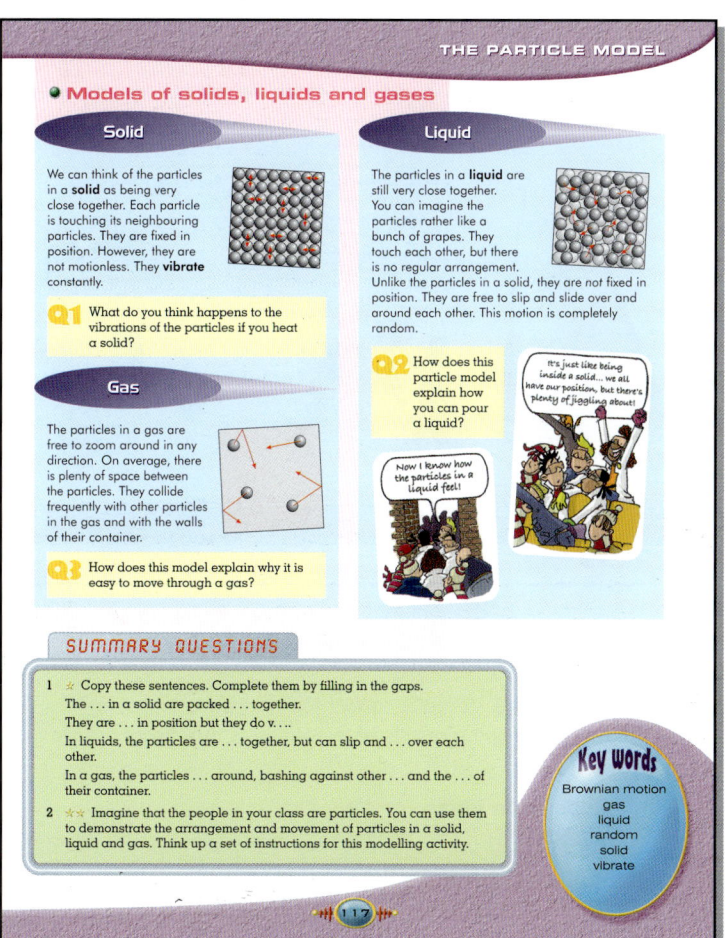

DURHAM HIGH SCHOOL
SCIENCE DEPARTMENT

7G4 Applying particle theory

National framework/QCA SoW references

National framework – Particles:
- Use the simple parrticle mode to explain:
  - why diffusion occurs in liquids and gases.
  - why air exerts a pressure.
  - why changes of state occur.

QCA SoW (7G)
How can the particle model explain other phenomena?
NC links: Sc3 1b.

Lesson starter suggestions

- **Does the particle theory help us to understand the behaviour of substances?** How does the motion of individual particles determine the properties of materials?
- **How does particle theory explain diffusion?** Movement and mixing of particles.
- **How does particle theory explain pressure?** Collision of particles with container.

Learning objectives

- To use particle theory to explain observations.
- To observe and explain diffusion in terms of the motion of particles.
- To explain pressure in terms of the movement of particles.

Gruesome science

Smelly gases do have their uses. In its pure form, North Sea gas is odourless. If there were a gas leak nobody would be able to smell it, and the results would be potentially catastrophic; so a odour is deliberately added in order to give it a characteristic smell that can be immediately recognised.

Teaching strategy

Key points to highlight

- Diffusion is the process by which particles of one gas mix with those of another.
- Particles of solute diffuse through a solvent when a solution is formed.
- In a gas the particles are constantly moving about.
- Pressure is the result of the particles colliding with the walls of their container.
- Pressure is the total force exerted over a unit area.
- We are surrounded by a 'sea of air'.
- Air exerts a pressure that we call 'atmospheric pressure'.
- A vacuum is a space that contains no particles.

Skills

Most of this lesson is concerned with observing and interpreting the results of experiments that demonstrate diffusion.

Extension ideas

Pupils could use particle theory to explain pressure, and what happens to the pressure inside a sealed tube (constant volume) if it is placed in a beaker of hot water.

Activity and technicians' notes

Looking at diffusion

Safety notes

Bromine is corrosive and very toxic. It must be handled with extreme care.

Potassium manganate(VII) is a powerful oxidising agent and is potentially harmful.

Equipment and materials required

- Potassium manganate(VII) crystals

For each group:
- A Petri dish of agar gel

For demonstration:
- Bromine
- Two gas jars

Learning styles

Kinaesthetic
- **Carrying out** a diffusion experiment with potassium manganate(VII) crystals.

Visual
- **Observing** the diffusion of bromine.
- **Drawing conclusions**.

Intrapersonal
- **Understanding** how diffusion and pressure can be in explained in terms of particles.

Answers to questions

Text

Q1 At the start the particles are fixed, but as the butter is heated the particles gain energy, vibrating more vigorously, and eventually they are able to move around each other and the butter melts.

Summary

1. stir, diffusion, gases, move, particles, container, force, pressure
2. The air pressure outside the can is greater than the air pressure inside and this forces the side of the can inwards.

Learning outcomes

Following this, most pupils will:
▸ Describe and explain observations using the particle model, and begin to use the particle model to explain phenomena.

Some pupils will also:
▸ Use the particle model to explain a range of phenomena, and evaluate whether evidence supports or refutes them.

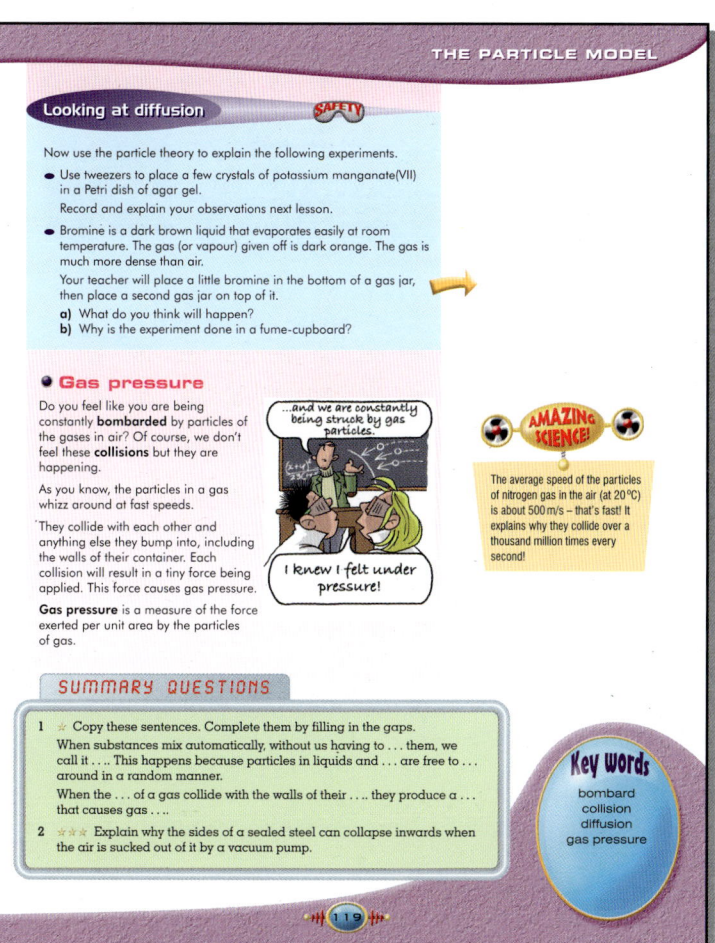

Suggestions for plenaries

- **Particle theory explains a range of phenomena:** Extend beyond simple properties.
- **What happens to the particles of a substance during diffusion?** Why do they spread out?
- **How do particles exert a pressure?** Collective force.

7G Read all about it!

Teaching strategy

Otto von Guericke and his amazing scientific demonstrations

Difficulties and misconceptions

▸ Pupils often talk about a vacuum as if it was some form of tangible entity. They should appreciate that a vacuum is what remains when everything has been removed.

Notes and tips

- Otto von Guericke's expertise was not limited to air pumps; he also investigated other fields of natural science. In 1672 he developed the first machine for producing an electric charge and in astronomy he worked on making predictions of the periodic return of comets.
- At the time of von Guericke there was a lot of important experimentation going on but it wasn't organised in the way scientific research is today. Early scientists were often rich people who regarded science as something of a hobby. They enjoyed performing grandiose experiments for the amusement of the audience. Pupils may be able to find details of experiments carried out by other scientists.

Skills

This unit has some opportunities for pupils to develop Sc1 skills of observing. More importantly, it provides an opportunity for pupils to appreciate that if a theory, in this case, the particle theory, is to be valid, it must be able to account for observations and the behaviour of materials.

Extension ideas

Pupils could work together to research in more detail one of the following topics:
- How is pressure measured?
- How does pressure vary with height in the atmosphere?
- How does pressure vary with depth in the oceans?
- What is the significance of isobars on a weather map?
- How is the weather linked to atmospheric pressure?
- Were people like Otto von Guericke serious scientists or were they just showmen?

Otto von Guericke and his amazing scientific demonstrations

The name of Otto von Guericke will always be linked to his beloved city of Magdeburg in Germany. His wealthy family had lived there for three centuries before Otto was born on 20 November, 1602. He went to university at the age of 15, and finally studied law at the Dutch university of Leiden for three years until he was 23. Whilst at Leiden, he also studied engineering and was especially interested in building fortresses.

His fortunes changed when Magdeburg was ransacked in the Thirty Years War, and he left Germany to work as an engineer in Sweden. However, he was able to return to Magdeburg in 1632, putting his engineering knowledge to good use in helping to rebuild the city. He became mayor of the city for over 25 years.

However, he will be best remembered for his famous demonstrations involving air pressure. Having invented the air pump in 1650, Otto was able to create a vacuum. He could remove the air from a container and show the great force that air pressure can produce. Otto really knew how to impress people; look at one of his experiments above:

In this experiment 20 men, using a pulley, tried to pull a piston upwards, as one man moved it downwards by sucking air out of the cylinder. People were amazed when the 'one man' (plus the help of air pressure) won the tug of war.

Then he organised an even more spectacular tug of war between two teams of six pack-horses.

Otto had two halves of a copper sphere made so that they fitted together perfectly. He had the local blacksmith, and some helpers, use a pump to suck the air out of the sphere. On his signal, the two teams of horses pulled and pulled, but could not separate the two halves of the sphere. The crowd of curious onlookers cheered. They were even more impressed when the horses had stopped and Otto returned to the copper sphere. He asked for silence as he released a valve and a hissing sound could be heard. As the air rushed back inside the sphere, it suddenly fell in two. Cue rapturous applause from the astounded crowd. What a demonstration!

Otto's demonstrations continued to be a great success as he toured around Europe, performing at several royal courts.

This technician is about to coat the silicon disc in a material that will form a tiny electrical circuit. The material used is heated in a vacuum, so that it evaporates easily. Then it condenses and coats the disc.

Questions

1. What evidence is there in Otto's education that he was very bright?
2. Why was Otto able to demonstrate the effects of air pressure?
3. Imagine that radio existed in the time of Otto. Write a radio commentary describing Otto's demonstration with the teams of horses.
4. Did Otto's demonstrations help science lose its public image as a form of magic? Why?

Answers

1. He went to university when he was only 15. Whilst at university, he studied both law and engineering.
2. He invented the air pump, which allowed him to create a vacuum.
4. probably so. His experiments were conducted out of doors in full view of his audience and there were no hidden devices or any attempt to hide what he did with the apparatus he used.

Danger – common errors

Pupils may regard the properties of solids, liquids and gases as isolated facts without fully appreciating how they can be explained in terms of the arrangement and motion of particles. They should appreciate that changes of state do not result in new substances but simply changes in the way in which particles are organised.

Learning styles

Visual
- **Observing** information on pressure and weather.

Interpersonal
- **Discussing** the contributions to science made by historical figures such as Otto von Guericke.

Intrapersonal
- **Understanding** the links between pressure and weather.
- **Reflecting** on the importance of early scientists to the evolution of science.

Learning outcomes

Following this unit, most pupils will:
- Describe the different properties of solids, liquids and gases.
- Classify materials as solid, liquid or gas and explain their classification of some 'difficult' materials.
- Appreciate that materials are made of particles and describe the movement and arrangement of these.
- Begin to use the particle model to explain phenomena.

Some pupils will also:
- Predict the properties of materials.
- Use the particle model to explain the properties of solids, liquids and gases.

In scientific enquiry:

Following this unit, most pupils will:
- Describe and explain observations on the different properties of solids, liquids and gases.
- Describe and explain observations using the particle model.

Some pupils will also:
- Describe and explain all of their observations using the particle model of matter.
- Use the particle model to explain a range of phenomena and evaluate whether evidence supports or refutes them.

7G Unit review

Answers to review questions

1 While the copper sphere was full of air, there were equal numbers of particle colliding with the inside wall and outside wall so the pressure inside and outside the sphere was the same.
   When the air was pumped out of the sphere, there were no longer any particles to collide with the inside wall, however, particles continued to collide with the outside wall. The resulting pressure on the outside wall was enough to keep the two halves of the sphere together, even when the teams of horses pulled their hardest.

2 When the flask is full of air, the bromine particles continually collide with air particles so it takes the bromine some time to spread from one side of the flask to the other. When all of the air has been removed from the flask, there is nothing for the bromine particles to collide with and the bromine spreads out more quickly.

3 The heat makes the tar soft. When the concrete sections get hot they expand, but because the tar is soft it can be easily compressed.

4 On a hot day, the wire would have expanded. If you fitted the wires too tightly, when the weather was cold the wires would contract and may break. When you fit the wires, you need to leave some slack to allow for the contraction.

5 Reduce the volume of the container: the same number of collisions would occur over a smaller area so the pressure would increase.
   Raise the temperature of the gas: the gas particles would collide more often and with greater force so the pressure would increase.
   Increase the amount of gas in the container: there would be more collisions over the same area so the pressure would increase.

9 28 800 000 000 000

10 $572 \, cm^3$

12 a $A = 19.2 \, g \, cm^{-3}$; $B = 0.003 \, g \, cm^{-3}$
    b A is the solid and B is the gas. The particles in a gas are much farther apart than in a solid. A small mass of gas has a large volume. Since density = mass ÷ volume, this means a gas has a low density.
    c volume expanded to = $2.7 \, cm^3$
       density of liquid = $18 \, g \, cm^{-3}$
    d i) More space between the particles in ice than in water.
       ii) If water in pipe freezes, it will expand and may split the pipe.

Answers to SAT-style questions

1 a i) A
   ii) C
   iii) B (1)
b gas (1)
c i) D = melting (1)
   E = freezing/solidifying (1)
   F = condensing (1)
   G = boiling/evaporating (1)
   ii) condensing and solidifying (2)
d Examples are: carry the experiment out in a fume cupboard, only warm the sulphur very gently, eye protection. (2)

2 a a stopwatch (1)
b Test 1 (1)
c Gas particles in air move much more quickly than solid particles in a liquid. There is more space between the particles of air than between the particles of liquid. (2)

3 a conductor of heat (1)
b Repeat the test.
   A and D, because the values are very close together. (2)
c e.g. rods given the same amount of heat/drawing pins same distance from heat source/same amount of grease used. (2)
d i) a bar chart (1)

ii) 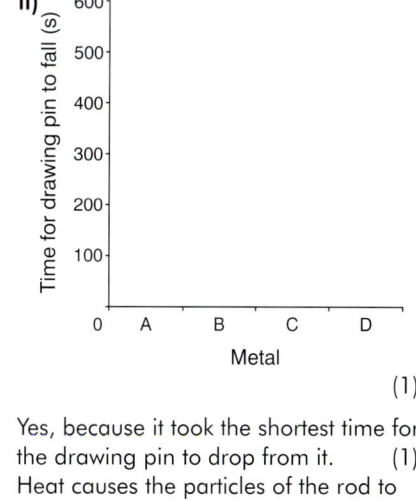 (1)

e Yes, because it took the shortest time for the drawing pin to drop from it. (1)
f Heat causes the particles of the rod to vibrate faster and faster. They take up more space and the rod expands. (2)

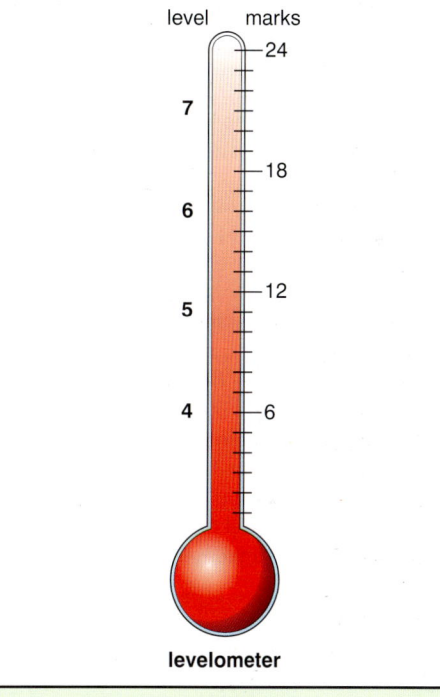

UNIT REVIEW QUESTIONS

d Most liquids contract when they freeze, but water is an exception.
   i) Find out why water is an exception.
   ii) Give one disadvantage that arises from this property of water.

SAT-STYLE QUESTIONS

1 Solid, liquid and gas are called the **three states of matter**.
The particles in a solid, liquid and gas are shown below.
The arrows represent changes of state:

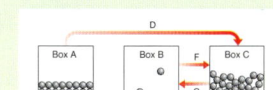
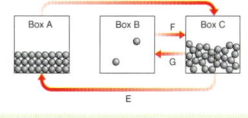

a Which box contains:
   i) a solid, **ii)** a liquid, **iii)** a gas? (1)
b Which state of matter is most easily compressed? (1)
c i) Identify the changes of state labelled D, E, F and G. (4)
   ii) Which changes of state require cooling down to take place? (2)
d Pip wanted to see how easily sulphur powder melts.
   Sulphur burns in air to form toxic sulphur dioxide gas.
   Give two safety precautions that Pip should take in her experiment. (2)

2 Mike and Pete are investigating diffusion.
In Test 1, they plan to time how long it takes a coloured gas to diffuse throughout a gas jar of air.
In Test 2, they want to fill the gas jar with water and time how long a coloured crystal takes to diffuse through the liquid.
a What measuring instrument will they need to judge which test is quicker? (1)
b Predict which test will happen more quickly. (1)
c Explain your answer to part **b** using particle theory. (2)

3 Benson and Molly were testing four metals (A, B, C and D). They set up the apparatus below and timed how long it took for each drawing pin to drop off the different metals:

Here are their results:

Metal	Time for drawing pin to fall (s)
A	550
B	470
C	360
D	545

a Finish off the question that Benson and Molly were investigating.
   Which metal is the best …? (1)
b How can they make their results more reliable? For which two metals is this particularly important? Why? (2)
c Give two ways in which Molly and Benson tried to make their investigation a fair test? (2)
d Benson wanted to display their results on a graph.
   i) What type of graph should he choose? (1)
   ii) Sketch the axes he should use. (1)
e Before their tests, Benson and Molly felt the metals.
   Molly predicted 'I think metal C will be best, because it feels coldest.'
   Do their results support her prediction? Explain your answer. (1)
f Use particle theory to explain what happens to the length of each rod in this investigation. (2)

Key words
Unscramble these:
dilos
riplatesc
offsinudi

7H Solutions

National framework/QCA SoW references

National framework – Scientific enquiry:
- Use scientific knowledge to decide how ideas and questions can be tested; make predictions of possible outcomes.
- Identify and control the key factors that are relevant to a particular situation.
- Use repeat measurements to reduce error and check reliability.
- Present and interpret experimental results through the routine use of tables, bar charts and simple graphs including line graphs.
- Describe and explain what their results show when drawing conclusions; begin to relate conclusions to scientific knowledge and understanding.

National framework – Particles:
- Use the simple particle model to explain:
  – why mass is conserved when substances dissolve to form solutions.
  – why temperature increases are likely to result in substances dissolving more quickly.
  – the formation of a saturated solution.

QCA SoW (7H)

How can we tell whether a liquid is a mixture? (7H1)
How much salt can we get from rock salt? (7H1)
What happens to the solute when a solution is made? (7H2)
How can we separate solvents from solutes? (7H3)
How can chromatography separate and identify substances in mixtures? (7H4)
Is there a limit to the amount of solid that will dissolve into a liquid? (7H5)
What else affects solubility? (7H5)
NC links: Sc1 2c, d, e, g, h, i, j, k, l, p; Sc3 1b, g, h; Sc3 2a, b.

Learning objectives

1.
- To understand that liquids may be pure or they may be mixtures in which one or more substances is dissolved in the liquid.
- To appreciate that salt is obtained from the sea and from under the ground.

2.
- To know that when a solute is added to a solvent, the mass of the solution formed is equal to the sum of the masses of the solute plus the solvent.
- To understand that when a solute dissolves in a solvent to form a solution, the solvent particles help to pull the solute particles apart and spread them throughout the solution.
- To know that it is possible to separate the solvent from the solute in a solution by a process called 'distillation', in which the solution is heated to the boiling point of the solvent.
- To know that a mixture of solutes can be separated by a process called 'chromatography', in which solutes are carried through absorbent paper at different speeds by a solvent to form a chromatograph.
- To appreciate that a solute may dissolve more readily in some solvents than others. A solute that is soluble in one solvent may be insoluble in another.
- To know that a graph showing how the solubility of a solute in a solvent changes with temperature is called 'a solubility curve'.

In scientific enquiry:
- To investigate the factors that affect how quickly a solute dissolves into a solvent.
- To understand the meaning of the terms 'independent variable', 'dependent variable' and 'control variable' in the context of a fair test.
- To use distillation to obtain pure water from a solution of food colouring.
- To use chromatography to investigate the food colourings used to colour sweets.
- To obtain data on the solubility of potassium nitrate by experiment and to use the data to draw a solubility curve.

Teaching strategy

This unit builds on the contents of units 4D 'Solids, liquids and how they can be separated', 5C 'Gases around us', 5D 'Changing state' and 6C 'More about dissolving' in the KS2 scheme of work. It also uses ideas that were developed in the KS2 programme of study.

The unit builds on ideas introduced in the previous unit 7G 'The particle model' (of solids, liquids and gases). It provides a foundation for work in later units that lend themselves to explanation in terms of particle theory, for example, 8A 'Food and digestion' (digestion), 8H 'The rock cycle' (crystal size related to rate of cooling), 8I 'Heating and cooling' (changes of state) and 9L 'Pressure and moments' (behaviour of gases).

Learning styles

Interpersonal
- **Reporting** results.

Intrapersonal
- **Interpreting**.
- **Making deductions**.

Kinaesthetic
- **Testing**.
- **Manipulating**.
- **Making**.

Visual
- **Making observations.**
- **Visualising**.

Notes to support 'What do you remember?'

- Pupils will already be familiar with the concept of dissolving and that some substances are soluble in water, whilst others are not from the programme of study from KS2.
- They will also have some knowledge of simple separating techniques like evaporation and filtration.
- The particle model has already been considered in some depth in the previous unit, 7G, so explanations given in terms of the behaviour of particles should present no problems.

Answers to questions

What do you remember?
1. chalk
2. filtration
3. petrol

Launch activity notes – Ideas about solutions

All these ideas are designed to encourage pupils to consolidate what they know and think about solutions thus far, and to formulate questions that they will go on to answer in the coming lessons.

Answers

a) Pupils are likely to realise that the salt has not disappeared in the sense that it has gone, but simply that it has dissolved rather like sugar in tea and coffee.
b) Pupils will already be familiar with some separation techniques such as filtration. They may suggest it would be possible to filter off all of the dirt.
c) Pupils will be able to observe the difference between a solution and a mixture of an insoluble solid in a liquid. They should be able to reason that since a solution contains no particles of solid, it would not be possible to separate solute and solvent by filtration.
d) It is obvious that there will come a time when it is not possible to add any more sugar to the solution.
e) Pupils will think of solution chemistry only in terms of water. However, they will be familiar with some other solvents such as ethanol. They could discuss whether the properties of all solvents will be the same or whether one solvent will remove a stain where another will not.
f) Initially pupils may think in terms of both processes forming a liquid. However they should consider differences such as the use of heat to melt a substance and the need for a solvent to form a solution.

7H1 Separating mixtures

National framework/QCA SoW references

QCA SoW (7H)
How can we tell whether a liquid is a mixture?
How much salt can we get from rock salt?
NC links: Sc3 1b, g.

Learning objectives

- To understand that liquids are either pure or mixtures, in which one or more substances are dissolved in the liquid.
- To appreciate that salt is obtained from the sea; the water evaporates leaving a mixture of salt and other solids.
- To appreciate that salt is also obtained from underground by mining, in the same way as we obtain coal and other minerals, or by solution mining.

Teaching strategy

Key points to highlight

- A solution is formed when a substance, often but not always a solid, dissolves in a liquid.
- In many of the solutions that will be familiar to pupils the liquid is water, but any liquid may form a solution.
- The substance that dissolves is called a solute.
- The liquid is called a solvent.
- A substance that dissolves in a liquid is said to be soluble.
- A substance that does not dissolve in a liquid is said to be insoluble.

Lesson starter suggestions

- **What solutions do we find in our homes?** Can you name something you have at home that is a solution of one substance dissolved in another?
- **Salt is an important substance:** Why is salt so important to our well being? What foods contain lots of salt? Can we have too much salt?
- **How can salt be taken from underground without digging?** Describe solution mining.

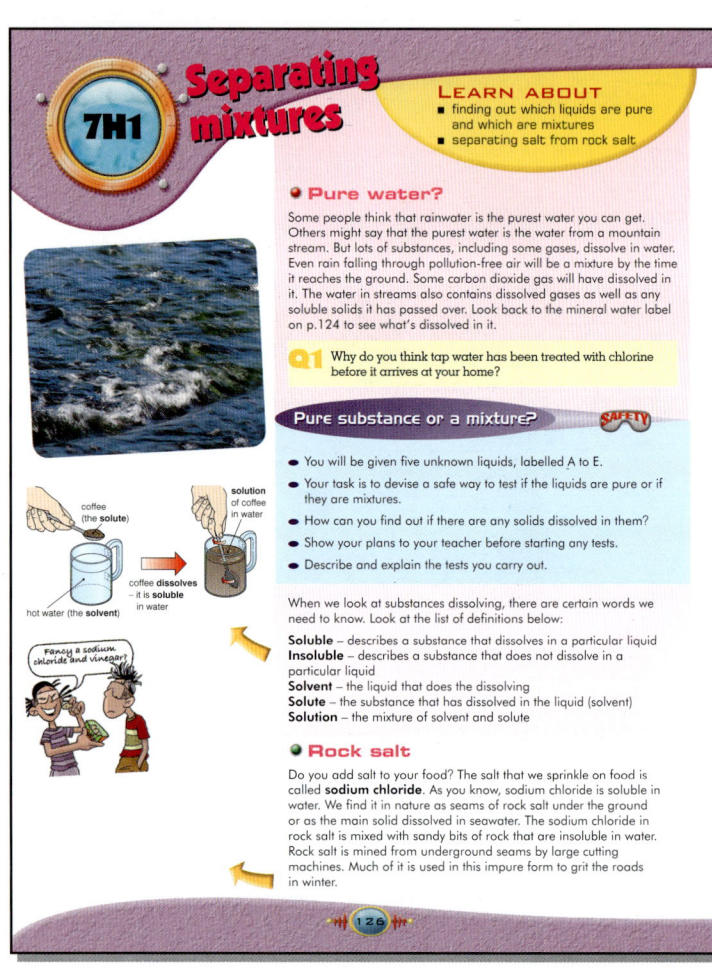

Activity and technicians' notes

Pure substance or a mixture?
Equipment and materials required

- Solution A is distilled water.
- Solution B is sodium chloride solution.
- Solution C is limewater.
- Solution D is ethanol.
- Solution E is copper(II) sulphate solution.

For each group:
- A set of five test tubes, labelled A to E, containing 1–2 cm³ of each solution.
- Eye protection

The activity requires pupils to devise methods of finding out whether each liquid is pure or a mixture. Methods are likely to involve driving off the liquid and seeing if any residue remains. The simplest method would be to spot the solutions on watch glasses or microscope slides, so a supply of these should be available. Dropper pipettes, or some other means of applying the liquids to the watch glass or slide, will also be needed. Other methods may require additional apparatus.

Safety notes

The solutions must not be heated directly with a flame, as ethanol is highly flammable. Copper(II) sulphate solution may be harmful, depending on concentration.

Teaching strategy contd.

Difficulties/Misconceptions
- Pupils often think that 'solution' relates only to a solid dissolved in a liquid. They should be made aware that it is also possible to have solutions of liquids and gases.
- Pupils may think that because a substance is soluble in one liquid, it is soluble in all liquids.

Skills
This lesson requires planning skills to devise suitable experiments, and then manipulative skills in carrying them out.

Notes and tips
Pupils may have started the year with the impression that if a liquid is colourless, it is pure. Hopefully, the work carried out on acids and alkalis in Unit 7E will have dispelled this notion. However, it is worth making the point that many substances dissolve in water to form colourless solutions, so lack of colour should not be seen as an indicator of purity.

Extension ideas
Pupils could find out how the halite, or rock salt, deposits under Cheshire were originally formed.

Activity and technicians' notes contd.

How much salt in rock salt?
Equipment and materials required
- Rock salt

For each group:
- A small beaker
- A filter funnel
- A filter paper
- An evaporating basin
- A glass rod
- A tripod and gauze
- A Bunsen burner
- Eye protection

Safety notes
The salt solution may spit if evaporated until it is nearly dry.

Access required to:
- A balance

Learning styles

Kinaesthetic
- **Manipulating** apparatus.

Interpersonal
- **Reporting** results to other pupils.

Answers to questions

Text
- **Q1** To kill any harmful micro-organisms that might be present.
- **Q2** They are insoluble in water.

Summary
1. dissolve, solution, solute, solvent

Learning outcomes

Following this, most pupils will:
- Classify some solids as soluble or insoluble, and be familiar with the terms solvent, solute and solution. They will use scientific knowledge and understanding to plan how to separate pure salt from rock salt.

Some pupils will also:
- Evaluate their method for finding the percentage of pure salt in rock salt.

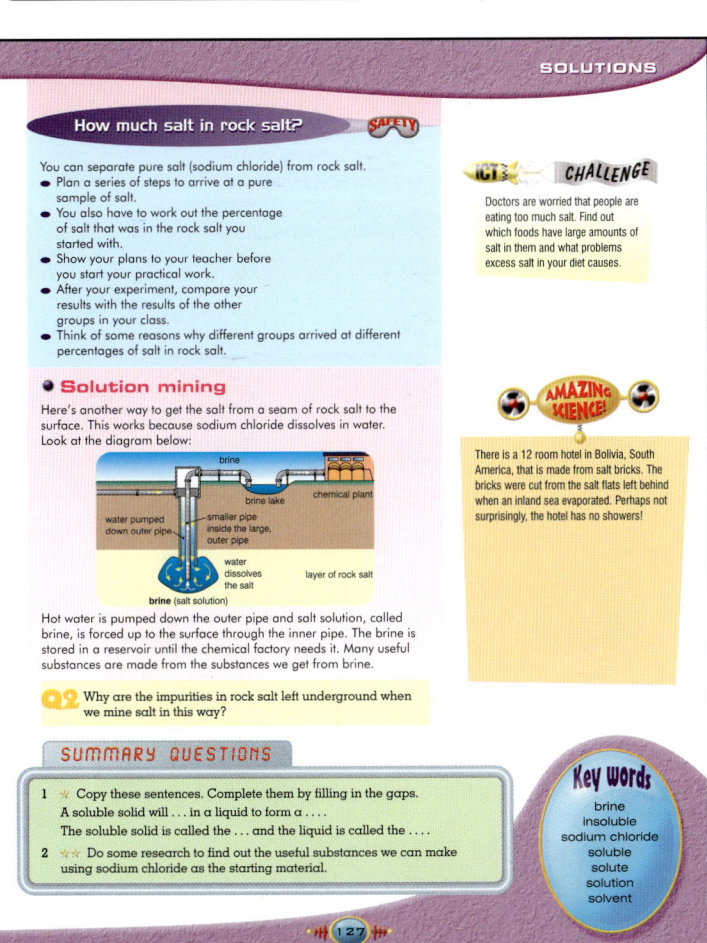

Suggestions for plenaries

- **Why do solutions leave a residue when heated?** The solute remains.
- **Obtaining salt from the sea by evaporation.** What happens to the water? Where does the energy come from?
- **Obtaining salt from underground by mining and solution mining.** What are the advantages of solution mining? Can you think of any disadvantages?

7H2 Particles in solution

National framework/QCA SoW references

National framework – Scientific enquiry:
- Use scientific knowledge to decide how ideas and questions can be tested; make predictions of possible outcomes.
- Identify and control the key factors that are relevant to a particular situation.
- Use repeated measurements to reduce error and check reliability.
- Present and interpret experimental results through the routine use of tables, bar charts and simple graphs, including line graphs.
- Describe and explain what their results show when drawing conclusions; begin to relate conclusions to scientific knowledge and understanding.

National framework – Particles:
- Use the simple particle model to explain:
  - why mass is conserved when substances dissolve to form solutions.
  - why temperature increases are likely to result in substances dissolving more quickly.

QCA SoW (7H)
7H Section 2: What happens to the solute when a solution is made?
NC links: Sc1 2c, d, e, g, h, i, j, k, l, p; Sc3 2a, b.

Learning objectives

- To know that when a solute is added to a solvent, the mass of the solution formed is equal to the sum of the masses of the solute plus the solvent.
- To understand that when a solute dissolves in a solvent to form a solution, the solvent particles help to pull the solute particles apart and spread them throughout the solution.

Teaching strategy

Key points to highlight

- When a solution forms, no mass is lost. The mass of the solvent increases by the mass of solute added.
- In a solid, particles are held in a regular arrangement by forces of attraction.
- When a solid dissolves, the solvent particles overcome these forces of attraction.
- When a solid dissolves, its particles move apart and spread throughout the solvent.
- A variable is something that can change and have different values.
- In an experiment, the factor that is to be investigated is called the independent variable.
- If an independent variable causes another variable to change, the other variable is a dependent variable.
- If the value of a variable is kept constant, then it is a control variable.

Lesson starter suggestions

- **When salt dissolves in water it forms a solution: is the mass of the solution less than, the same as, or greater than the mass of the water?** How much greater?
- **Imagine you could shrink in size, like Alice in Wonderland, so small that you could watch particles of salt and particles of water mix together. What do you think you would see?** How would the particles be arranged? How would they be moving?
- **Why do some solids dissolve in water but others don't?** Solubility and insolubility.

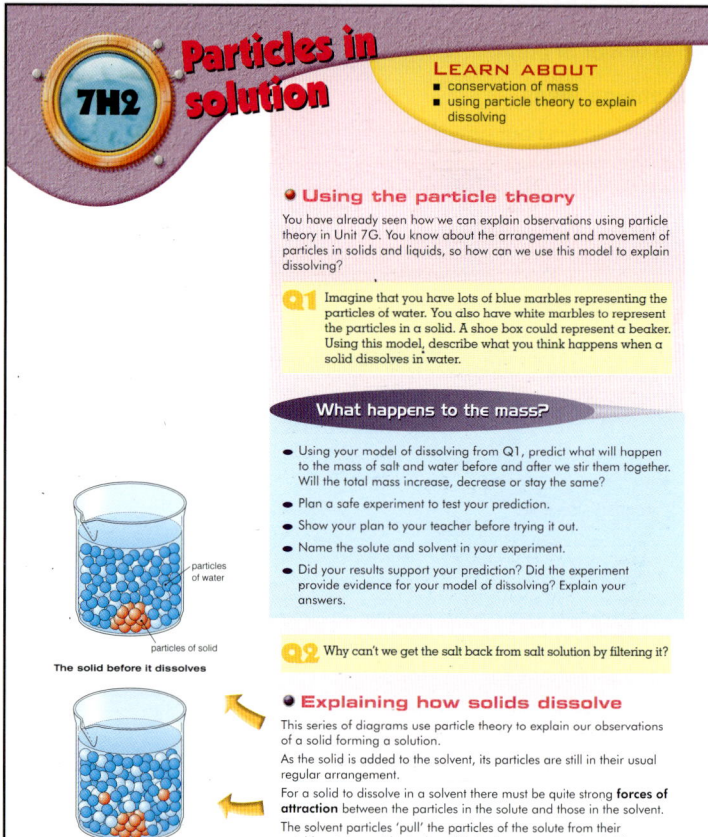

Activity and technicians' notes

What happens to the mass?
Equipment and materials required
- Salt

For each group:
- A small beaker
- A glass rod

Access required to:
- Water
- A balance

Difficulties/Misconceptions

Pupils may find the concept of a variable difficult to understand. Discuss some examples of variables in an everyday context. A simple example might be buying chocolate bars. The independent variable is the number of bars, the dependent variable is the cost of the bars and the control variable is the price per bar.

Activity and technicians' notes contd.

Investigating the rate of dissolving

Safety notes

Copper(II) sulphate solutiion may be harmful depending on concentration.

Equipment and materials required

- Copper(II) sulphate

For each group:
- A small beaker
- A glass rod
- A stopwatch or access to a clock reading to the nearest second
- A 100 cm³ measuring cylinder

If particle size is investigated then pupils will also need:
- A mortar and pestle

If temperature is investigated then pupils will also need:
- A thermometer
- A tripod and gauze
- A Bunsen burner

Access required to:

- If mass is investigated, pupils will need access to a balance.

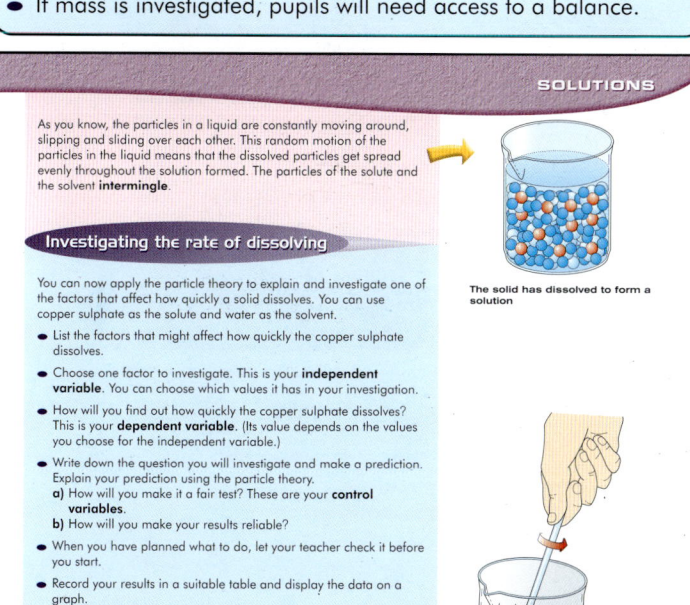

Teaching strategy contd.

Skills

This lesson requires planning skills to devise suitable experiments, and then manipulative skills in carrying them out. It will also require the display and interpretation of data.

Extension ideas

Pupils could use the particle theory to explain how each of the following factors affects the rate at which copper sulphate dissolves in water: mass of copper sulphate, volume of water, particle size of copper sulphate, temperature and stirring.

Learning styles

Visual
- **Making observations** on dissolving.
- **Imagining** the movement of particles during dissolving.
- **Displaying data**.

Kinaesthetic
- **Manipulating** apparatus.

Intrapersonal
- **Making deductions** from the results of experiments.

Answers to questions

Text

Q1 The different-coloured marbles would mix so that each blue marble is surrounded by white marbles.

Q2 Salt is soluble and would pass through the filter paper in the solution.

Summary

1 forces, particles, solute, solvent, solvent, spread.
2 Water particles don't attract the sand particles enough to pull them from each other.

Suggestions for plenaries

- **What conditions affect how quickly a solute dissolves in a solvent?** Concentration, particle size, temperature, stirring.
- **In the context of an experiment, what are independent, dependent and control variables?** Explain in context of an example.
- **What makes an experiment a fair test?** Control of all but one variable – use an example.

Learning outcomes

Following this unit, most pupils will:
▶ Use the particle model to explain what happens when a solid dissolves in water, explaining why mass is conserved.
▶ Make measurements of temperature and mass.
▶ Describe and explain observations.
▶ Identify patterns in solubility data and make predictions from them.

Some pupils will also:
▶ Use the particle model to explain why solids dissolve at different rates under different conditions.

7H3 Distilling mixtures

National framework/QCA SoW references

QCA SoW (7H)

7H Section 2: How can we separate solvents from solutes?
NC links: Sc3 1h

Learning objectives

▶ To know that it is possible to separate the solvent from the solute in a solution by a process called distillation.

▶ To understand that in distillation, the solution is heated to the boiling point of the solvent. The solvent becomes a gas and passes into a condenser (where it is cooled to give pure solvent) leaving pure solute.

Lesson starter suggestions

- **How can a solid be obtained from a solution?** The water is allowed to evaporate; how can the process be speeded up?
- **What happens to the particles when a liquid boils?** How does the speed at which they move and the distances between them change?
- **What is distillation?** Name something that is made in a distillery.

Teaching strategy

Key points to highlight

- When a liquid is heated, it boils and becomes a gas.
- When a gas is cooled, it condenses and becomes a liquid again.
- Energy is taken in when a liquid boils, and is given out when a gas condenses.
- The solvent can be separated from the solute in a solution by distillation.
- In a distillation apparatus, liquid turns to gas in the flask and the gas turns back to a liquid in the condenser.

Difficulties/Misconceptions

The terms 'gas' and 'vapour' and when to use them often confuse pupils. In essence they mean the same thing; however there is a tendency to use the term 'gas' for a substance that is gaseous at room temperature, such as oxygen, and the term 'vapour' when a substance that is a liquid at room temperature is turned into a gas by heating, such as water vapour.

Skills

This lesson requires manipulative skills to carry out a small-scale distillation, and observational and interpretational skills to appreciate the operation of a simple distillation apparatus.

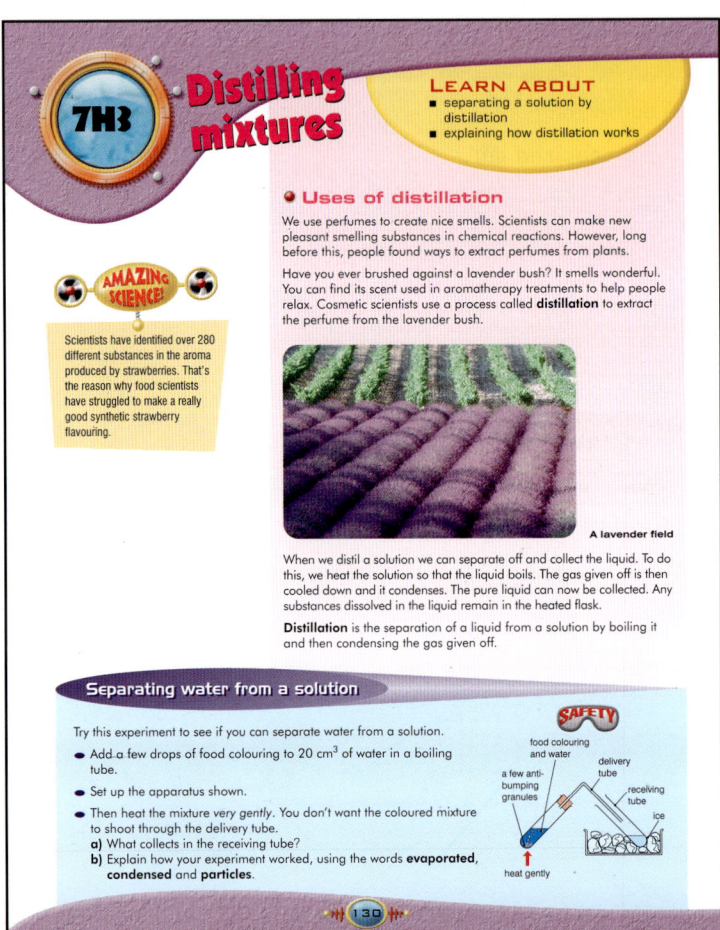

Activity and technicians' notes

Separating water from a solution

Safety notes

- Safety glasses must be worn.

Equipment and materials required

- Solution of water containing a few drops of food colouring
- Crushed ice
- A boiling tube containing 20 cm³ of the coloured solution
- A bung fitted with a delivery tube
- A test tube
- A stand and clamp
- A Bunsen burner
- A small beaker
  a) water
  b) The water particles evaporated from the mixture when it was heated and condensed in the cool receiving tube.

Teaching strategy contd.

Notes and tips
- For maximum efficiency, it is important that cold water is passed through the outer jacket in the correct direction i.e. from bottom to top. If it is connected from top to bottom, it is possible for cold water to only pass down one side of the jacket. This is a good opportunity to show pupils a simple distillation apparatus and explain the problem.
- Pupils should also take note of the position of the thermometer. The bulb should be level with the side arm and not immediately above the boiling water, as superheating can give inaccurate readings.
- Boiling and condensing provide a good opportunity to introduce the idea of a reversible change.

Extension ideas
Pupils could find out about the role of distillation in the manufacture of a spirit, such as whisky or brandy.

Learning styles

Visual
- **Observing** distillation apparatus and visualising how it works.

Kinaesthetic
- **Manipulating** apparatus.

Answers to questions

Text
Q1 The blue spheres gain sufficient energy to become a gas and pass into the condenser where they are cooled and turn back into liquid. The white spheres remain in the flask.

Q2 The cold water flowing through the outer jacket ensures that the tube inside the condenser is always cold. Hot vapour passing into the condenser is cooled and turns from a gas to a liquid.

Learning outcomes

Following this, most pupils will:
▶ Describe how mixtures can be separated by distillation.

Some pupils will also:
▶ Use the particle model to explain distillation.

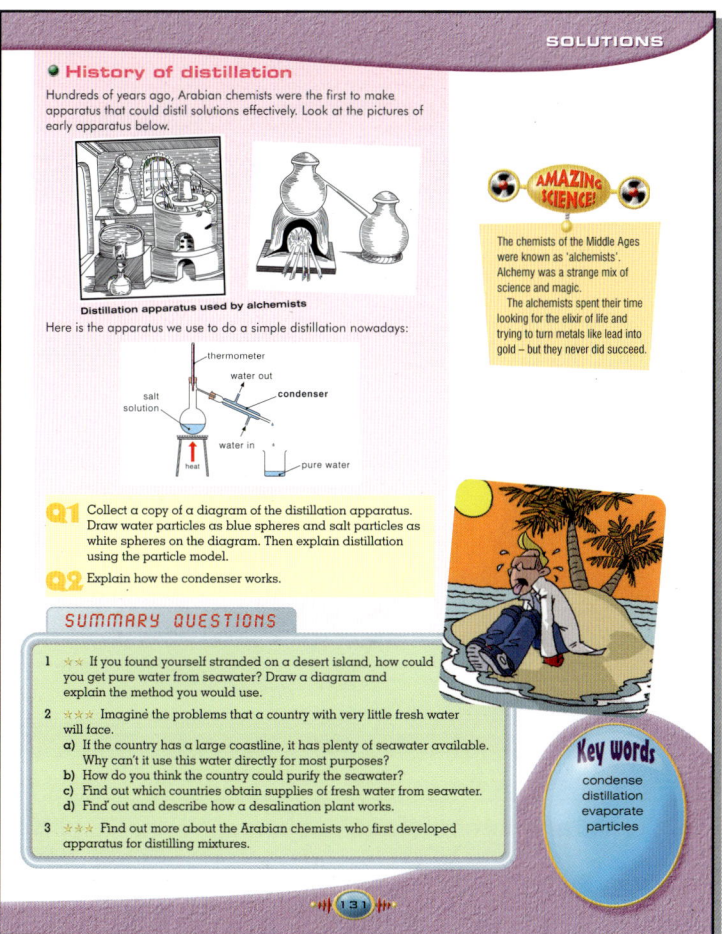

Suggestions for plenaries

- **What happens to a solution during distillation?** Explanation of separation.
- **How does the movement of the particles in a liquid change when the liquid is heated to its boiling point?** Effect of increasing kinetic energy.
- **How does distillation separate a mixture?** Boiling points of solvents.

7H4 Chromatography revealed

National framework/QCA SoW references

QCA SoW (7H)

7H Section 3: How can chromatography separate and identify substances in mixtures? NC links: Sc3 1.

Lesson starter suggestions

- **What forces exist between particles of solute and particles of solvent?** Attractive forces.
- **What happens when a tube of absorbent paper is stood in a solvent?** Explain absorption.
- **How can a mixture of solutes in the same solution be separated?** Retention on the absorbent paper.

Learning objectives

▸ To know that a mixture of solutes can be separated by a process called 'chromatography'.
▸ To understand that, in chromatography, solutes are carried through absorbent paper at different speeds by a solvent.
▸ To know that the resulting pattern, in which the solutes reach different places on the absorbent paper, is called a 'chromatogram'.

Teaching strategy

Key points to highlight

- Chromatography is a relatively new method of separating substances.
- Chromatography uses only very small amounts of substances.
- Ink must not be used to mark the base line on the chromatogram.
- The base line must be above the level of solvent in the container.
- Care must be taken when applying the sample.
- Solvents other than water can be used.

Difficulties/Misconceptions

Pupils must be told to draw the base line for their chromatogram in pencil and not in ink. If the line is drawn in ink, the dye(s) in the ink will move up the paper with the solvent.

Skills

This lesson is mainly given over to practical activities involving making chromatograms. Pupils are required to make deductions from their results.

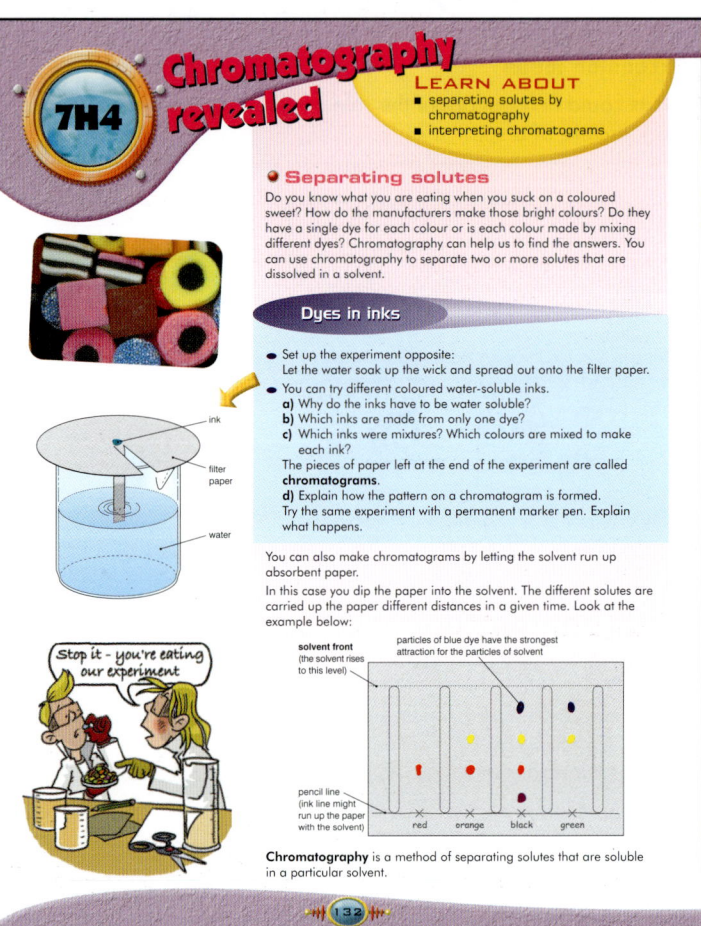

Activity and technicians' notes

Dyes in inks

Equipment and materials required

- Pens containing different coloured water-soluble inks

For each group:
- Several filter papers
- A small beaker

Activity and technicians' notes contd.

Testing sweets
Equipment and materials required
- Colour-coated sweets such as 'Smarties'

For each group:
- A thin paintbrush
- A small beaker
- A piece of chromatography paper
- A paper clip

Teaching strategy contd.

Notes and tips
In order to obtain clear chromatographs, pupils need to take care when applying the sample to the chromatography paper. If the colour solution is too weak, or the spots are too small, the results may be difficult to see. If the spots are too large, they will spread into each other.

Extension ideas
Introduce pupils to the R_f value as a quantitative aspect of chromatography.

Learning styles

Kinaesthetic
- **Making** chromatograms.

Visual
- **Observing** the separation brought about by chromatography.

Intrapersonal
- **Interpreting** chromatograms.

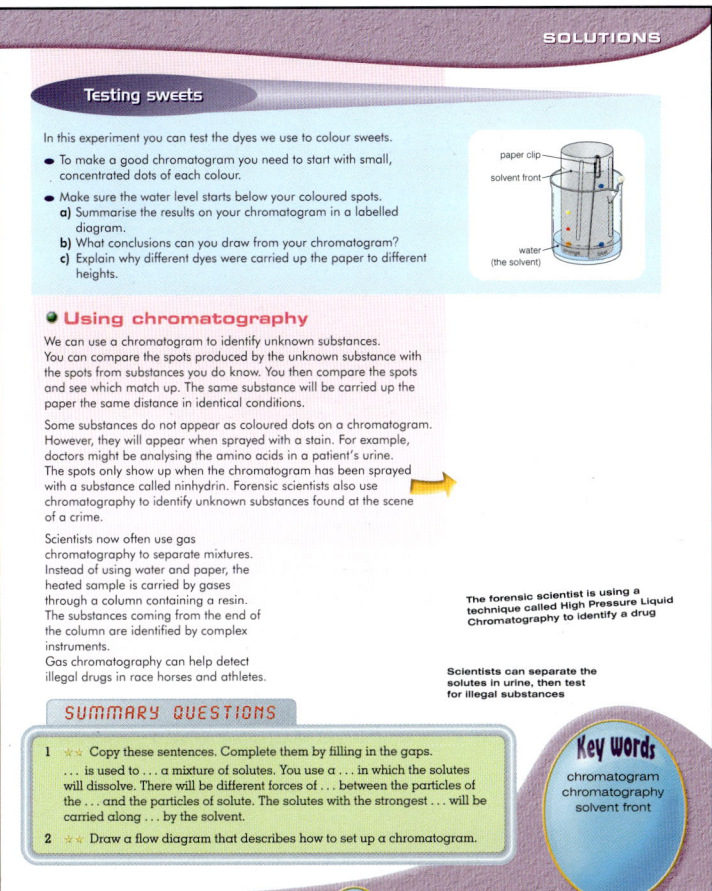

Answers to questions

Summary
1. Chromatography, separate, solvent, attraction, solvent, attraction, farthest
2. Draw a pencil line at the bottom of the chromatography paper. → Mark the starting positions of each spot. → Place a small spot of each mixture on its starting position. → Stand the chromatography paper in a dish of solvent so that the spots are above the level of solvent. → Leave the chromatography paper until the solvent has risen up the paper.

Learning outcomes

Following this unit, most pupils will:
▶ Describe how mixtures can be separated by chromatography and interpret data from chromatograms.

Some pupils will also:
▶ Use the particle model to explain chromatograms.

Suggestions for plenaries

- **Why is chromatography such a useful separation technique?** Do you need a large amount of the sample?
- **How do we make a chromatogram?** What happened when the sample was placed in the absorbent paper?
- **For what purposes might chromatography be used?** How could you check if an athlete had been taking performance enhancing drugs?

7H5 Finding solubility

National framework/QCA SoW references

National framework – Particles:
- Use the simple particle model to explain:
  - the formation of a saturated solution.

QCA SoW
7H Section 4: What else affects solubility?
NC links: Sc1 2i, j; Sc3 2b.

Learning objectives

- To appreciate that a solute may dissolve more readily in some solvents than others. A solute that is soluble in one solvent may be insoluble in another.
- To know that a graph showing how the solubility of a solute in a solvent changes with temperature is called a 'solubility curve'.

Teaching strategy

Key points to highlight

- There are many solvents other than water.
- There is a limit to the amount of solute that will dissolve in a given volume of solvent.
- When no more solute will dissolve, a solution is said to be 'saturated'.
- Solubility of solids in liquids increases with temperature.
- A graph showing solubility against time is called a 'solubility curve'.
- Solubility curves for water run from 0 °C, the freezing point of water, to 100 °C, the boiling point of water.
- Solubility is often given as grams of solute dissolved in 100 g of solvent.
- When giving the solubility of a solute, the temperature must be stated.

Difficulties/Misconceptions

When checking the solubility of solutes in different solvents, careful observation is needed. Pupils may find it difficult to see when some but not all of the solute dissolves.

Skills

This lesson is mainly given over to practical activities involving testing solubility in two different solvents, and in collecting data on the solubility of potassium nitrate at different temperatures. Pupils also need to consider how best to tabulate the data and use it to draw a solubility curve.

Lesson starter suggestions

- **If a solute is soluble in one solvent, does that mean it is soluble in all solvents?** Are all solvents the same?
- **When making a solution, what happens if you keep on adding more and more solute to the solvent?** Explain a saturated solution.
- **Is the solubility of a solute in a solvent the same at all temperatures?** Discuss effects of temperature.

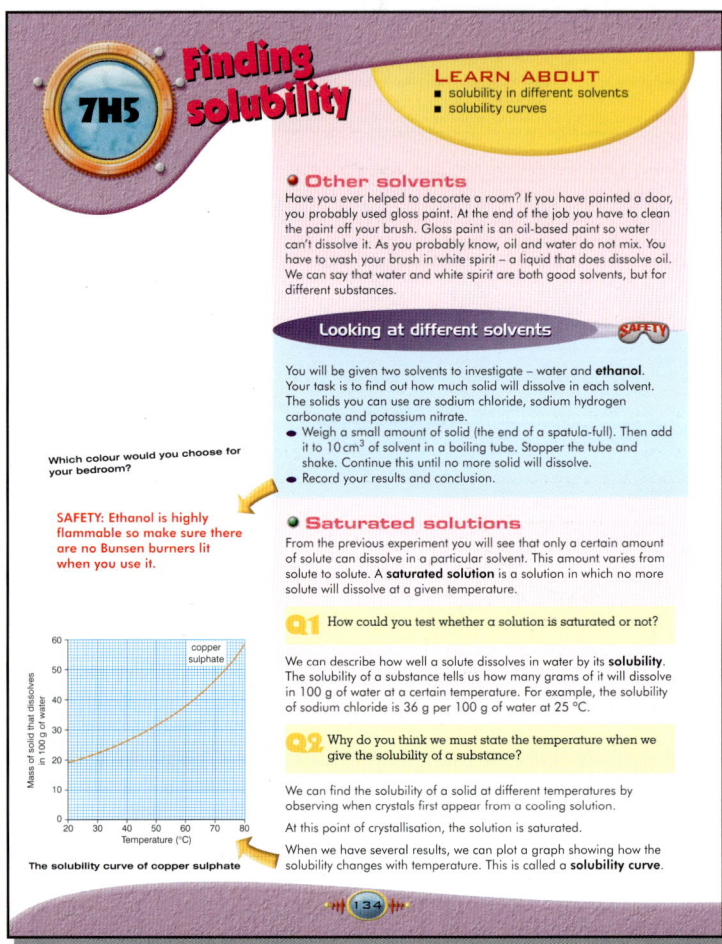

Activity and technicians' notes

Looking at different solvents

Safety notes
- Ethanol is highly flammable, so make sure there are no Bunsen burners lit when you use it.
- Potassium nitrate is a powerful oxidising agent and is potentially hazardous.
- Safety glasses must be worn.

Equipment and materials required
- Ethanol
- Sodium chloride
- Sodium hydrogencarbonate
- Potassium nitrate

For each group:
- A measuring cylinder
- A boiling tube fitted with bung
- A spatula

Accesss required to:
- A balance

Learning styles

Kinaesthetic
- **Testing** the solubilities of solutes in different solvents.
- **Finding** the solubility of potassium nitrate at different temperatures.

Visual
- **Making observations** on solubility.
- **Presenting data** in the form of a cooling curve.

Intrapersonal
- **Interpreting** cooling curves.

Activity and technicians' notes contd.

Solubility curve of potassium nitrate
Safety notes
- Potassium nitrate is a powerful oxidising agent and is potentially hazardous.
- Safety glasses must be worn.

Equipment and materials required
- Potassium nitrate

For each group:
- A small beaker
- A Bunsen burner
- A stand and clamp
- A thermometer
- A graduated pipette
- A pipette filler
- A boiling tube fitted with a bung

Access required to:
- Water
- A balance

Answers to questions

Text
Q1 Add more solute and see if it will dissolve.
Q2 The solubility of a solute in a solvent varies with temperature.

Summary
1 solubility, temperature, solubility, water, curve
2 a)

Temperature (°C)	10	30	50	70	90
Solubility (g/100g of water)	9.3	13.0	16.5	19.8	22.9

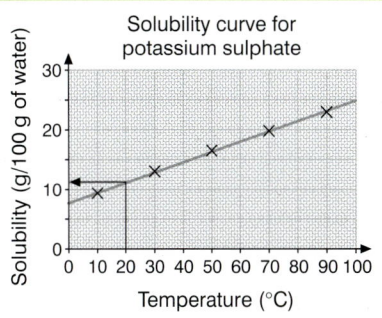

b) 11.5 g/100 g of water
c) 18 ÷ 4 = 4.5 g

Learning outcomes

Following this unit, most pupils will:
▶ Explain the meaning of the term 'saturated solution'.
▶ Make measurements of temperature and mass.
▶ Present experimental results as line graphs and point out patterns.

Some pupils will also:
▶ Describe and explain their observations about solubility and make predictions from them.

Suggestions for plenaries

- **How does solubility change with temperature?** Does it increase, decrease or stay the same?
- **Why won't any more solid dissolve in a saturated solution?** Describe saturation in terms of no more room available in the solvent.
- **Do ethanol and water dissolve solids equally well?** Explain different properties of different solvents.

7H Read all about it!

Teaching strategy

Dopey horses

Difficulties and misconceptions

- At this stage, any scientific experimental work conducted by pupils will have been limited to the manipulation of traditional apparatus such as test tubes, and direct observations of results. Pupils may find it a little suspicious of processes like gas chromatography where a sample is injected into one part of a machine and results appear from another part, as if by magic. They will need reassurance that the principle behind gas chromatography is exactly the same as the paper chromatography they will have carried out in the laboratory.

Notes and tips

- Most pupils will be aware of the use and abuse of drugs in sport. It seems there is scarcely a week that goes by without some new drug-related revelation.
- Detecting the presence of a drug in body fluid may not be quite as straight forward as it seems. Sometimes it is not the drug itself which is detected but the products formed when the drug is metabolized by the body. Additional substances may be taken which screen the drug.
- There are huge profits to be made from performance enhancing drugs. Companies design drugs so that they cannot be detected.

Skills

This unit provides a wealth of opportunities for pupils to develop Sc1 skills of manipulating apparatus, making observations and recording and displaying results. It also underpins an understanding of the particle theory developed in the previous unit by considering solubility in terms of the behaviour of particles.

Extension ideas

Pupils could research together or individually to find out more about one of the following topics:

- The scientific contributions of Friedlib F Runge, David T Day and Mikhail Tsvet to chromatography
- Column chromatography
- Thin-layer chromatography
- GCMS – combined gas chromatography and mass spectrometry

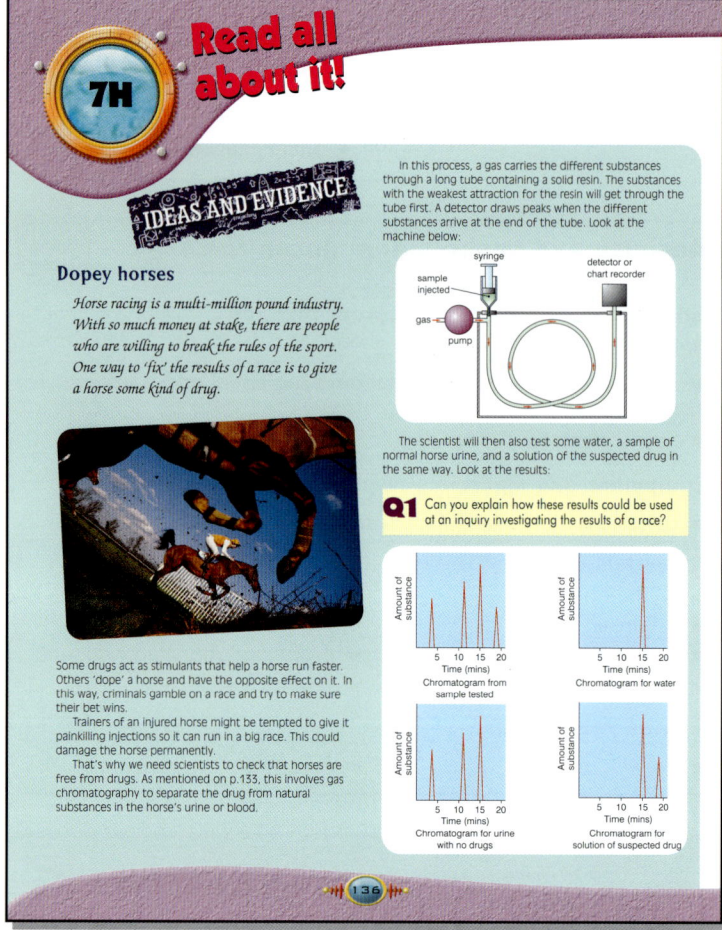

Questions

1. What effects can drugs have on race horses?
2. Why is gas chromatography used to analyse samples taken in drugs tests?
3. Explain how gas chromatography works.
4. Write about the similarities between gas chromatography and paper chromatography.
5. Explain how the results of the sample shown could be used at an enquiry into horse doping.

Answers

1. Some drugs help a horse to run faster while others slow it down.
2. Gas chromatography separates the components urine or blood so that drugs can be identified.
3. A gas carries the different substances in a sample through a long tube containing a solid resin. The resin attracts substances by different amounts and so the substances take different amounts of time to pass through the tube.
4. Both techniques are able to separate complex mixtures of substances by the differing attractions they have for the medium through which they travel, either resin or paper.
5. The pattern shown on the chromatogram can be compared to those for prohibited drugs. If a peak on the sample chromatogram is in the same position as that of a drug, then the drug must be present in the sample.

Danger – common errors

There are two common misconceptions about making solutions. One is that the solute somehow disappears in the solvent and the other that it is possible to add any amount of solute to the solvent. These can both be addressed by discussing the formation of a solution in terms of the solute particles being broken down and becoming surrounded by solvent particles. Once the solute is broken down into particles it is no longer visible. The solution becomes saturated when all of the solvent particles are associated with solute particles.

Learning styles

Visual
- **Researching** around chromatography.

Interpersonal
- **Discussing** the contributions made by different scientists to the development of chromatography.

Intrapersonal
- **Understanding** how different forms of chromatography work.
- **Appreciating** the underlying principle behind all forms of chromatography.

Learning outcomes

Following this unit, most pupils will:
- Classify solids as soluble or insoluble and be familiar with the terms solvent, solute and solution.
- Use scientific knowledge and understanding to plan how to separate pure salt from rock salt.
- Use the particle model to explain what happens when a solid dissolves in water, explaining why mass is conserved.
- Describe how mixtures can be separated by distillation.
- Describe how mixtures can be separated by chromatography.
- Most pupils will be able to explain the meaning of the term 'saturated solution'.

Some pupils will also:
- Use the particle model to explain why solids dissolve at different rates under different conditions.
- Use the particle model to explain distillation and chromatography.
- Describe observations about solubility.

In scientific enquiry:

Following this unit, most pupils will:
- Make measurements of temperature and mass.
- Interpret data from chromatograms.
- Make measurements of temperature and mass; present experimental results as line graphs and point out patterns.

Some pupils will also:
- Evaluate their method for obtaining pure salt in terms of the mass obtained.
- Explain observations about solubility and make predictions from them.

7H Unit review

Answers to review questions

1 a The colliding water particles are able to overcome the forces of attraction between the sugar particles; each sugar particle becomes surrounded by water particles. The sugar particles spread.

  b The forces of attraction between the dyes and the solvent are different, so the dyes are pulled at different rates by the solvent.

  c When the seawater is boiled, there is enough energy to overcome the forces of attraction between the water particles; the water boils, becoming a gas. In the condenser the gas turns back to a liquid and water is collected. The salt particles do not change into a gas and remain in the flask.

2 a They could place a few drops on a glass slide and leave it until the liquid evaporates. If a solid is dissolved in the liquid, a solid residue will be left on the slide.

  b

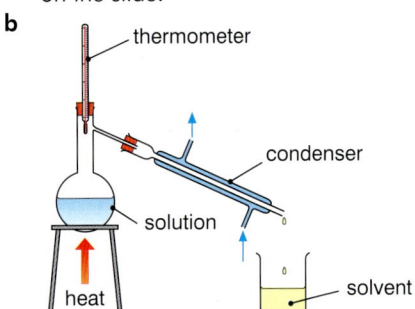

  c they could use chromatography

3 a the orange squash contains two dyes; one safe dye and one of the banned dyes

  b B

5 a

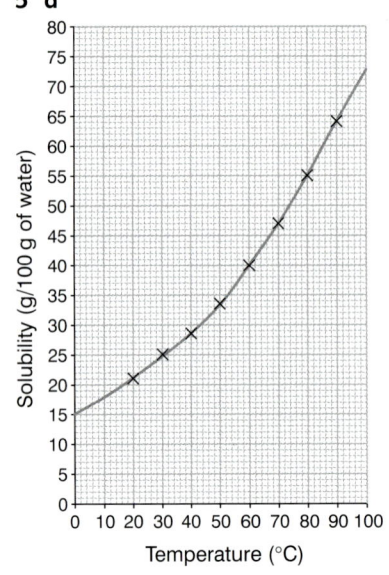

  b i) 37.0 g/100 g water
       ii) 16.5 g/100 g water
       iii) 72.0 g/100 g water
  c 22.5 g
  d 22.5 g

6 a 1.5 g
  b 11.25 g

7 36 g/100 g water

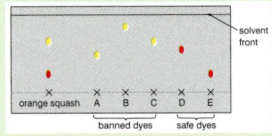

UNIT REVIEW QUESTIONS

Thinking skills

6 The solubility of potassium chlorate at 20 °C is 7.5 g/100g of water.
 a How much potassium chlorate will dissolve in 20 g of water at 20°C?
 b A student added 15 g of potassium chlorate to 50 g of water at 20°C. How much potassium chlorate remained undissolved after stirring?

7 If a maximum of 9 g of sodium chloride dissolve in 25 g of water at 20 °C, what is the solubility of sodium chloride?

SAT-STYLE QUESTIONS

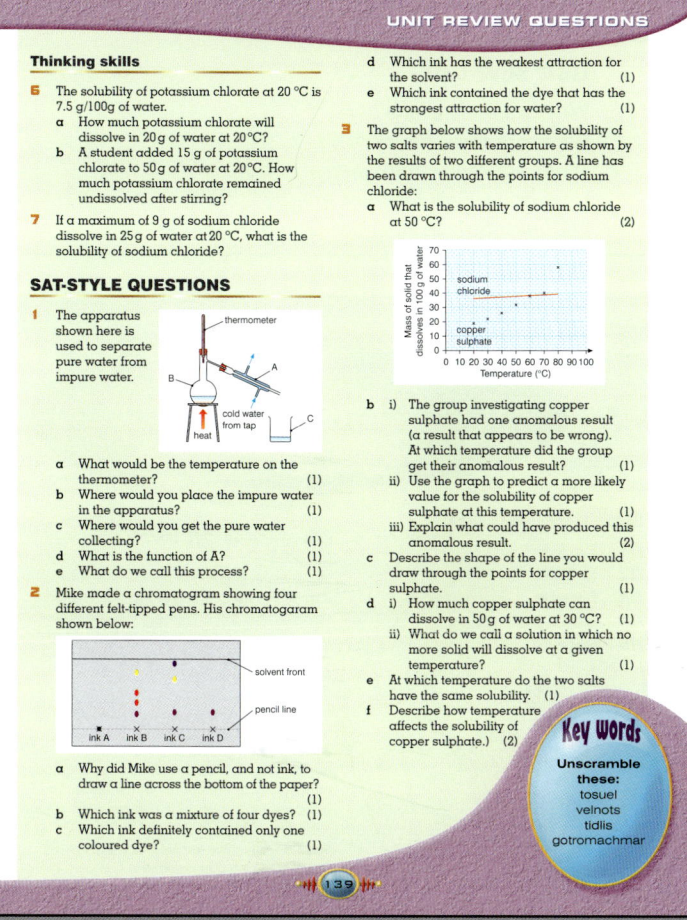

1 The apparatus shown here is used to separate pure water from impure water.
 a What would be the temperature on the thermometer? (1)
 b Where would you place the impure water in the apparatus? (1)
 c Where would you get the pure water collecting? (1)
 d What is the function of A? (1)
 e What do we call this process? (1)

2 Mike made a chromatogram showing four different felt-tipped pens. His chromatogaram shown below:

 a Why did Mike use a pencil, and not ink, to draw a line across the bottom of the paper? (1)
 b Which ink was a mixture of four dyes? (1)
 c Which ink definitely contained only one coloured dye? (1)
 d Which ink has the weakest attraction for the solvent? (1)
 e Which ink contained the dye that has the strongest attraction for water? (1)

3 The graph below shows how the solubility of two salts varies with temperature as shown by the results of two different groups. A line has been drawn through the points for sodium chloride:
 a What is the solubility of sodium chloride at 50 °C? (2)

 b i) The group investigating copper sulphate had one anomalous result (a result that appears to be wrong). At which temperature did the group get their anomalous result? (1)
  ii) Use the graph to predict a more likely value for the solubility of copper sulphate at this temperature. (1)
  iii) Explain what could have produced this anomalous result. (2)
 c Describe the shape of the line you would draw through the points for copper sulphate. (1)
 d i) How much copper sulphate can dissolve in 50 g of water at 30 °C? (1)
  ii) What do we call a solution in which no more solid will dissolve at a given temperature? (1)
 e At which temperature do the two salts have the same solubility. (1)
 f Describe how temperature affects the solubility of copper sulphate.) (2)

Key words
Unscramble these:
tosuel
velnots
tidlis
gotromachmar

Answers to SAT-style questions

1 a 100 °C (1)
 b B (1)
 c C (1)
 d cools water vapour, so it changes back into liquid (1)
 e distillation (1)

2 a The dyes in ink would run up the paper. (1)
 b B (1)
 c A or D (1)
 d A (1)
 e C (1)

3 a 37–39 g/100 g water (2)
 b i) 70 °C (1)
  ii) 43–47 g/100 g water (1)
  iii) e.g. may have used less than 100 g of water/temperature of the water may have been hotter than 70 °C (2)
 c curved (1)
 d i) 10–12 g (1)
  ii) saturated (1)
 e 60 °C (1)
 f The solubility of copper sulphate increases as the temperature increases. (2)

levelometer

7I Energy

National framework/QCA SoW references

National framework – Energy:
Identify a range of fuels and explain their uses.
Explain that the Sun is the original energy source and explain how its energy is transferred to a range of living and non-living things.
Explain why conservation of fuels is important.

QCA SoW (7I)
Why are fuels useful? (7I1)
What are fossil fuels? (7I2)
What are renewable energy resources? (7I3, 7I4)
How do living things use energy? (7I5)
NC links: Sc1 1a,b, c; Sc1 2 a, d, f; Sc4 5a, b, c.

Learning objectives

- To know that fuels are materials that are burned to release energy.
- To state that coal, mineral oil and natural gas are fossil fuels, and renewable energy resources include solar power, wind power, moving water and tidal water, biomass and some geothermal sources.
- To explain how fossil fuels formed, describe some of their uses, state that supplies are limited and explain why energy conservation is important.
- To describe some methods of energy conservation.
- To describe how renewable energy resources can be used to generate electricity and how a device works using a renewable energy resource.
- To know that all living things need energy for all activities, that animals obtain their energy from food and green plants obtain their energy from sunlight.
- To know that, in science, energy is measured in joules or kilojoules.

In scientific enquiry:

- To design an investigation to compare the energy in different fuels or different foods, controlling relevant variables.

Answers to questions

What do you remember?

1. all of them
2. oil
3. Examples are wood, paper, wax, fuels, etc. Oxygen from the air is needed for burning. It is not a reversible process.
4. An insulator, e.g. wool, bubble wrap, polystyrene, anything with trapped air.

Notes to support 'What do you remember?'

This unit builds on work from KS2. Pupils will have learned that some materials burn better than others and that when materials burn the change is irreversible because new materials are made. They will have learned that people need food for growth and food for activity, and that green plants need water, warmth, light and air to grow. Before beginning this unit, it will be useful if pupils:

- Can name some materials that are good at 'keeping warmth in'.
- Have had experience of burning materials.
- Know that animals need food to grow and be active.
- Know that plants need light, water, warmth, and healthy leaves to grow.

Teaching strategy

This unit builds on the KS2 units 4C 'Keeping warm', 6A 'Interdependence and adaptation' and 6D 'Reversible and irreversible changes' and leads into units 8I 'Heating and cooling' and 9I 'Energy and electricity'.

Difficulties/Misconceptions

- Pupils often think of energy as 'stuff' with substance. Discussion of energy resources as things that supply energy encourages pupils to begin to distinguish energy from 'stuff' like fuel and from the effects of energy, like activity, forces and power.
- A common misconception is that activity gives you energy, because it makes you healthier and more active. Discuss how you need food (an energy input) before you can do activities.

Notes and tips

- Time spent developing the idea of energy, in familiar contexts, is time well spent, as it 'prepares the ground' for the concept of energy transfers and energy transformations that pupils will meet in units in Years 8 and 9.
- Pupils do not need to know details about electricity generation, just some of the energy resources that can be used.

Learning styles

Visual
- **Visualising**.
- **Imagining**.
- **Obtaining** information.

Auditory
- **Explaining**.

Kinaesthetic
- **Practical** activities involving comparisons.
- **Making** models.
- **Designing** solar panels.

Interpersonal
- **Discussing and collaborating** with other groups.

Launch activity notes – Young technologists

This activity can be used as the starting point to encourage pupils to discuss their own ideas about 'making things go'. Many pupils will be very original and inventive with their ideas, but this activity allows you to spot basic misconceptions about 'what causes what'.

Answers

a) Pupils will be aware that most cars use petrol or diesel. Some may be keen viewers of programmes such as 'Tomorrow's World' and may be aware of electric vehicles, solar-powered vehicles and vehicles using hydrogen fuel cells.

b) Beyond 'they have a battery', most pupils will not have considered this. At this stage, you could discuss that the battery is charged from the mains, and discuss batteries running down and the limitations this puts on the use of the car. You may wish to return to this topic when you discuss energy conservation and the use of fossil fuels, or renewable energy resources, to generate electricity later in the unit.

c) Pupils will be aware – though they may need prompting – of the use of wind power to do useful things, as in old windmills. They may be aware of wind farms using windmills to generate electricity. They will probably be aware that, in principle, this car should work. They may also have an intuitive feel for 'petrol works better' in terms of 'it makes the car go faster' or 'it doesn't stop working when the wind stops blowing'.

d) Sunlight is freely available. Most pupils will feel this would be a good idea in really sunny countries, 'but wouldn't work here because we don't have enough sunshine', showing a basic concept of energy and energy use.

7I1 Fuels on fire

National framework/QCA SoW references

National framework – Energy:
Identify a range of fuels.
Explain that fuels are valuable resources and explain how they are used.

QCA SoW
7I Section 1: Why are fuels useful?
NC links: Sc4 5a; Sc1 2f.

Learning objectives

▸ To know that fuels are materials that are burned to release energy.
▸ To decide how to make a fair comparison between fuels, considering what factors have to be controlled.

Teaching strategy

Key points to highlight

- Fuels are materials that are burned to release the energy stored in them.
- The range of different types of fuel and what they are used for.
- All fuels store energy. Some fuels release more energy than others, when they are burned.
- **Safety:** How to use a Bunsen burner safely.

Difficulties/Misconceptions

Some pupils may confuse fuels and energy resources. A fuel is an energy resource that must be burned to release its energy.

Skills

- Scientific enquiry skills to identify and control key factors.

Lesson starter suggestions

- **Can you identify what all the samples have in common?** Pass round samples of fuels, such as coal, charcoal, wood, empty gas canisters, wax, sealed samples of artificial crude oil. Ask pupils to identify them all.
- **What is the reason for burning?** Pupils work in pairs to identify ten situations where people burn things.
- **Burning and safety:** Ask pupils: 'What safety rules about burning do you have in the laboratory? Do you know the reason for each of these rules? Make up a suitable set of safety rules for people burning things at home, for things like barbeques, bonfires, fireworks.'

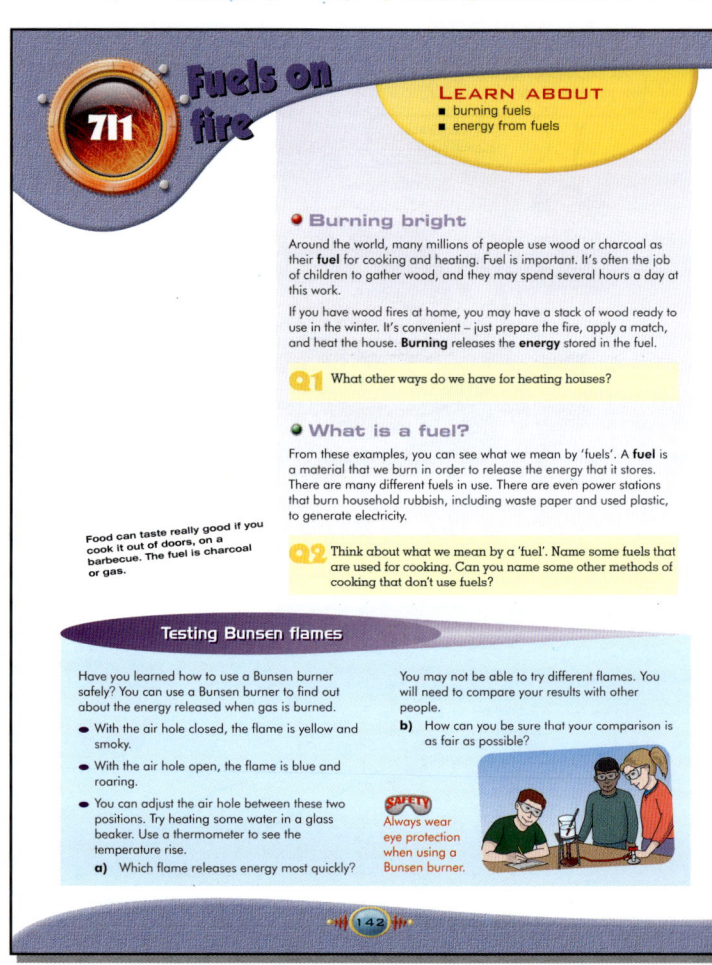

Activity and technicians' notes

Testing Bunsen flames
Safety notes

- Pupils must always wear safety goggles when using Bunsen burners.
- Stress the need for care near flames; particularly blue flames that are not clearly visible.
- Ensure pupils know not to put cold water into hot glassware.

Equipment required

For each group:
- Bunsen burner, tripod and gauze
- Small beaker
- Thermometer
- Safety goggles for each pupil

Technician's notes

- Beakers should have graduations on the side so they can be filled with water to the same level each time.

Access required to:
- Water

Teaching strategy contd.

Notes and tips
- If you burn different fuels to compare the energy output from them, follow your employer's risk assessments on the use of fuels. Discuss with pupils the hazards of fuel use and the precautions that must be taken.
- Do not allow pupils to use any fuels unsupervised.
- Do not use petrol.
- Only use tiny quantities of fuels and use these well away from stock supplies of fuel.
- Always wear eye protection when burning fuels.

Extension ideas
Pupils could be asked to consider the efficiency and convenience of different fuels, discussing reasons why particular fuels are chosen for particular purposes. Questions to prompt discussion on this topic might include: 'Why don't we put petrol on a coal fire in our house?' 'Why do you think people changed from burning wood to burning coal on the fires in their homes?'

Learning styles

Kinaesthetic
- **Practical activity** to compare Bunsen flames.

Interpersonal
- **Discussing and collaborating** with other groups to make fair comparisons of results.

ICT challenge

Access required to:
- Computers
- Internet
- CD-ROM encyclopaedias

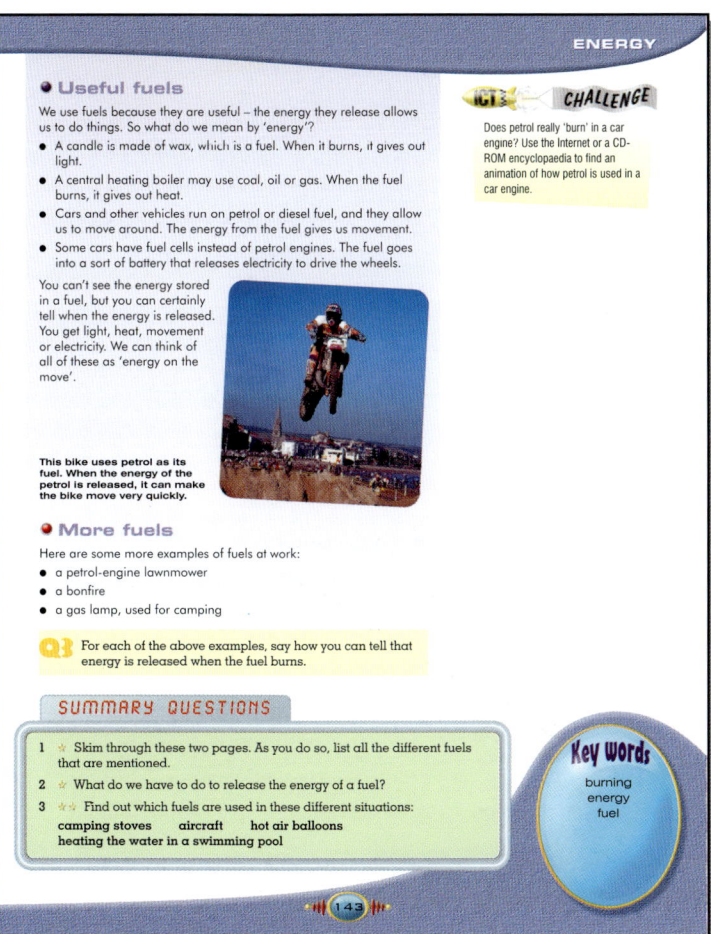

Answers to questions

Text
Q1 coal fires, central heating radiators or 'hot air' central heating fuelled by mineral oil, natural gas or calor gas

Q2 fuels used for cooking: natural gas, bottled gas, biomass; cooking without fuels: electricity (but see electricity generation later)

Q3 a petrol-engine lawnmower: get movement out; a bonfire: get heat out; a camping gas lamp: get light out

Summary
1 wood, charcoal, gas, household rubbish, wax, coal, oil, petrol, diesel, fuel cells
2 burn the fuel
3 camping stoves: butane or propane gas, methylated spirits, 'Coleman's fuel' or other similar liquid fuel;
aircraft: kerosene fuel;
hot air balloons: propane, or a propane and hydrogen mix;
heating the water in a swimming pool: fossil fuels or solar power

Learning outcomes

Following this, most pupils will:
▶ Use a Bunsen burner safely, producing rules for its safe operation.
▶ State that fuels release energy when they are burned.
▶ Plan how to compare the energy output of different fuels, making a fair comparison.

Some pupils will also:
▶ Use secondary sources to find information about the use of fuels.

Suggestions for plenaries

- **What do you know about fuels?** Pupils write four sentences beginning with F, U, E, L.
- **Fuels in the future:** Write on the board a list of tasks for which fuels are used. Pupils to think what fuels might have been used for each task 100 years ago, what fuels are used now, and then guess what fuels might be used 100 years from now. Discuss ideas in class discussion.
- **Red, amber, green cards:** Read out a series of sentences containing the word 'fuel'. Pupils to decide whether 'fuel' is being used in a scientific sense or not.

7I2 Fossil fuels

National framework/QCA SoW references

National framework – Energy:
Identify a range of fossil fuels.
Explain that they are a valuable resource and explain how they are used.

QCA SoW
7I Section 2: What are fossil fuels?
NC links: Sc1 1a; Sc4 5a.

Learning objectives

- To know that coal, mineral oil and natural gas are fossil fuels and to explain how they formed.
- To describe some of the uses of fossil fuels and to explain why conservation of fossil fuels is important.

Teaching strategy

Key points to highlight

- Coal, mineral oil and gas are fossil fuels. They formed millions of years ago from plant and animal remains.
- We each use a lot of 'hidden' fossil fuels – the fossil fuels used to generate the electricity we use.
- Fossil fuels are irreplaceable – they will run out. It is important to conserve them so they last as long as possible.
- Burning fossil fuels give off harmful gases that damage the environment, causing global warming and acid rain.

Difficulties/Misconceptions

A few pupils may believe that it is our industrial lifestyle that causes problems for the environment. Encourage all pupils to consider which of our activities damage the environment and to look at ways in which technology is solving problems as well as creating them.

Skills

Both the activity 'Running low' and the 'Link up to Citizenship' are activities where pupils will benefit from working together to discuss ideas, particularly as some may have strong views about some of the topics discussed. Encourage 'active listening', where pupils reflect back what they have heard. Groups could be asked to present a summary of the results of their discussions to the rest of the class.

Lesson starter suggestions

- **What fossil is this?** Pass round samples, or pictures, of coal containing fossils. Ask pupils to suggest what the fossil is and how it got there. Go over ideas in class discussion.
- **How much fuel do we use?** Pupils work together to identify all the fuel they have used so far today. Encourage them to look for hidden fuels, such as the fuel used to bake the bread they made their breakfast toast from.
- **Anagrams:** Give pupils anagrams for a range of different fossil fuels. They have to work out the names of the fuels, and then suggest a reason why traditional fuels such as wood and animal dung are not on the list. (Hint: Where are the different fuels found?)

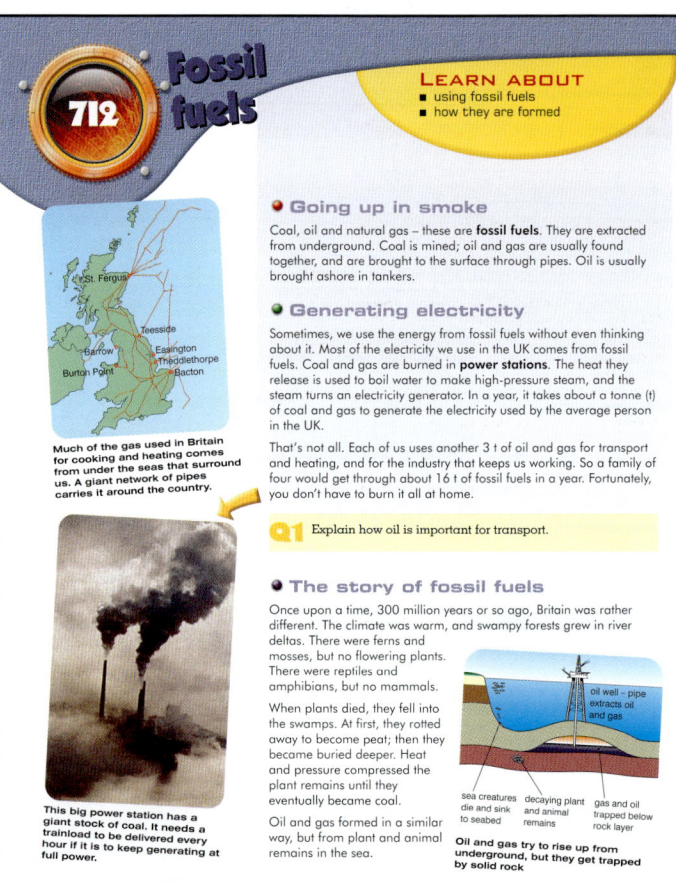

Teaching strategy contd.

Notes and tips
Many websites and other resources have information about problems caused by burning fossil fuels. As many of these also contain information about a range of renewable energy resources, you may wish to consider the environmental impact of both fossil fuels and renewable energy resources at the same time, after completing section 7I3 'Renewables'.

Extension ideas
Pupils could look at how coal forms and is extracted. They can consider factors that affect figures: stating how long coal supplies will last, discussing our use of coal and whether we can ask other countries to restrict their use of coal.

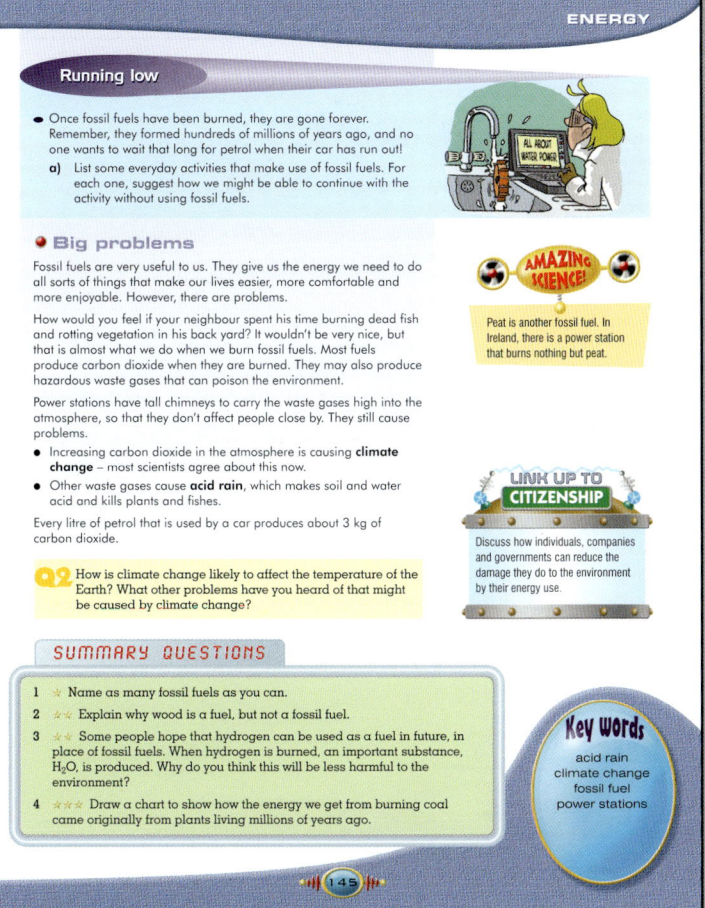

Learning outcomes

Following this, most pupils will:
▸ State that fuels release energy when they are burned.
▸ Explain why conservation of fuels is important.

Some pupils will also:
▸ Describe energy transfer links between the Sun, fossil fuels and themselves.

Learning styles

Visual
- **Imagining** the waste products produced from burning fossil fuels.

Interpersonal
- **Discussing and collaborating** with others about use and conservation of fossil fuels.

Answers to questions

Text
Q1 Most motor vehicles run on fuels made from refining crude oil.
Q2 Global warming will make the Earth warmer. This may cause: flooding of extensive areas of low lying land, as polar ice caps melt; extinction of plants and animal species poorly adapted for warmer temperatures; more extreme weather patterns such as tornadoes, droughts and floods.

Summary
1 coal, mineral oil (crude oil, petrol, diesel, aircraft fuel, paraffin, central heating fuel oil), natural gas
2 Wood has not been 'processed' by millions of years of heat and pressure.
3 H_2O is water. The environment has evolved to cope with water in it!
4 Chart should show energy in sunlight falling on plants, then plants being compressed to form coal.

Suggestions for plenaries

- **Verbal quiz about fuels and fossil fuels:** Ask all pupils to stand up. Ask them questions in turn. They sit down when they answer correctly. Provide questions of differing complexity so all pupils have an equal chance of answering correctly.
- **Written questions and answers:** Each pupil writes an answer to a question about fuels or fossil fuels. They swap answers with a partner and the partner has to decide on a suitable question that has the given answer.
- **Cognitive map:** Provide pupils with a partially completed cognitive map about the types and uses of fuels, including fossil fuels. They then complete the map.

7I3 Renewables

National framework/QCA SoW references

National framework – Energy: the Sun is a source of energy.

QCA SoW

7I Section 3: What are renewable energy resources?
NC links: Sc4 5a, c.

Learning objectives

- To know that renewable energy resources include solar power, wind power, moving water and tidal water, biomass and some geothermal sources.
- To know how renewable energy resources can be used to generate electricity.
- To explain how a device works using a renewable energy resource.

Teaching strategy

Key points to highlight

- Energy resources are any resources that provide energy for activities.
- Renewable energy resources are those that can be replaced as they are used.
- Renewable energy resources can be used to generate electricity.
- Even renewable energy resources will run out if we use them faster than they can be replaced.
- Renewable energy resources do not produce polluting gases, but they still affect the environment.

Difficulties/Misconceptions

Pupils often think that renewable energy resources do not affect the environment. Encourage them to look in detail at the impact of a chosen renewable energy resource on people and the environment, then to list good and bad features of that resource.

Lesson starter suggestions

- **Different sources of energy:** Ask pupils to think of as many different sources of energy as they can. Question to prompt ideas: 'If you were making a toy car or toy go-kart, how could you make it move?'
- **Pedal power:** Show picture of person lighting light bulb and running fan using pedal power from a bicycle. Work in pairs to suggest advantages and disadvantages of generating electricity this way, instead of using fossil fuels.
- **Alternatives to fossil fuels:** Show pictures of different activities. Pupils to state what fossil fuel/s might be used in the activity, and then to decide if there is any other way to make the activity work, e.g. sails on a boat instead of a diesel engine.

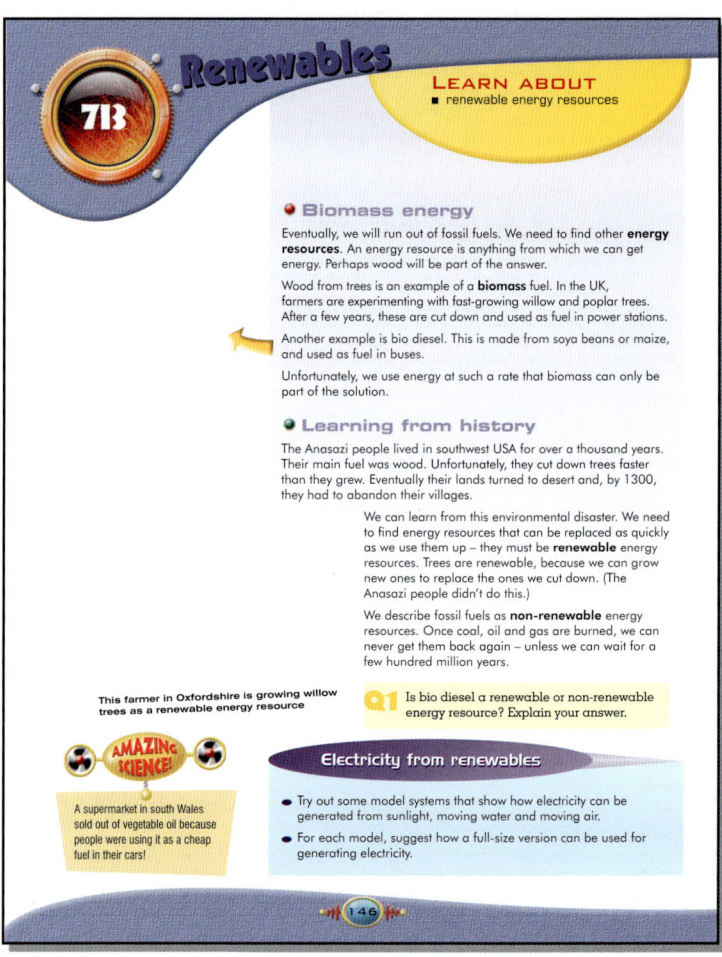

Activity and technicians' notes

Electricity from renewables

Equipment required
- Electric motor kits
- Wires, bulbs, small current ammeters
- Solar cells
- Windmill and water mill construction kits

Technician's notes

Solar cells, windmill and water mill kits can be connected to small electric motor kits, to make the electric motor work 'in reverse' to give out a current. Often this current is too small to light a bulb, so pupils will need to connect a sensitive ammeter to the motor to measure current output.

Access required to:
- Water
- Hair dryers or similar source of moving air

Teaching strategy contd.

Skills
This lesson provides plenty of opportunities for open-ended discussion about the impact of different energy resources, about questions of energy use and lifestyle. There are often no 'right answers' to these issues, so encourage pupils to listen to evidence, and then make up their own minds.

Notes and tips
The activity 'Electricity from renewables' involves using an electric motor 'in reverse' to generate electricity. Draw pupils' attention to this as it is valuable to look back on when they study the Year 8 unit 'Magnets and electromagnets'.

Extension ideas
Different types of power station need different locations. Pupils could look at the requirements of a range of power stations, deciding where they would site them, what effect the power station would have on the environment, and the advantages and disadvantages of generating electricity this way.

Learning styles

Kinaesthetic
- **Practical activity** to make electricity generating models.

Interpersonal
- **Discussing** electricity generation using renewable energy resources.

Answers to questions

Text
Q1 Renewable: more maize or soya beans can be grown.

Q2 Hydroelectric power: refer to water cycle.
Wind power: moving air, driven by the Sun's energy.
Wave power: waves created by the wind.
Solar power: the Sun gives out radiation continuously, so it will not run out.
Geothermal power: rocks will take hundreds of millions of years to cool down.

Q3 Yes, if some of the electricity generated is used to charge batteries.

Summary
1. wood, biomass (such as willow, poplar), bio diesel (from maize or soya), vegetable oil, sunlight, moving water, moving air, wave power, geothermal power
2. Non-renewable; it is a fossil fuel.
3. Solar power is available in space and reduces the need to carry fuel.
4. Nuclear fuel is a non-renewable energy resource. Eventually the supplies of uranium will run out.

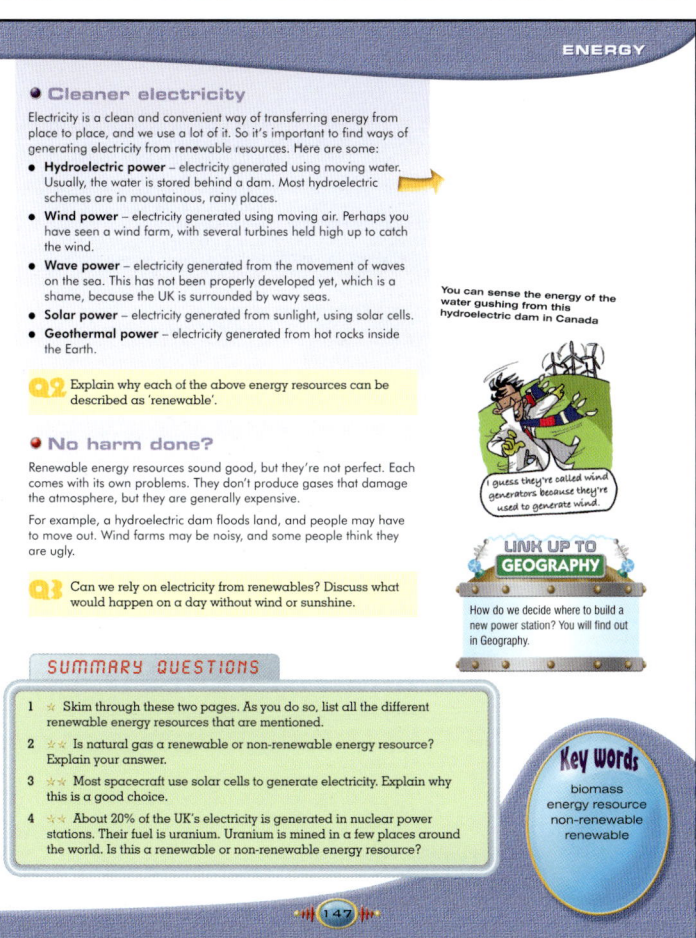

Learning outcomes

Following this, most pupils will:
▶ Use secondary sources provided to find information about energy resources.
▶ Describe how electricity is generated using renewable energy resources.

Some pupils will also:
▶ Compare the effectiveness of different energy transforming devices.
▶ Compare the advantages and limitations of various energy resources.

Suggestions for plenaries

- **Energy resources:** Provide the jumbled up letters of the words 'energy resources'. Pupils to try to make as many words as they can from these letters. Can they make scientific words? Can they work out the two words that use all the letters?
- **Connectives:** Ask pupils to complete the sentence 'Renewable energy resources are important . . .' using one of these words: 'and, because, however, as, but, although'.
- **Sorting energy uses:** Give pupils a set of card with different activities that use energy on them. Pupils to sort them into activities that 'can only use fossil fuels', 'can only use renewable energy resources', 'can use both'. Ask them to justify their choices.

7I4 Energy – use with care!

National framework/QCA SoW references

National framework – Energy:
Recognise that fuels are a valuable resource and explain why it is important to conserve fuels.
Recognise the Sun as an energy source.

QCA SoW
7I Section 2: What are fossil fuels?
7I Section 3: What are renewable energy resources?
NC links: Sc4 5a, b, c; Sc1 2a, d.

Learning objectives

▶ To know that fossil fuel supplies are limited.
▶ To understand the need for energy conservation.
▶ To discuss and describe methods of energy conservation.

Teaching strategy

Key points to highlight

- Saving energy is good for us and good for the environment.
- Our knowledge of the damage burning fuels do to the environment has improved, so modern buildings are built to waste less energy.
- Sunlight is free. Solar panels and solar cells trap and use the energy in sunlight.
- We can save energy by stopping warmth escaping from our houses.

Difficulties/Misconceptions

Pupils may confuse solar panels and solar cells. Solar cells make electricity, just like the cells they will have used in electrical circuits.

Skills

The activity 'Designing solar panels' is intended as a quick comparison only, although the activity could be extended into a full Sc1 investigation or, with cooperation with the design and technology department, into a full design project. If done as a full-scale project, pupils will need to predict, test, make appropriate observations and modify their designs in response to findings. The main factors they will need to consider include the shape, size and insulating properties of the tray (the volume of water being heated and the amount of heat 'leaking out' from the tray), and the colour (they will find that black absorbs heat best, but also loses heat fastest when the tray is no longer being heated).

Lesson starter suggestions

- **Using energy:** Draw two sketch diagrams on the board, one of a person and one of a house. Ask pupils to copy the diagrams and write: around the person all the ways they can think of that they use energy; around the house all the ways their home uses energy.
- **Keeping warm:** Ask the pupils to imagine their Mum says they have to pay the fuel bills this year! What things can they do to keep the bills as low as possible? Pupils to discuss and to decide on two or three good ideas to share with the class.
- **True or false:** Give pupils a series of statements about fuels and energy resources. Pupils to identify which statements are true, and which are false and then to correct the false ones.

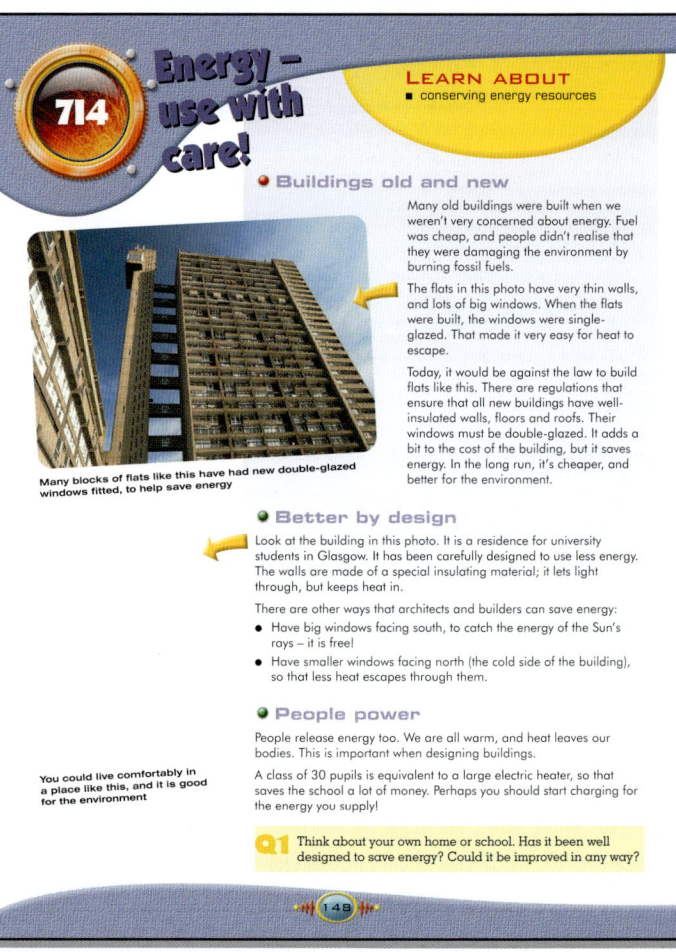

Notes and tips

There are so many aspects to energy saving and sensible energy use, that it may not be practical to ask pupils to look at them all. You may prefer to give groups a single aspect, such as solar panels or double-glazing, to look at, and then ask each group to report back to the class on what they have found. Their work could be presented as part of a class display.

Extension ideas

- Pupils could look at devices using solar power to provide electricity, discussing the benefits they bring and how practical they are.
- Some pupils may extend their research to look at devices using a renewable energy resource other than solar power.

Activity and technicians' notes
Designing solar panels
Equipment required
For each group:
- Metal and plastic trays of different sizes, shapes, colours and depths
- Thermometers
- Cling film/clear plastic
- Range of insulating materials

Technician's notes
All trays should be relatively shallow; very deep trays will show very little warming in the time for one lesson, even on very hot days.

Access required to:
- Water
- A sunny day or, failing that, a high wattage ordinary light bulb to simulate heat from the Sun!

Learning styles
Visual
- **Visualising** buildings.
- **Imagining** characteristics of buildings that are good and bad for energy conservation.

Kinaesthetic
- **Practical activity** to design solar panels.

Answers to questions
Text
Q1 A well-designed home or school would have cavity wall insulation, a thick layer of loft insulation, double glazed windows, large south-facing windows, small north-facing windows, well fitting doors to exclude draughts and possibly solar panels.

Q2 So energy from the sunlight can be used at night, and on days when it is not very sunny.

Summary
1. Energy resources are limited: fossil fuels will run out. Using less energy saves us money. Using less energy often protects the environment.
2. large south-facing windows, solar panels and solar cells on roof or walls
3. Electricity generating machinery still has to be bought and maintained. Electricity still has to be transported from where it is generated to where it is used. Electricity generation using wind power is less efficient than electricity generation using fossil fuels: less of the energy in the wind is transformed into electrical energy.

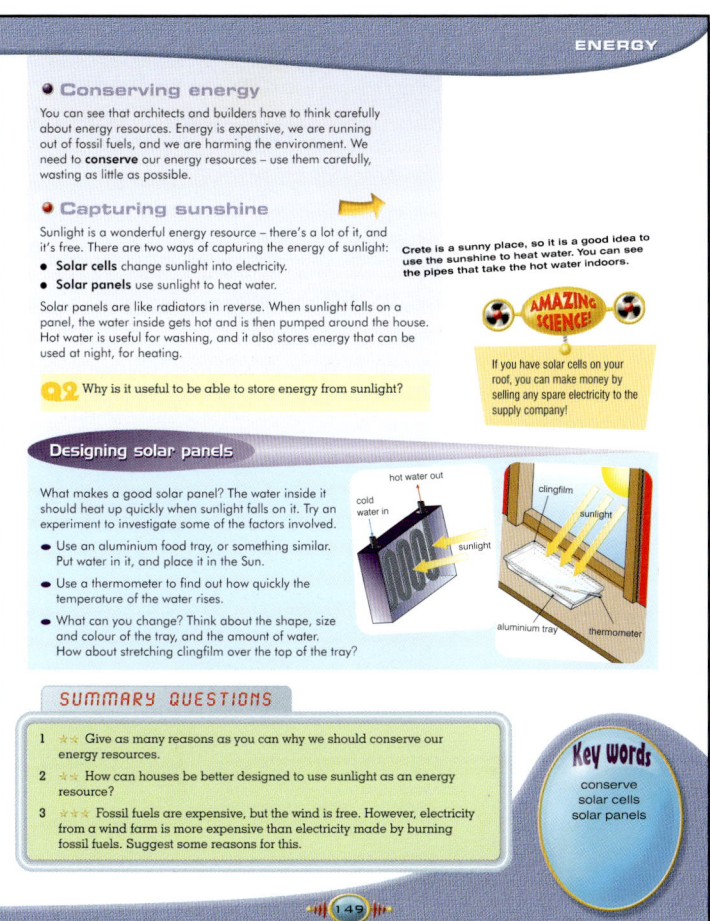

Suggestions for plenaries
- **What is energy?** Working in pairs, give pupils one or two minutes to write an answer to the question: 'What is energy?' Use class discussion to agree a class definition of energy.
- **Cognitive map:** Working alone, or in pairs, ask pupils to construct either a spider diagram or a concept map to summarise what they know about energy resources.
- **'Taboo cards':** You will need at least six cards with words related to fuels and energy. Pupils to work in threes: a describer, a listener, and a referee. The describer describes the scientific word on their card, without using any of the 'taboo' words that are written on the card. The listener guesses what the word is. The referee checks they don't cheat.

Learning outcomes
Following this, most pupils will:
▸ Use secondary sources provided to find out information about energy devices.
▸ State how renewable energy resources can provide heating and electricity.

Some pupils will also:
▸ Select appropriate secondary sources to find information about energy devices.
▸ Give examples of how to conserve fuels and energy resources.

715 Food as fuel

National framework/QCA SoW references

National framework – Energy: the uses of food by living things, the Sun is the original source of energy for other energy resources.

QCA SoW

7I Section 4: How do living things use energy?
NC links: Sc1 1b, c; Sc1 2d; Sc4 5b.

Learning objectives

▸ To know that humans, and all other living things, need energy for all activities.
▸ To know that animals obtain their energy from food, and green plants obtain their energy from sunlight.
▸ To know that, in science, energy is measured in joules or kilojoules.
▸ To design an investigation to compare the energy in different foods, controlling relevant variables.

Teaching strategy

Key points to highlight

- All living things need energy for all activities, including growing and living processes such as breathing and keeping warm.
- Animals obtain their energy from the food they eat.
- Green plants make their own energy from sunlight.
- The scientific unit of energy is the joule or kilojoule.

Difficulties/Misconceptions

Some pupils may think that exercise gives them energy because 'exercise is healthy'. Discussion about feeling tired and hungry after strenuous exercise should help resolve this misconception.

Lesson starter suggestions

- **Food:** Make a class spider diagram, with the word 'food' in the centre, to establish pupils' prior knowledge about food and why they need to eat. Keep the spider diagram to refer back to in the plenary session.
- **Energy foods:** Pass round samples or packaging from 'energy foods'. Ask 'Who knows what these are?' 'Who might eat them?' 'Why?' Record class ideas.
- **Food for energy:** Show a picture of a person lying on the beach surrounded by mounds of biscuits, chocolates, chips and other 'fattening' foods. Show a second picture of a person in hiking gear, with a mountain in the background, carrying only a bunch of celery. Pupils to decide 'Have these people chosen their food sensibly?' Pupils to explain why and how they could improve their choices of food.

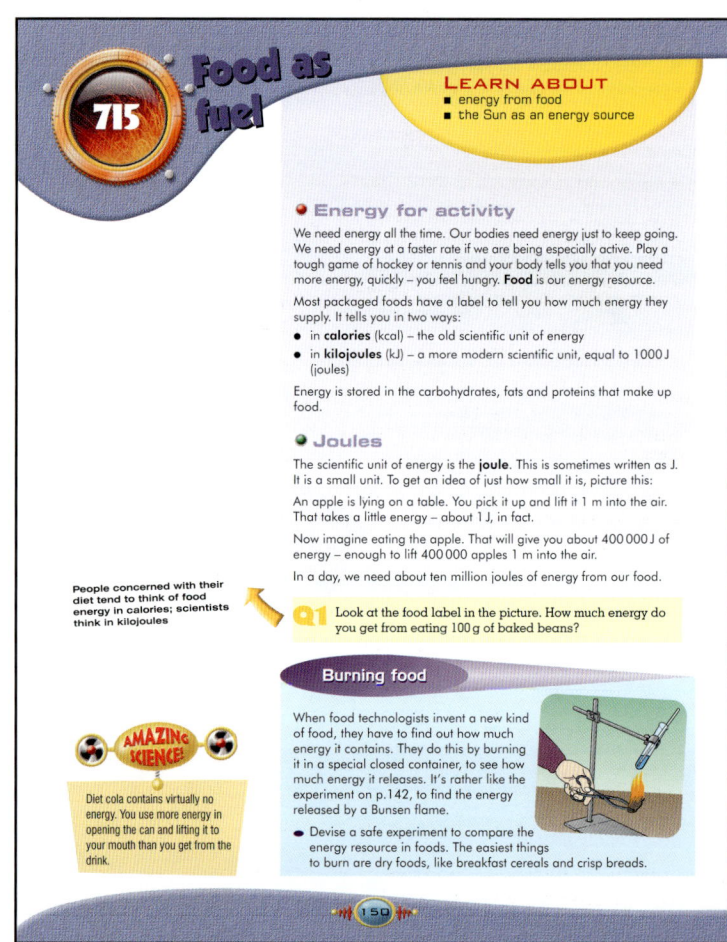

Activity and technicians' notes

Burning food

Safety notes

- Employer's risk assessment should be followed.
- Discuss safety hazards and precautions with pupils before carrying out this activity.
- Safety goggles must be worn.
- Ensure good ventilation as this activity is smoky and smelly.
- Avoid peanuts – they burn well but some pupils may have serious allergic reactions to them.
- Remind pupils of the hazards of putting cold water in hot glassware.

Equipment required

- Safety goggles for each pupil

For each group:
- Bunsen burner
- Metal spike or other method of holding food in flame
- Test tubes, clamps, thermometers

Technician's notes

Clamps should enable pupils to hold test tube securely above pieces of burning food.

Access required to:
- Water

Teaching strategy

Skills

The activity to compare the energy content of different foods by burning them provides an opportunity to:
- Devise an experiment.
- Consider factors that will affect the result.
- Consider ways to make the results as reliable as possible.
- Cooperate as a class, since it is likely that pupils will not have time to burn all the different types of food and they will need to share results.

Notes and tips

- Sensitivity is needed when discussing slimming, health and diet. Pupils will learn about the need for a balanced diet in Unit 8A.

- Pupils do not need to know in detail how photosynthesis works, as this is covered in Unit 9C. In this lesson, simply remind pupils of Key Stage 2 work in which they learned that plants need light, water and healthy leaves to make their own food.

Extension ideas

Some pupils could consider why different people have different energy requirements, discussing the things that affect the energy needed, and how and why the amount of energy a person needs changes throughout their lifetime.

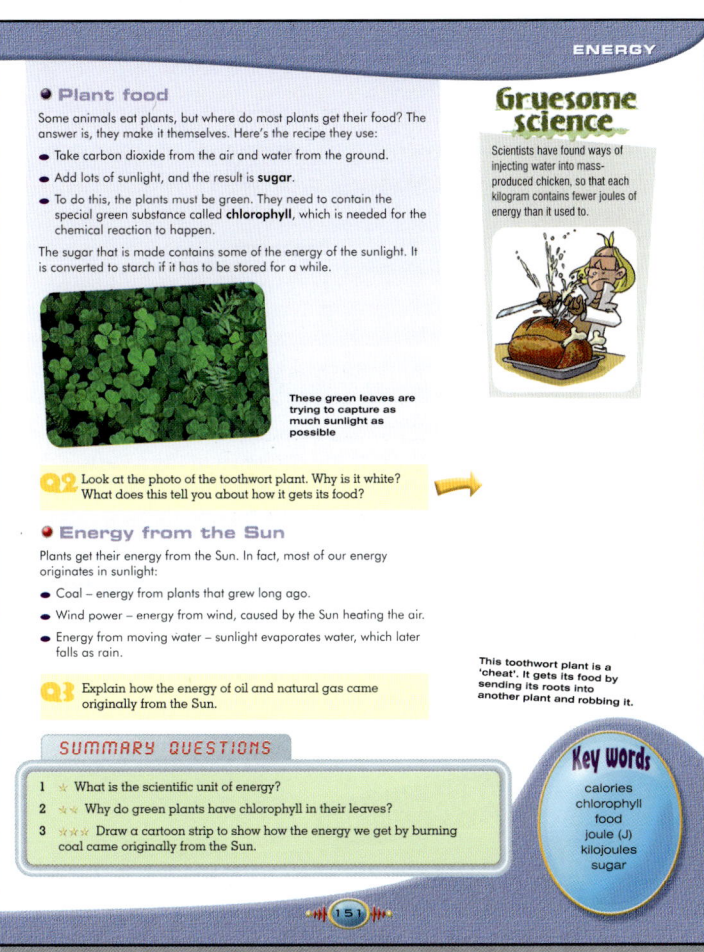

Learning styles

Visual
- **Viewing** food labels and photographs to obtain information.

Auditory
- **Explaining** how oil and gas store energy that came originally from the Sun.

Kinaesthetic
- **Practical activity** to burn foods and compare their energy content.

Answers to questions

Text

Q1 308 kJ (73 kcal)

Q2 Toothwort has no chlorophyll, so it does not make food from sunlight.

Q3 Energy in sunlight was absorbed by green plants in the sea; the plants were eaten by marine animals; the plants and animals fell to the seabed; they were converted to oil and gas; the energy was stored in the oil or gas.

Summary

1 joule, or kilojoule
2 Chlorophyll is needed for the chemical reaction that turns water, carbon dioxide and the energy in sunlight into sugar.

Suggestions for plenaries

- **Food:** Revisit the spider diagram created in the Lesson starter 'Food'. Ask pupils how they can add to it or modify it to show what they have learned this lesson. Ensure that the energy links between sunlight, food and themselves are shown.
- **Words and meanings:** Group work: each pupil writes a scientific word connected with food or energy, and writes its meaning on a separate piece of paper. The groups swap papers, then pupils in each group work together to match words and meanings.
- **Energy route:** Pupils use a word story about an egg eaten for breakfast to construct a complete energy diagram, showing how energy was transferred from sunlight to them. They also consider other things 'en route' that some of the energy was used for.

Learning outcomes

Following this, most pupils will:
▸ Compare the energy output of a range of foods, using a fair comparison and controlling relevant variables.
▸ Identify energy transfers in living things.

Some pupils will also:
▸ Describe energy transfer links including the Sun, living things and themselves.

71 Read all about it!

Teaching strategy

Fair shares for all?/Going under?/Predicting climate change

Difficulties/Misconceptions

- Most pupils will have heard of climate change and many will be aware that scientists think it is caused by our modern lifestyles. However many will expect changes in temperature to be dramatic if climate change is happening, and will not be aware that, in global terms, an increase in average temperature of just a few degrees is considered huge.
- Nor will many be aware that the problems are not caused directly by plants and animals dying because they are too hot, but because of the disruption the change in climate causes in breeding cycles, weather patterns and water flow patterns.

Notes and tips

- Energy consumption and climate change are both huge topics, that could be developed into class project work. Each group of pupils could find out about one aspect of the topic and report back.
- The topic of energy consumption allows plenty of scope for pupils to develop their own opinions – encourage them to do so – but also to listen to the opinions of others.
- Pupils should be aware that energy consumption figures for different countries can be written in two different ways:
  1. Energy consumption per individual in a country
  2. Energy consumption for the whole country
- If they are looking at the total energy consumption for a country, they will need to take into account how many people live in that country.
- Some pupils may also wish to consider the environmental effect of technologies. Old technology is usually more damaging to the environment than new technology.
- This opens up a whole range of topics linking with geology and citizenship issues.
- The topic of climate change provides opportunity for pupils to look at how scientific ideas about global warming have changed in the last decade. Most scientists now agree that global warming really is happening and are discussing ways to solve the problem, but only a decade ago scientists were involved in heated debate over whether or not global warming existed, or whether changes recorded were just part of a natural cycle of changes that had been happening for millennia.

Skills

This unit has many opportunities for pupils to develop the Sc1 skills of research and evaluating evidence. It also provides opportunities for pupils to practise working cooperatively together, looking at citizenship, history and geography issues about the effect that people have on each other and on the world around them.

Extension ideas

Pupils could work together to research in more detail one of the following topics:
- How does a device using a renewable energy resource work?
- How is electricity generated using a renewable energy resource?
- How does a particular method of conserving energy work?
- How is a particular country/group of people affected by energy use/climate change?
- How do scientists record weather patterns, and predict how they are going to change?

Questions

1. Which areas of the world use most energy?
2. Give a benefit of using lots of energy, and a problem that it causes.
3. What type of energy use is causing climate change?
4. What did scientists do, to help them decide whether climate change really was happening?
5. What are the biggest and smallest temperature changes predicted by 2100?

Answers

1. North America and Europe
2. gives us a high standard of living, harms the environment
3. burning fossil fuels
4. gathered new scientific evidence in the 1990s
5. 5.5 °C, 1.5 °C

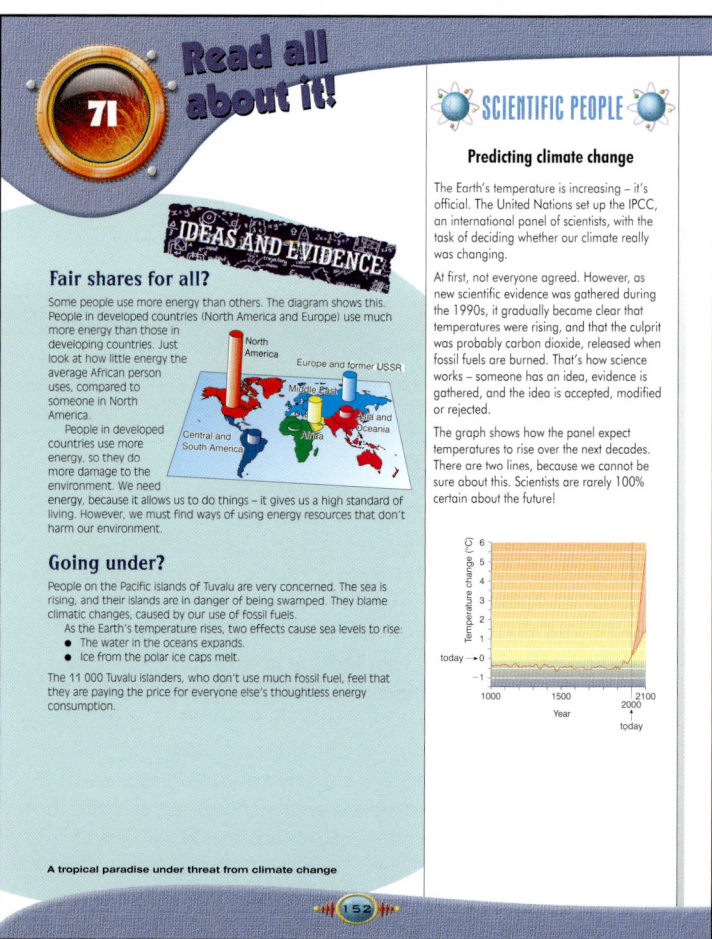

Danger – common errors

The concept of energy causes problems for some pupils because energy is a very abstract concept, and some pupils are not mentally ready to deal with abstract concepts at this age. Although eventually pupils will need to learn a scientific definition of energy, it may be better at this stage to concentrate on everyday situations where they have used the word, such as 'When I have lots of energy I like to run around a lot', as these will help pupils become familiar with the concept that 'when something has energy it can do something'.

Learning styles

Visual
- **Visualising, and imagining** lifestyles around the world.

Kinaesthetic
- **Role play** of different characters in debate on energy use or climate change.

Interpersonal
- **Discussing** energy use or climate change.

Learning outcomes

Following this unit, most pupils will:
- State that fuels release energy when they are burned.
- Explain why conservation of fuels is important.
- Describe how renewable energy resources are used to provide heating and to generate electricity.
- Identify energy transfers in living things.

Some pupils will also:
- Describe energy transfer links between the Sun, fossil fuels and themselves, and between the Sun, living things and themselves.
- Compare the advantages and limitations of various energy resources.
- Give examples of how to conserve fuels and energy resources.

In scientific enquiry:

Following this unit, most pupils will:
- Use a Bunsen burner safely.
- Plan how to compare the energy output of different fuels and foods, making a fair comparison and controlling relevant variables.
- Use secondary sources provided to find information about energy resources and energy devices.

Some pupils will also:
- Select appropriate secondary sources to find information about energy devices and the use of fuels.
- Compare the effectiveness of different energy transforming devices.

7.1 Unit review

Answers to review questions

1. **a** biomass: it eats green plants
   b fossil fuels, occasionally renewables (hydrogen fuel cells, solar power)
   c fossil fuels (diesel and in power stations), renewables (to generate some of electricity for electric trains)
   d sunlight

3. **a** A museum keeper conserves old things, protects and restores them so they don't decay.
   b A wildlife ranger conserves wildlife, protects its habitat so it doesn't become extinct.
   c an athlete conserves his/her energy; he/she doesn't use it up so he/she can save it for the big event

4. **a** a flow chart
   b energy at every stage, except the first stage, would increase

5. **a** mineral oil, natural gas, biomass, geothermal, wave power, nuclear, electricity

6. **a** natural gas
   b the renewable energy resources part of the diagram will be larger; this change is important, as fossil fuels will run out and their burning damages the environment
   c more renewable, less non renewable; first column will be larger, most or all of the other columns will be shorter

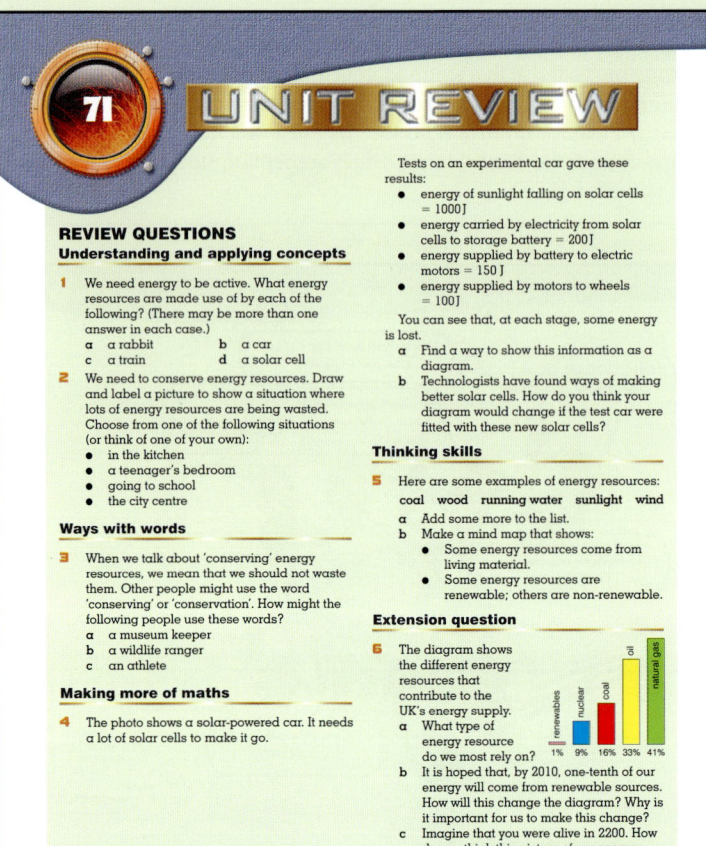

UNIT REVIEW QUESTIONS

SAT-STYLE QUESTIONS

1. Pip's house is on a remote island. Her parents have installed solar cells on the roof, and a wind turbine. These generate electricity.
   The house also has a diesel generator that can generate electricity when the solar cells and wind turbine are not working.

   a For each method of generating electricity, draw a line to the energy resource which it makes use of: (3)

Method	Energy resource
diesel generator	sunlight
solar cells	running water
wind turbine	chemicals
	moving air

   b The solar cells do not work at night. Explain why not. (1)
   c The wind turbine does not always supply electricity. Explain why not. (1)
   d Explain why it is important for Pips family to have a petrol generator. (1)

2. Year 7 have been studying energy resources. They have made a list of different resources, which their teacher shows on the whiteboard:

   geothermal   biomass   oil   coal
   moving air   tidal   running water
   nuclear   solar   natural gas

   a From the list, name three fossil fuels. (3)
   b From the list, name three renewable energy resources. (3)
   c The class has been reading about a new tidal power station to be built near their town. The local newspaper published this picture of how it will work:

The boxes below show the stages in generating electricity.

A	The generator produces electricity.
B	The gates shut, trapping the water.
C	Water flows out past the turbine, making it turn.
D	As the tide comes in, the water level behind the dam rises.
E	The turbine turns the generator.

Put the boxes in the correct order. The first one has been done for you.

| D | | | | | (4) |

3. Reese and Pete are investigating a small electric heater. They are using it to heat 200 cm³ of water in a beaker. They measure the temperature of the water every minute.

The graph shows their results.
a Explain why Reese stirred the water each time before taking its temperature. (1)
b Pete wrote this prediction:
'If we wait twice as long, the temperature of the water will go up twice as much.' (2)
Do the results of the experiment support Pete's prediction?
c Copy the graph. Add a line to show the results you would expect if the experiment were repeated with 400 cm³ of water in the beaker. (2)

Key words

Unscramble these:
greeny
correuse
near belew
nescover
metalic ganche

Answers to SAT-style questions

1. **a** diesel generator: chemicals;
   solar cells: sunlight;
   wind turbine: moving air (3)
   b There is no sunlight at night. (1)
   c Sometimes it is not windy. (1)
   d A petrol generator generates electricity when the renewable resources do not work. (1)
2. **a** oil, coal, natural gas (3)
   b moving air, running water, solar (3)
   c D, B, C, E, A (4)
3. **a** so all the water was at the same temperature (1)
   b No, the graph is a curve, not linear. (2)
   c Graph is less steep, temperature rise is slower. (2)

levelometer

7J Electric circuits

National framework/QCA SoW references

National framework – Energy: Cells provide energy in an electrical circuit; the electric current transfers the energy to all components in series and parallel circuits.

QCA SoW (7J)

How do electrical circuits work? (7J1, 7J2)
What happens in a circuit? (7J2, 7J3, 7J4)
How can we explain what happens in electrical circuits? (7J4, 7J5)
What kinds of circuits are useful and what are the hazards? (7J5, 7J6)
NC links: Sc1 1b; Sc4 1a, b, c.

Learning objectives

- To learn about complete circuits and use circuit symbols to represent simple circuits.
- To describe how conductors and insulators are used in electrical circuits.
- To use a water flow model to visualise and explain electric current flow and energy transfer in circuits.
- To explain how batteries are made from cells.
- To explain the meaning of voltage.
- To describe how current and voltage in a circuit are related.
- To understand the chemical origin of the energy in electrical circuits.

In scientific enquiry:

- To use an ammeter to measure current and describe how the current in a circuit can be changed.
- To identify series and parallel circuits and investigate the current flow round them.
- To describe electrical dangers and explain some electrical safety features.

Safety notes

You will find relevant safety information in CLEAPPS *Laboratory Handbook* 12.2. and 12.3.

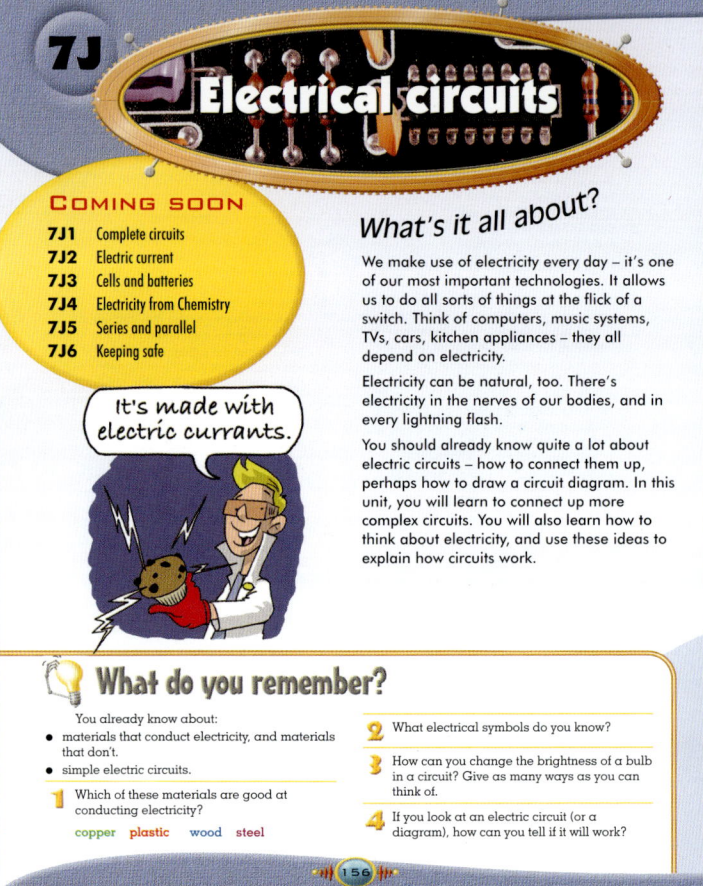

Answers to questions

What do you remember?

1. copper, steel. (Some pupils may know water is a good conductor, and copper conducts electricity very well.)
2. symbols for a bulb, a cell (or battery), a switch and a wire
3. Use more cells (or batteries) or less bulbs. (Some pupils may mention using a 'bigger' battery – though they may not know that this really means a cell with a higher voltage.)
4. Look for a complete circuit. (Some pupils may mention that it could look like it would work and still not work, because components may not work or connections may be bad.)

Notes to support 'What do you remember?'

This unit builds on work from KS2. Pupils will have drawn and built a range of simple series circuits. They will know how switches work and will probably be able to describe some ways to make bulbs brighter or dimmer. Before beginning this unit, it will be useful if pupils can:

- Remember that electrical components only work if there is a complete circuit.
- Connect up simple circuits.
- Recognise and draw standard symbols for a cell, a bulb and a switch.

Teaching strategy

This unit builds on the KS2 units 4F 'Circuits and conductors' and 6G 'Changing circuits' and leads into units 8J 'Magnets and electromagnets' and 9I 'Energy and electricity'.

Difficulties/Misconceptions

- Common misconceptions are the 'clashing current' view of current flow, where current is supposed to flow from both sides of cells, meeting in the components, and the 'diminishing current' view where current is supposed to be used up as it passes through components. Discussion and repetition will help clear these misconceptions.

- Ideas of 'voltage pushing' and a range of different flow models will help clarify the difference between current and voltage, another common area of confusion.

Notes and tips

This unit should concentrate on enabling pupils to understand two abstract concepts.

- The first is that electricity, or electrical current, is 'stuff' that flows round circuits, through conductors.
- The second is that electrical current carries, or transfers, energy around a circuit, from the cells to the electrical components, and that while electrical energy is used up in the components, current is not used up.

Learning styles

Visual
- **Visualising** and tracing current flow.
- **Matching** circuits to diagrams.
- **Imagining** thought models of current flow.
- **Predicting** behaviour of circuits.
- **Visualising** electrical hazards.

Auditory
- Explaining.

Kinaesthetic
- **Building** circuits and measuring current.
- **Investigating** combinations of cells.
- **Investigating** fuse wires.

Intrapersonal
- **Evaluating** circuit models.

Launch activity notes – Young electricians

This activity can be used as the starting point to explore pupils' knowledge, and misconceptions, about electrical circuits. Discussing the misconceptions of the Scientifica characters provides a non-threatening way for pupils to identify their own misconceptions.

- Pupils will know from KS2 that working batteries and bulbs are necessary if a circuit is to work.
- Pupils may have used a red wire to connect the + side of the battery to one side of a bulb and a black wire to connect the – side of the battery to the other side of the bulb. Here they could discuss what is inside the wires and whether the colour of the plastic insulation matters.
- Many pupils will already have a clear idea of which materials conduct and which are insulators. This can be linked to their knowledge about using electricity and building circuits safely.

- Some pupils may already be using statements such as 'Close the switch and the light goes on.' to help them explain the way current flows around a circuit, even though they may not be aware of the term 'current' at this stage.

Answers

a) Mike Roscope and Molly Kewell.
b) Reese Cycle and Pete Ridish – it's getting hot!
c) No.
d) It can't – there are metal contacts.

7J1 Complete circuits

National framework/QCA SoW references

National framework – Energy: The purpose of cells in an electric circuit.

QCA SoW
7J Section 1: How do electrical circuits work?
NC links: Sc4 1a.

Learning objectives

▸ To learn about complete circuits.
▸ To use circuit symbols to represent simple circuits.
▸ To describe how conductors and insulators are used in electrical circuits.

Teaching strategy

Key points to highlight

- There has to be a 'complete loop' of conducting metal all the way round a circuit.
- Each component has to have two connections, so electricity has a 'way in' and a 'way out'.
- Metal is a conductor, that electricity can flow through. Electricity cannot flow through insulators.

Difficulties/Misconceptions

Pupils readily realise that the wires allow electricity to flow from battery or cells to the components. Some pupils may not realise that the electricity needs to flow back again. Emphasising the two connections for each component – a 'way in' and a 'way out' helps here.

Skills

- In the 'circuit challenge', pupils use problem solving skills and visualisation skills to match real circuits to the appropriate circuit diagrams.
- Then they predict how they would behave when the switches are closed.
- As pupils progress to slightly more complicated, parallel circuits, this ability to identify whether or not a circuit is the same as a circuit diagram becomes very important.

Lesson starter suggestions

- **Red, amber, green:** Checking pupils' prior knowledge. Read out a series of names of electrical components. Pupils hold up a red card if they have not heard of it, an amber card if they have heard of it but don't know what it is for, and a green card if they can name where it might be used.
- **What is it?** In pairs: give pupils cards with circuit descriptions on. Pupils take it in turns to read out the card, their partner tries to name an electrical appliance that it could be.
- **Component descriptions and symbols:** Pupils match component names with descriptions and circuit symbols. Can they name a different electrical appliance where each is used?

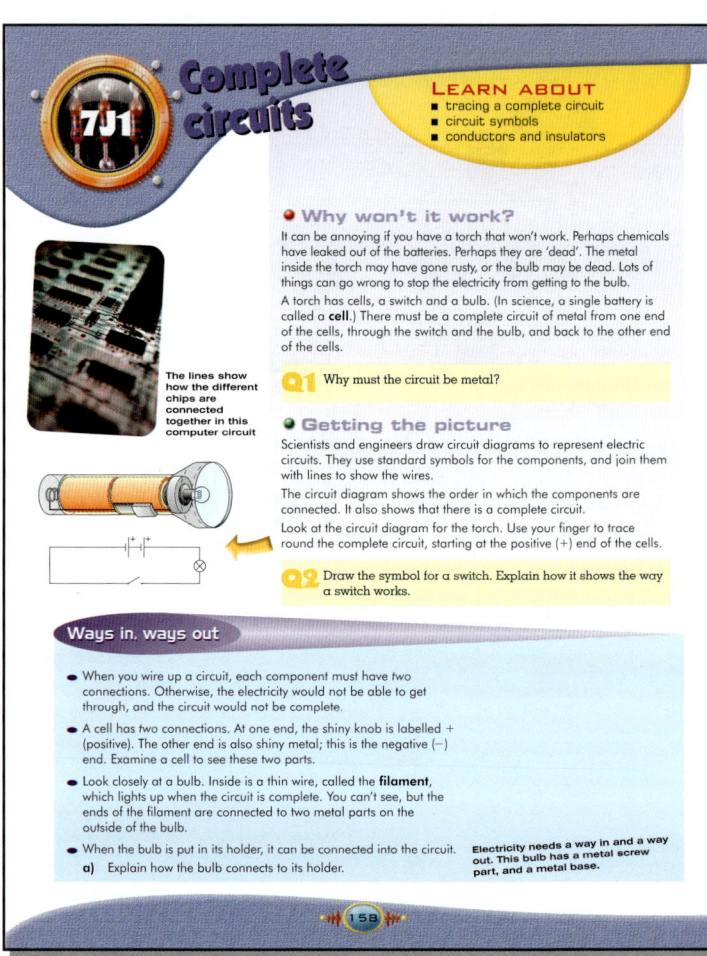

Activity and technicians' notes

Ways in, ways out and Circuit challenge

Equipment required
- Cells
- Bulbs, bulb holders
- Switches
- Wires

Technician's notes

- Ensure all bulbs and cells are working.
- Ensure enough equipment for a minimum of two bulbs, two cells, two switches and six wires per group.

Teaching strategy contd.

Notes and tips
Pupils often approach Key Stage 3 electricity with a wide range of levels of understanding from Key Stage 2. Some pupils may find work in this section easy, whilst for others it will be challenging. Using circuits of differing complexity enables this section to provide challenge for pupils at all levels. Pupils can also be challenged to find out why different circuits do not work, with huge differentiation in the faults provided for pupils to detect.

Extension ideas
Pupils consider the many different electrical circuits used in cars. More able pupils could be asked why the metal body of the car is not used for both halves of the circuits (out from the battery and back to the battery again), leading to consideration of the role of conductors and insulators in different places, and possibly what might happen if the insulation wore out or was damaged.

Learning styles

Visual
- **Visualising** and tracing current flow around circuits.
- **Matching** circuits to circuit diagrams.

Kinaesthetic
- **Building** circuits.

Answers to questions

Text
Q1 because metal conducts electricity

Q2 When the switch is drawn open, the wire is broken, there is no complete circuit and electricity cannot flow. When the switch is closed, there is a complete circuit and electricity can flow.

Q3 So it is easy to tell which wire is which, without having to follow it all the way along the cable.

Summary
3. Use a simple series circuit of bulb, switch and cell. Check the cell is working first, before testing the bulbs.
4. It makes it easy for scientists/engineers to communicate circuits to other scientists/engineers; it avoids confusion and helps with safety issues.

Suggestions for plenaries

- **Making it work:** Pupils work in small groups to write a list of things that must be present for a circuit to work, such as 'each component must have two connections'. Groups then contribute to a class list.
- **Questions and answers:** Give pupils a series of answers for which they have to write suitable questions.
- **Card dominoes:** Give pupils a set of card dominoes, with descriptions and circuit diagrams for a range of circuits. Pupils play dominoes with the cards.

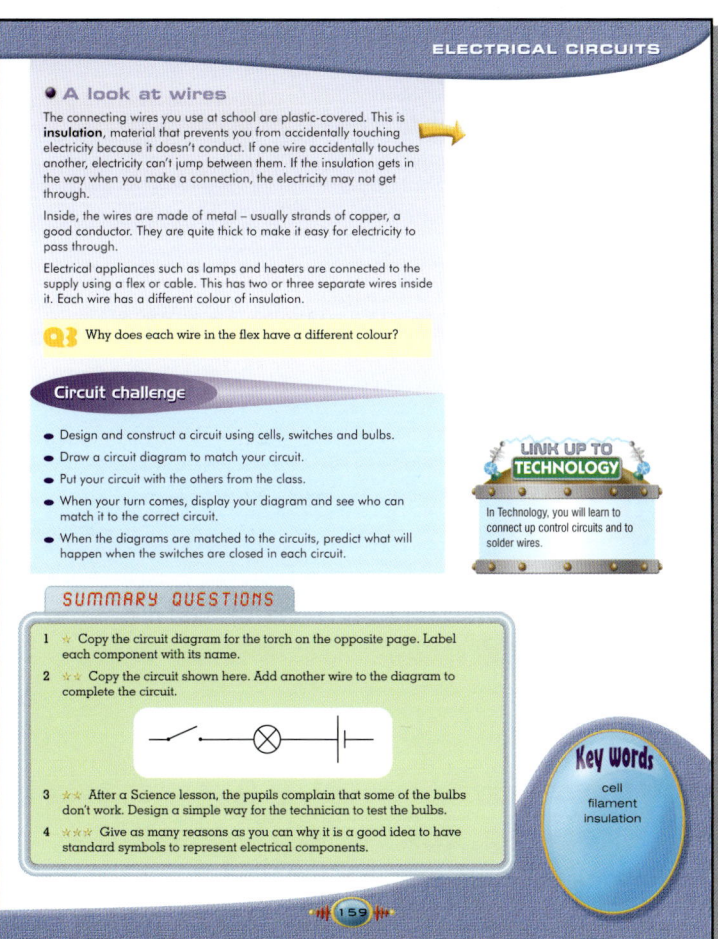

Learning outcomes

Following this, most pupils will:
▶ Build working electrical circuits and represent them using circuit diagrams.

7J2 Electrical current

National framework/QCA SoW references

National framework – Energy: The purpose of cells in an electric circuit.

QCA SoW

7J Section 1: How do electrical circuits work?

7J Section 2: What happens in a circuit?

NC links: Sc1 1b; Sc4 1a, b.

Learning objectives

- To use a model to help visualise the electric current in circuits.
- To use an ammeter to measure current.
- To describe how the current in a circuit can be changed.

Teaching strategy

Key points to highlight

- Current flows from positive to negative.
- The current is the same at all points around a series circuit. Current is not 'used up' by components.
- Resistance makes it harder for current to flow, so current is less.

Difficulties/Misconceptions

The most common misconceptions about current flow are:

- The 'current used up' model: some current is used up as it passes through each component. Correct this by using an ammeter to measure the current at different points around a series circuit.
- The 'clashing current' model: current flows out of both sides of the cell and meets in the component. Use the water flow model and 'Where does the current go then?' to show that this idea is incorrect.

Skills

In this section, pupils use scientific enquiry skills to build and evaluate thought models about current flow, assessing how well evidence from observations supports each model.

Notes and tips

Electricity is a very abstract concept, which many pupils find difficult. The more ideas and models they can test practically the better. This helps ensure pupils have as wide a base of practical experience as possible to draw on when trying to predict how circuits will behave.

Lesson starter suggestions

- **What's wrong?** Draw a circuit diagram of a simple series circuit with cell, bulb, switch, wires, on the board. Tell pupils that when the circuit is connected the bulb doesn't light. In groups, can they identify all the faults that might stop the bulb lighting?
- **Symbol check:** Working in pairs, one pupil names an electrical component, their partner draws the correct symbol. Some pupils may know components not met in class yet – ask them to explain what these components are.
- **Circuit descriptions:** Give pupils descriptions of different circuits. They then draw a circuit diagram for each circuit described, using correct circuit symbols. In groups, they compare diagrams. Are they all identical?

Activity and technicians' notes

Using an ammeter

Equipment required

- Cells
- Bulbs, bulb holders
- Ammeters
- Wires

Technician's notes

- Ensure all bulbs and cells are working.
- Ensure one cell, one bulb, one ammeter and three wires per group. You may wish to extend to more bulbs or cells per group.

Teaching strategy contd.

Extension ideas

'Electronics' and 'electronic chips' are possible because scientists can control how relatively small numbers of electrons (the tiny, freely moving, charged particles in metals and semi-conductors) move in tiny pieces of material. Ask pupils to consider how electronics and 'computer chips' affect their every day lives. Electronics are found in almost all modern appliances, from washing machines to telephones to clocks.

Learning styles

Visual
- **Visualising** current flow around circuits.
- **Imagining** different thought models of electric current flow.

Kinaesthetic
- **Building** circuits and measuring current.

Answers to questions

Text
Q1 the 'flow along the red wire only' model
Q2 Water flows around a complete circuit, similar to electricity flowing round a complete electrical circuit. The 'current' is water, the 'cell' is the pump.

Summary
1. current, positive, negative, ammeter, less
2. 1A, 1A
3. Increasing the resistance decreases the current flowing, so the electric motor will run slower.

Suggestions for plenaries

- **'20 questions'**: In pairs: first pupil draws a circuit. Their partner has to ask questions to find out what the circuit does and then draw the correct diagram. Answers can be only 'yes' or 'no', with a limit given of 20 questions. Then swap roles.
- **Verbal quiz:** Ask all pupils to stand. In turn ask questions about electricity, current and resistance. Pupils sit when they answer a question correctly. It does not matter if questions are repeated.
- **True or false:** Give pupils a list of statements about electric current, circuit representations and resistance. They decide if statements are true or false.

ELECTRICAL CIRCUITS

● From positive to negative

Electric current flows all the way round a circuit, without getting used up. We imagine it in the wires, flowing steadily round a circuit. It flows out of the positive end of the cell, and back in at the negative end.

Current can't escape from wires, because it needs to stay inside a conductor. It can't leak out through the insulation.

It's hard to picture electric current. Even if you had a very powerful microscope, you wouldn't be able to see an electric current in a wire. This is because the current is actually made up of vast numbers of tiny particles, even smaller than atoms, called **electrons**. Particles can't just disappear, so they flow all the way round the circuit.

Q2 Look at the picture of the garden fountain. In what ways is it similar to an electric circuit? What is its 'current' made of? What is its 'cell'?

Some gardens have fountains like this. The pump makes water flow round and round, like current going around a circuit. This elastic pump is specially designed to work under water.

● Changing current

What happens if you connect two bulbs in a circuit, one after the other? You don't need an ammeter to tell you. The bulbs are equally bright, because each has the same current flowing through it. The bulbs are dimmer than before, showing that the current is less. Why is this?

With two bulbs, it is harder for current to flow around the circuit. It has to push its way through one bulb, and then through the other. We say that there is more **resistance** in the circuit.

Resistance is a bit like friction. Friction is a force that makes it difficult for things to move past each other. Resistance makes it difficult for electric current to flow.

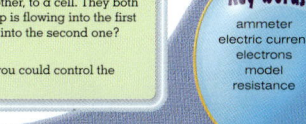

The volume control on this radio works by changing the resistance in an electric circuit

SUMMARY QUESTIONS

1. ★ Copy these sentences. Complete them by filling in the gaps.
   Electric . . . flows all the way round a circuit.
   Current flows from . . . to . . . .
   Current is measured using an . . . .
   If there is more resistance in a circuit, the current flowing will be . . . .
2. ★★ Two light bulbs are connected, one after the other, to a cell. They both light up. An ammeter shows that a current of 1 amp is flowing into the first bulb. How much current flows out of the first bulb, into the second one? How much flows out of the second bulb?
3. ★★★ Use the idea of 'resistance' to explain how you could control the speed of an electric motor.

Key words
ammeter
electric current
electrons
model
resistance

Learning outcomes

Following this, most pupils will:
▶ Investigate simple electrical circuits, choosing appropriate equipment.
▶ Use an ammeter to measure current.
▶ Use circuit diagrams to represent electrical circuits.
▶ State that a series circuit has the same electric current flowing at all points.

7J3 Cells and batteries

National framework/QCA SoW references

National framework – Energy: The purpose of cells in an electric circuit.

QCA SoW

7J Section 2: What happens in a circuit?
NC links: Sc1 1b; Sc4 1b.

Learning objectives

- To explain how batteries are made from cells.
- To explain the meaning of voltage.
- To describe how current and voltage in a circuit are related.

Teaching strategy

Key points to highlight

- Cells push electric current round the circuit.
- The voltage tells the amount of 'push' of the cell.
- Bigger voltages can be made by connecting cells together so that their voltages add up.

Difficulties/Misconceptions

Pupils do not always realise that the polarity of cells (which way round the cells are) matters. Help pupils understand this by explaining that the voltage of the cell pushes the current out of the positive terminal of the cell, around the circuit and back into the negative terminal of the cell.

Skills

In the activities in this lesson, pupils use problem solving skills to demonstrate for themselves: the connection between voltage and brightness of bulbs; the connection between polarity, voltage and brightness of bulbs.

Notes and tips

With some pupils this would be an appropriate point to introduce ways of measuring voltage and discussions of energy, though pupils will all use a voltmeter to measure voltage in unit 9I 'Energy and electricity'. At this stage, they only need to know that voltage is what makes the current flow.

Gruesome science

An electric eel is like a giant battery, with up to 6000 cells along its body. It has to touch its head and tail against you at the same time to electrocute you.

Lesson starter suggestions

- **Using batteries:** Ask pupils to list 10 different things that use batteries. How many different types of batteries can they describe or draw?
- **Free writing:** Pupils write for 1 minute on 'batteries'. Discuss briefly. Then write for a further 1 minute on either 'What batteries are for' or 'How batteries vary'. (Choose a second topic, depending on outcome of earlier class discussion.)
- **True or false:** Check pupils' present understanding of batteries and cells, by asking them to decide which of a series of statements about cells and batteries are true and which are false.

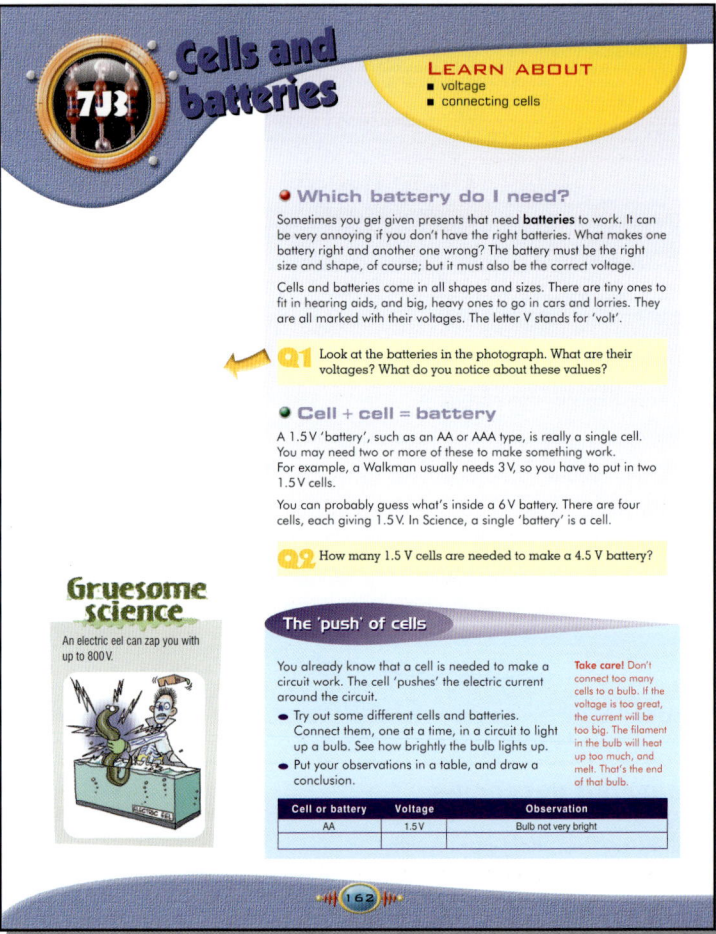

Activity and technicians' notes

PUPIL SHEET

The 'push' of cells, Adding up, cancelling out

Safety notes

Don't connect too many cells to a bulb. If the voltage is too great, the current will be too big. The filament in the bulb will heat up too much, and melt.

Equipment required

- Cells
- Bulbs, bulb holders
- Wires

Teaching strategy contd.

Extension ideas

Some pupils could be introduced to the correct circuit symbol for a battery (two cells separated by a dotted line). Ask them to look at a range of 'battery powered' devices, finding out what voltage they need. (For some devices they may be able to tell if the device uses a single cell, or several cells connected to form a battery.) Pupils could also find out why it is important for devices to be supplied with the correct voltage.

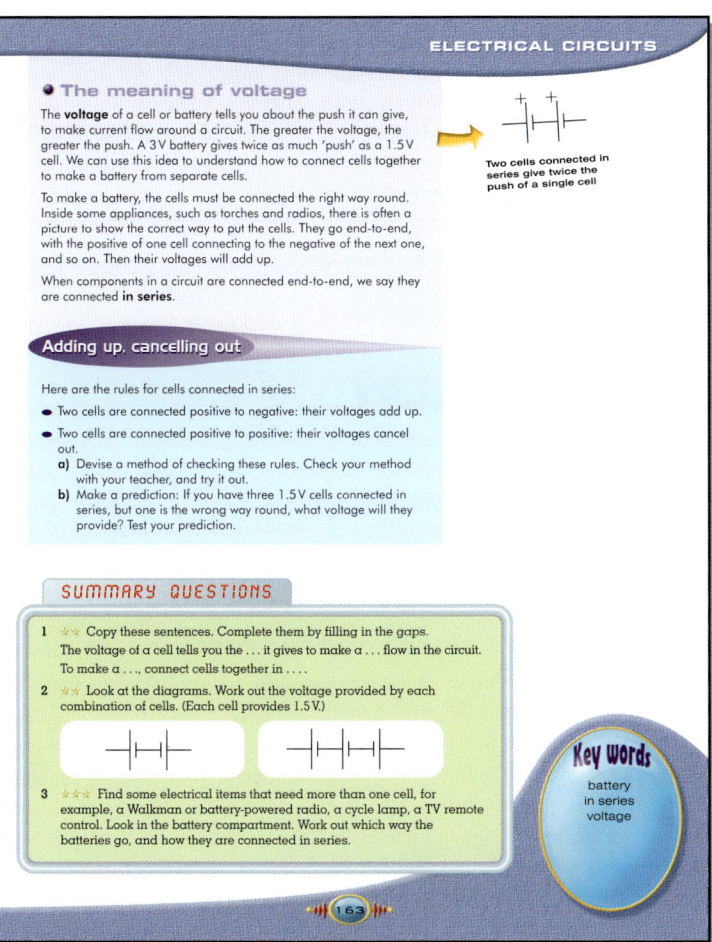

Activity and technicians' notes contd.

Technician's notes

- Provide enough cells for up to four per group, with battery holders or other appropriate way of connecting cells together.
- Ensure all cells are equally new. The voltage output of cells changes as they age, giving potentially confusing results.
- Have voltmeters available. Teachers may wish to introduce pupils to measuring voltages with a voltmeter.

Learning styles

Visual
- **Imagining** and predicting the result of different combinations of cells.

Kinaesthetic
- **Practical activity** to investigate the effect of combining cells in different ways.

Answers to questions

Text
Q1 All the voltages are multiples of 1.5 V.
Q2 three connected positive to negative

Summary
1 push, current, battery, series
2 0 V, 1.5 V.

Learning outcomes

Following this, most pupils will:
▶ Investigate the effect of changing cells in simple electrical circuits.

Suggestions for plenaries

- **Acrostic:** Use the letters of the word VOLTAGE to begin seven facts or statements about electrical circuits.
- **Red, amber, green:** Read out a series of sentences about electrical circuits. Pupils to hold up cards to show if they agree (green), disagree (red) or don't know (amber).
- **Anagrams:** Pupils solve clues to un-jumble anagrams for a range of key words associated with electricity and electrical circuits.

7J4 Electricity from Chemistry

National framework/QCA SoW references

National framework – Energy: An electric current carries energy to all the components in an electric circuit.

QCA SoW
7J Section 2: What happens in a circuit?
7J Section 3: How can we explain what happens in electrical circuits?
NC links: Sc4 1c.

Learning objectives

- To explain how current transfers energy round a circuit.
- To understand the chemical origin of the energy in electrical circuits.
- To explain current flow and energy transfer using a water flow model.

Teaching strategy

Key points to highlight

- Energy is stored in a cell as chemical energy.
- The current is caused by chemical reactions in the cell.
- The current transfers energy from the cell to the components in the circuit.

Difficulties/Misconceptions

Many pupils find the water flow model helpful in understanding electrical circuits, but it is not perfect. (Water flows out of cut pipes, electricity does not leak out of cut wires. Making a wire thinner increases the resistance and decreases the current, making a pipe thinner makes the water flow faster.) For this reason, encourage pupils to make up and explain their own thought models of electricity, if they find these easier than the water model. Ensure, however, that any model includes all the elements needed to represent simple circuits.

Skills

Pupils use scientific enquiry skills to visualise different models of electrical circuits and to assess how well different models represent real circuits.

Lesson starter suggestions

- **Spot the difference:** Show pupils pictures/examples of two lights, both on. One should be a small bulb, such as a torch bulb, the other a bright lamp such as a spotlight or floodlight. Pupils should identify as many differences as they can between the two circuits.
- **Question loops:** On separate pieces of paper, each pupil writes a question about electricity or circuits and the corresponding answer. Use these for a 'question loop' activity to check pupils' understanding so far of electrical circuits.
- **Current or voltage:** Pupils look at a series of statements and decide which are about current, which are about voltage, and which could be about either.

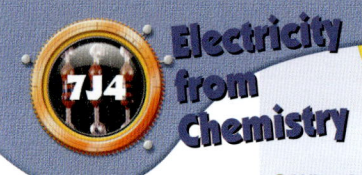

LEARN ABOUT
- energy from a cell
- the difference between current and voltage

What's going on?

Batteries are an expensive way of buying electricity. We buy them because they are convenient, but they soon run out. Worse than that, they sometimes leak chemicals, which can cause damage to the appliance they are being used in.

Inside a cell, there are chemicals. When you connect the cell into a complete circuit, the chemicals start to react with each other, and this pushes the current around the circuit. Disconnect the circuit and the reaction stops.

Because the chemicals used in a cell are hazardous, you should not open one up. If the chemicals leak, wipe them up and throw away the cell. Don't get the chemicals on your hands.

Running down

Look at the picture of the circuit. How can we describe what's going on?

There is a chemical reaction going on in the cell. At the same time, the bulb is giving out light. We can describe this by thinking about **energy**. We say that the chemicals in the cell are a store of energy; when the circuit is complete, the energy from the chemicals is transferred to the bulb, and comes out as light.

Eventually, the chemicals run out, and the battery is 'dead'. Its store of energy has been used up, and the bulb won't shine.

Q1 An AA cell is bigger than an AAA cell. Both have a voltage of 1.5 V. Use the idea of 'energy' to explain why the AA cell will keep a bulb lit for longer. Why is the bulb the same brightness whichever cell is used?

There are other ways to light a bulb. You could plug a reading lamp into the mains. Electricity is coming from a power station, where coal or gas is being burned. That's another chemical reaction that provides the energy needed to light the bulb.

Some cycle lights work from a **dynamo**. When the dynamo is working, you have to pedal a little harder. You are providing the energy needed to light the bulb.

Q2 Explain why the lights go out when the bicycle stops. Some lights keep shining for a while after the cycle stops. How can this happen?

Answers to questions

Text

Q1 The AA cell is larger, so contains more stored energy. The voltages are the same, so the bulb is equally brightly lit by both cells.

Q2 When the bicycle stops, you are no longer supplying energy. Sometimes some of the energy is stored in a battery, so the light continues shining for a short while after you stop pedalling.

Q3 Breaking the circuit at any point stops the current flow and so stops energy getting to the bulb.

Q4 The flowing water represents current. The pump represents the cell or battery. The energy is transferred from pump to wheel by the flowing water.

Summary

1 current, voltage, energy, energy, current
2 a) Electric current is represented by the pupil messenger.
   b) The cell is represented by the office.
   c) Energy is represented by the messages.

Teaching strategy contd.

Notes and tips

- Spend plenty of time on ideas of current transferring energy, as pupils with a thorough understanding of this find later work on electricity much easier.
- Ensure pupils understand the connection between voltage and energy: a cell with a high voltage pushes the current hard so the current carries lots of energy around the circuit, just as kicking a ball hard means the ball transfers lots of energy across the football pitch.

Extension ideas

Pupils could consider the use of rechargeable batteries, and how a battery can be made rechargeable; this links in with chemistry work about reversible and irreversible chemical reactions. Looking at the advantages of rechargeable batteries could also link in with citizenship topics about disposal of waste materials, pollution and concern for the environment.

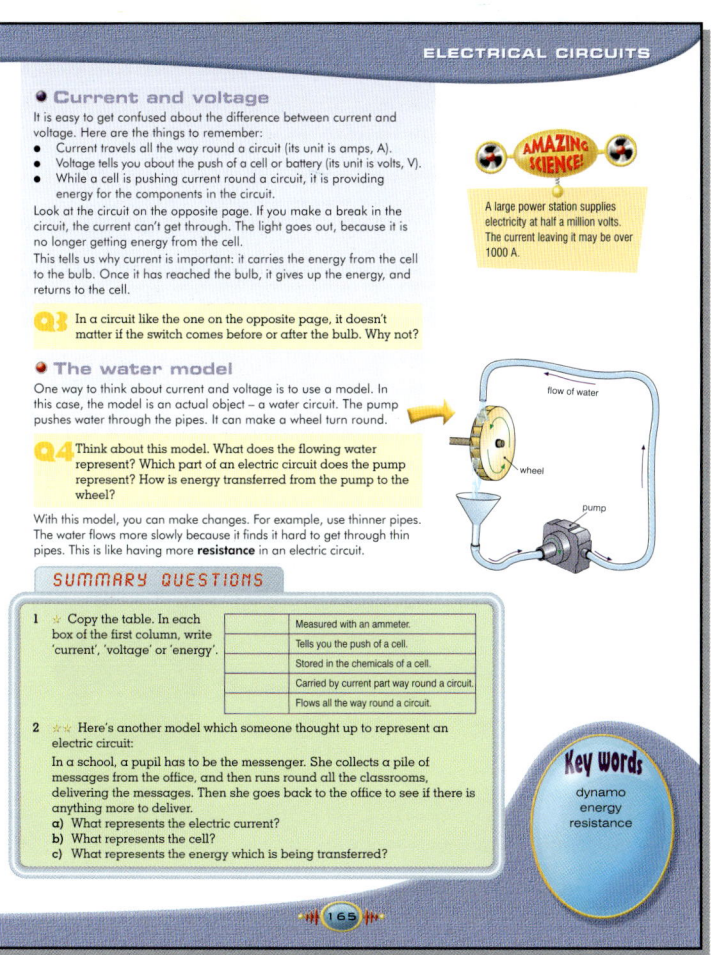

Learning styles

Visual

- **Visualising/imagining** different models to represent electrical circuits.

Intrapersonal

- **Understanding and evaluating** models to represent electrical circuits.

Suggestions for plenaries

- **Thought models:** Pupils to draw or describe a thought model to show an electrical circuit with bulb, cell, resistor and switch. Go over some thought models in a class discussion.
- **Connectives:** Pupils to complete the sentence 'A water flow model can be used for an electric circuit . . .' continuing with one of the following words: 'so, but, and, although, because, however, therefore, if'.
- **Energy in circuits:** Pupils describe, using words and/or pictures, what is happening to the energy in a series of given circuits. Where does the energy come from? Where does it go? What forms is it in?

Learning outcomes

Following this, most pupils will:
▶ Describe the relationship between current flow and energy transfer in an electrical circuit, using a flow model.
▶ Describe the effects of resistance using a flow model.

Some pupils will also:
▶ Relate voltage to energy transfer in electrical circuits.
▶ Explain the difference between current flow and energy transfer, using a flow model.

7J5 Series and parallel

National framework/QCA SoW references

National framework – Energy: Electric current transfers energy to components in series and parallel circuits.

QCA SoW

7J Section 3: How can we explain what happens in electrical circuits?
7J Section 4: What kinds of circuits are useful and what are the hazards?
NC links: Sc4 1a.

Lesson starter suggestions

- **Car electrics:** Give pupils a list of electrical appliances from a car, including headlights. Ask 'How can you tell that these are not wired in one complete loop?' Ask pupils to discuss reasons.
- **Problem solving:** Pupils work together to solve this problem: Sadie's little brother has been given a new toy car. One switch turns on the headlights; a second switch turns on the siren. What circuits might be inside the car? Can you make the headlights and siren work using only one battery?
- **Flow diagram:** Draw a flow diagram to show what happens when a switch is closed in series circuit (circuit diagram given). The pieces of flow diagram are in the wrong order; pupils are to put them into the correct order.

Learning objectives

▶ To identify series and parallel circuits.
▶ To describe current flow round series and parallel circuits.

Teaching strategy

Key points to highlight

- Parallel circuits have more than one possible path for the current flow.
- Parallel circuits allow components to be controlled separately.
- Current in a parallel circuit splits, some going along each path.

Difficulties/Misconceptions

Most pupils expect two bulbs in parallel to be dimmer than a single bulb, just as two bulbs in series are dimmer than a single bulb. The 'obstacle course' analogy of two obstacles side-by-side being easier for a group of people to get through, than two obstacles one after the other, often helps here. Some pupils may consider energy, and a cell going flat faster if it lights two bulbs in parallel than if it lights just one bulb.

Skills

- In this section pupils use mental 'model building' skills, as they must build for themselves a mental picture of the current flow in different circuits.
- They also practise applying their mental model in a range of different situations, comparing what the model predicts with what they observe.

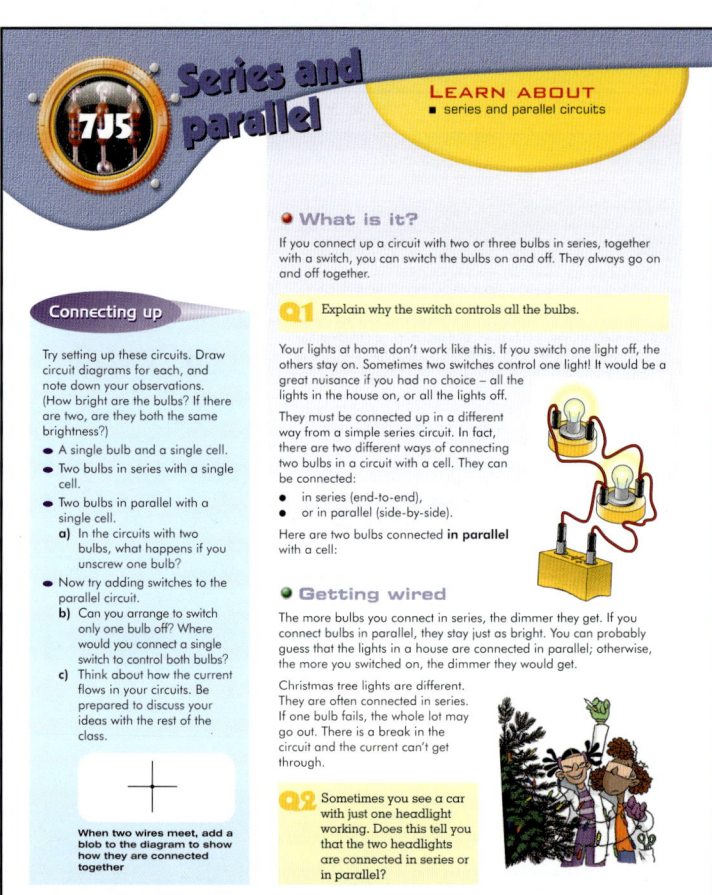

Activity and technicians' notes

Connecting up

Equipment required

- Cells
- Bulbs, bulb holders
- Switches
- Wires

Technician's notes

- Ensure all bulbs and cells are working.
- Ensure one cell, two bulbs, one switch and six wires per group.
- Teachers may wish to have ammeters available so pupils can measure current.

Teaching strategy contd.

Notes and tips
- Many pupils will assess and consolidate their mental models of current flow far more easily if they discuss their model with other pupils, explaining how their model works.
- Giving pupils a range of different models of current flow will enable them to choose the model that they find easiest to 'picture'.

Extension ideas
Pupils could use the Internet or a multimedia encyclopedia to find out how modern Christmas tree lights work. The bulbs are wired in series with each other, but a resistor is wired in parallel with each individual bulb. Pupils could be asked to explain the purpose of these resistors, by imagining what would happen if one bulb 'blew'.

Learning styles

Visual
- **Visualising** current flow around series and parallel circuits.

Auditory
- **Explaining** ideas about current flow to other pupils.

Answers to questions

Text
Q1 Opening the switch breaks the current that flows through all the bulbs.

Q2 Headlights are in parallel. If they were in series they would both go out when one broke.

Q3 Because each bulb is equally bright; if one got more current, it would be brighter (if the bulbs are identical).

Summary
1 series, parallel, parallel, parallel

Learning outcomes

Following this, most pupils will:
▶ Measure current in series and parallel circuits.
▶ Identify patterns in the results.
▶ State that current divides in a parallel circuit, some flowing along each possible path.
▶ Compare series and parallel circuits.

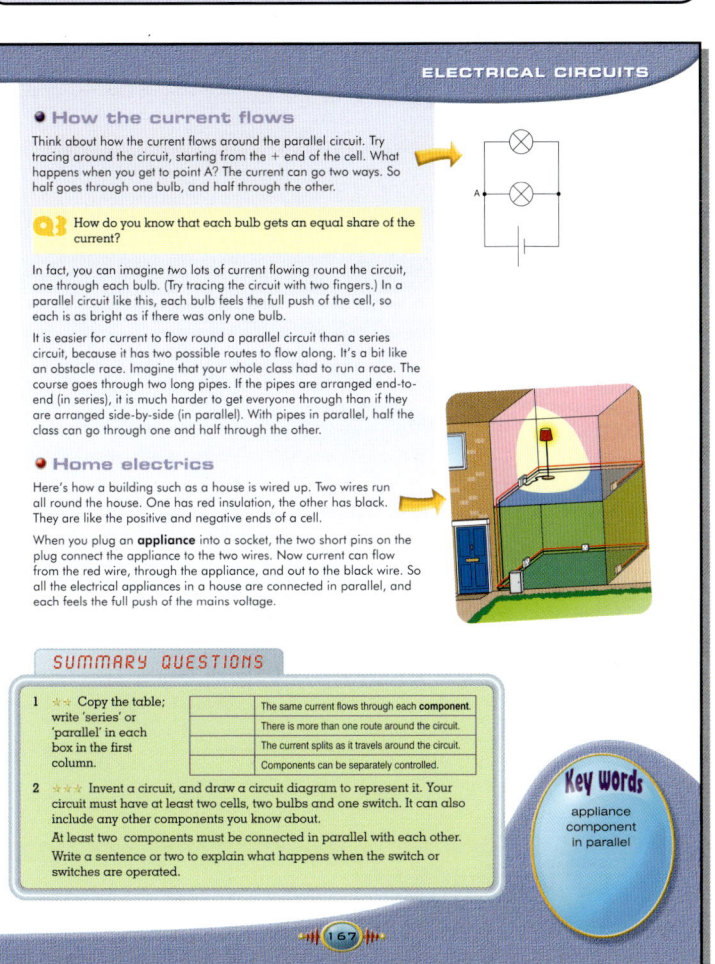

Suggestions for plenaries

- **Concept map:** Working in pairs or small groups, pupils make a concept map to show what they know about electrical circuits. Then they contribute ideas to a class concept map.
- **Series or parallel:** Give pupils a list of statements about circuits. Pupils to decide if statements apply to series circuits or to parallel circuits.
- **Acrostic puzzle with clues:** Pupils use clues to complete the across clues for an acrostic puzzle, to find a word downwards.

7J6 Keeping safe

National framework/QCA SoW references

National framework – Energy: Electric current transfers energy to components in electric circuits.

QCA SoW

7J Section 4: What kinds of circuits are useful and what are the hazards?
NC links: Sc4 1c.

Learning objectives

▸ To describe electrical dangers.
▸ To explain some electrical safety features.

Teaching strategy

Key points to highlight

- Mains electricity is dangerous because it has a much greater energy than small cells.
- A fuse is a safety device that switches off the current if it becomes too large.
- A fuse works by a thin piece of fuse wire melting because it becomes hot when a large current flows. Thicker fuse wires melt for larger currents.
- Fuses must be connected in series with the components they are to protect.
- People can be killed by very small currents flowing through them.
- Circuit breakers protect people by switching off the current before it can flow for long enough to cause damage.

Difficulties/Misconceptions

Pupils may be confused about whether it is currents or voltages that are dangerous. Potentially it is both, and pupils are best taught at this stage that mains electricity is hazardous because of its high energy.

- High voltages are generally dangerous because they drive a high current, which causes serious burns and tissue damage, but high static voltages (such as Van der Graaff generators) can be harmless because they do not have enough energy to drive enough current through the skin to do any damage.
- Very small currents – a few milliamps – can kill, because they interfere with the electrical signals controlling the heart's rhythm and cause fatal heart attacks. Relatively small voltages can cause these currents, if connections to the moist tissue under the skin are good enough.

Lesson starter suggestions

- **Spot the hazard:** Give pupils pictures showing electrical hazards, such as pylons, sub-stations, frayed cables, warning signs about removing covers from electrical equipment, etc. Use class discussion to establish what's known about electrical safety.
- **Free writing:** Ask pupils to write for 1 minute on 'electrocution'. Go over in class discussion.
- **Red, amber, green:** Read out a series of statements about electrical safety. Pupils to decide if they agree with each statement and hold up the appropriate card: green (agree), red (disagree) or amber (don't know).

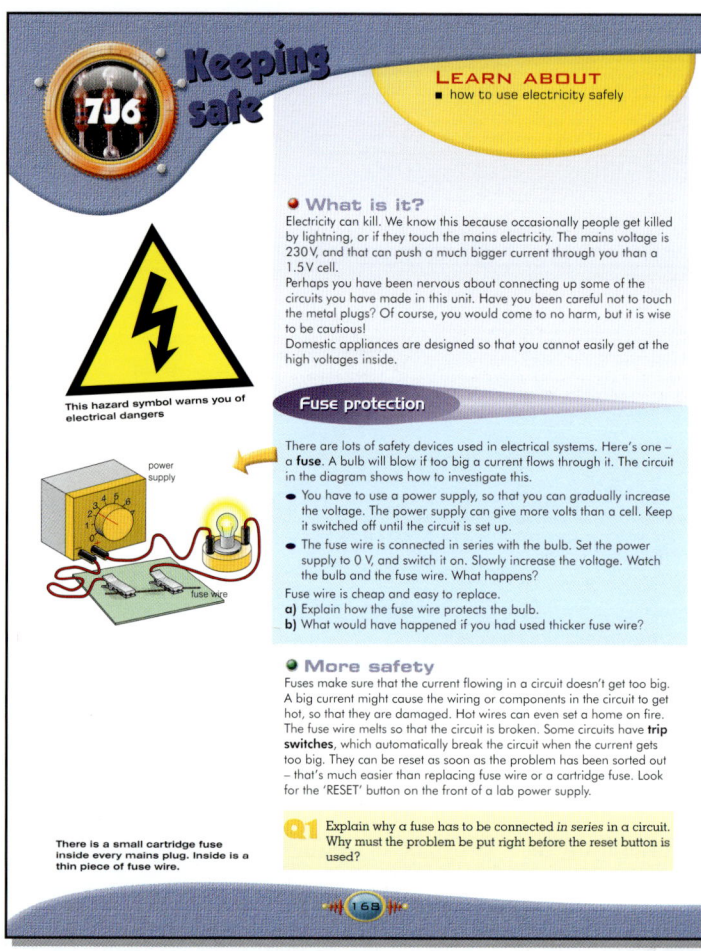

Activity and technicians' notes

Fuse protection

Safety notes

Fuse wires must be tested on heatproof mats. Pupils must be made aware that fuse wire will get very hot.

Equipment required

- Variable voltage power supplies
- Bulbs, bulb holders
- Wires
- Crocodile clips
- Heatproof mats
- Fuse wire

Teaching strategy contd.

Skills
- Research skills.

Notes and tips
Many pupils will already have a clear idea about electrical safety, knowing how to keep themselves safe when using electrical appliances, and near power sub-stations and overhead cables. This section provides an opportunity to develop understanding of when and why electricity is dangerous, and to explain how safety devices (which pupils may well already have met or heard of) work.

Extension ideas
Pupils extend their knowledge of electrical safety into the community, considering where they see parts of the electricity supply system and how the public is protected from its dangers.

Activity and technicians' notes contd.

Technician's notes
- Protect bulbs by using fuse wire that blows at a lower current than the bulbs in the circuit.
- A low voltage power unit may trip if the current is too large.
- Teachers may wish to have a range of fuse wires with different current ratings.

Access required to:
- Mains electricity supply

Learning styles

Visual
- **Visualising** current, voltage and energy in electrical hazard situations.

Kinaesthetic
- **Practical activity** to investigate fuse wires.

Answers to questions

Text
Q1 The fuse wire melts, breaking the circuit, before the current gets large enough to damage the bulb. All the current must flow through the fuse. Otherwise the trip switch will just 'trip' again.

Q2 Water conducts electricity, so you may get electrocuted with wet hands.

Q3 Your body has a higher resistance. The current still flows through the bulb, if you touch the circuit, not through you.

Summary
1. 5 A; this is large enough to allow the normal current to flow through the bulb, but not large enough to allow a damaging current to flow.
2. If there is a fault in one of the circuits, all the other circuits will continue to work. So if there is a fault in the upstairs power sockets, for example, you can still use the electric lights to look for the fault, instead of possibly having to do so in darkness.

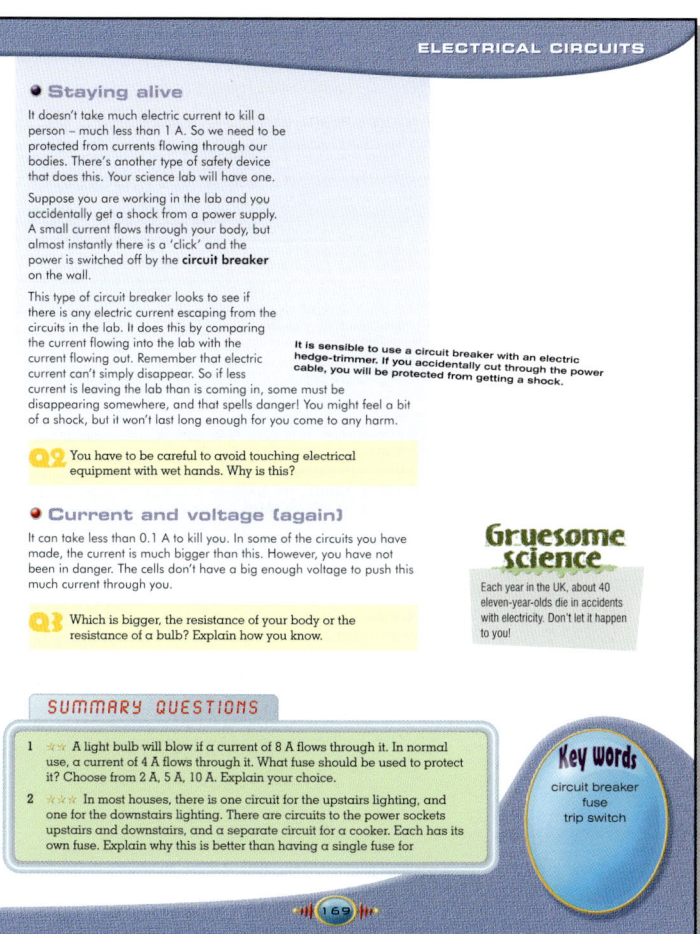

Suggestions for plenaries

- **Question loop:** Ask each pupil to write a question and answer about electrical safety on separate pieces of paper. Shuffle and distribute questions and answers. One pupil reads a question, the pupil with the appropriate answer then reads their question and so on.
- **Safety poster:** Ask pupils to work together to design a poster or leaflet about electrical safety. Or this could be homework.
- **Acrostic puzzle with clues:** Pupils use clues to complete the across clues for an acrostic puzzle, to find the word 'electrocution' downwards.

Learning outcomes

Following this, most pupils will:
▶ Describe some electrical hazards, and how to use electricity safely.

Some pupils will also:
▶ Explain how electricity can be hazardous to people.
▶ Use the idea of nerves as electrical conductors to explain some electrical hazards.

7J Read all about it!

Teaching strategy

Nervous about electricity?/Shocking stuff

Difficulties/Misconceptions

Pupils may find the idea of our bodies being controlled by electric currents flowing through our nerves confusing, when they know that currents cause electrocution and can be fatal. They need to understand that the currents in nerves are really tiny, and that even small electric currents from outside the body can disrupt these 'nerve currents' so that the heart does not know when to beat.

Skills

- This section gives an opportunity for pupils to practise scientific enquiry skills, finding out how scientists have worked in the past, and how they have used experiments to test ideas.
- There are also many opportunities for pupils to practise research skills, to find out about other early electricity discoveries.

Notes and tips

Some pupils may enjoy research to find out about animals that use electricity, such as the hammer head shark that detects the minute electrical signals given off by prey creatures under the sand, or electric eels that create large voltages to stun fish.

Safety notes

Some pupils may have heard of the idea of testing batteries by placing them against the tongue. If pupils mention this, they should be warned not to do so, as some pupils may find this can give them a painful shock.

Extension ideas

- Pupils could find out more about earlier experimenters with electricity, such as Benjamin Franklin or Alessandro Volta. They could also find out more about modern ideas, such as how defibrillators work, or why birds can sit on power cables but it is fatal to touch them with a fishing line.

Questions

1. Explain how our nerves work by electricity.
2. Who first discovered the connection between our bodies and electricity?
3. What important invention did the work on 'animal electricity' lead to?
4. How do electric fences work?
5. Explain how a defibrillator can help someone having a heart attack.

Answers

1. Tiny currents flow along our nerves, so our brain can control organs.
2. Luigi Galvani
3. cells and batteries
4. Electric current flows through them, giving a mild electric shock to animals (and people) that touch the fence.
5. It gives an electric shock to the heart, restarting its natural rhythm.

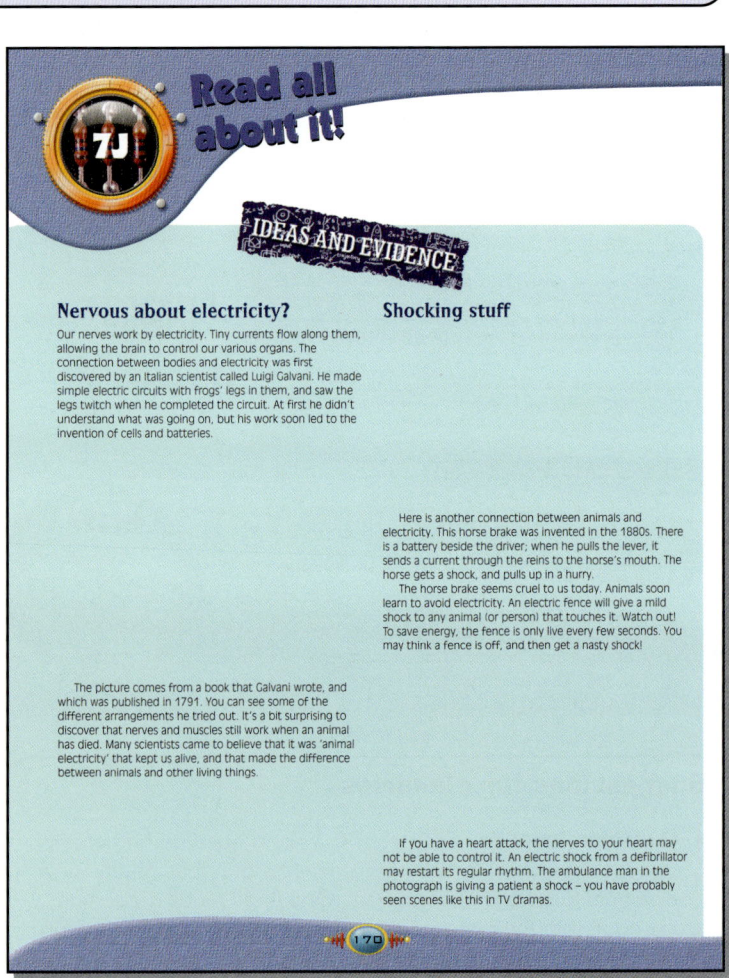

Danger – common errors

- The difference between current and voltage is something pupils often find difficult. It is worth spending time on here, since a clear understanding of the difference makes later work on electricity much easier. It is important to emphasise that voltage is the force that pushes current round: introduce the idea of current being a flow of tiny particles, if your pupils will find this helpful.
- Pupils should be given the opportunity to explore and discuss several different flow models of electric current, and encouraged to explain them to each other. Provide a mixture of good and not so good models, so that pupils can begin to assess when a model is a good representation of reality, and when it has a lot of limitations.

Learning styles

Interpersonal

- **Conversation and feedback** about 'Ideas and Evidence'
- **Conversation and feedback** about cartoon characters' discussions.

Learning outcomes

Following this unit, most pupils will:

▸ Build working electrical circuits and represent them using circuit diagrams.
▸ State that a series circuit has the same electric current flowing at all points and that current divides in a parallel circuit, some flowing along each possible path.
▸ Compare series and parallel circuits.
▸ Describe the relationship between current flow and energy transfer in an electrical circuit, using a flow model.
▸ Describe the effects of resistance, using a flow model.

Some pupils will also:

▸ Relate voltage to energy transfer in electrical circuits.
▸ Explain the difference between current flow and energy transfer, using a flow model.
▸ Use the idea of nerves as electrical conductors to explain some electrical hazards.

In scientific enquiry:

Following this unit, most pupils will:

▸ Investigate simple electrical circuits, choosing appropriate equipment.
▸ Use an ammeter to measure current in series and parallel circuits, and identify patterns in the results.
▸ Investigate the effect of changing cells in simple electrical circuits.
▸ Describe some electrical hazards, and how to use electricity safely.

Some pupils will also:

▸ Explain how electricity can be hazardous to people.

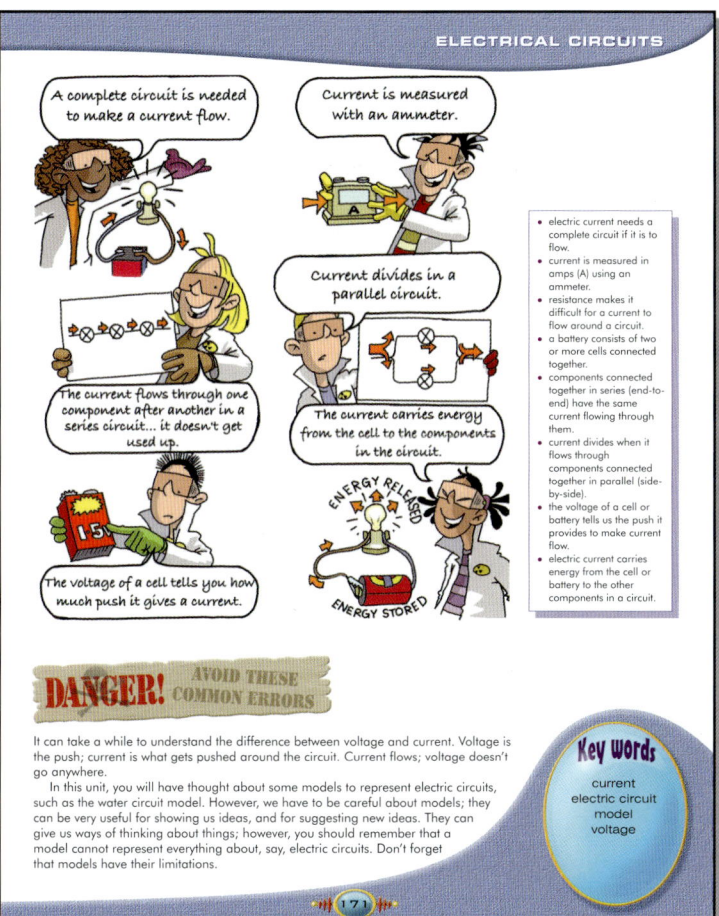

7J Unit review

Answers to review questions

1 a

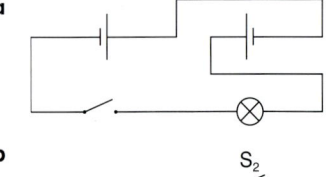

b

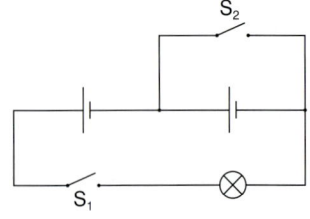

S1	S2	Comments
open	either	bulb off
closed	open	bulb bright – 2 cells in series
closed	closed	bulb dim – 1 cell short circuited

2 a a bulb
  b a switch
3 melts; this explains what happens to the fuse wire
4 a motor
  b motor, buzzer, bulb, heater
6 a 0.2 A
  b current of 0.2 A through each bulb, total current for circuit of 0.4 A; two bulbs are as bright in parallel as each bulb is on its own
  c current of 0.1 A; it is harder for current to flow through two bulbs (resistance is doubled, current halved)

REVIEW QUESTIONS
Understanding and applying concepts

1 The diagram shows four electrical components.

  a Make an exact copy of the diagram. Add connecting wires to make this circuit:
  • When the switch is open, the bulb is off.
  • When the switch is closed, the bulb is on.
  b Now draw a second circuit using the same components and an extra switch. By changing the switches, the bulb can be off, dimly lit or brightly lit. Explain how the switches are used to change the brightness of the bulb.

2 What electrical devices are described here?
  a This device includes a thin wire, which heats up and glows when an electric current flows through it.
  b This device has two positions. In one, it allows an electric current to flow through it.

Ways with words

3 A fuse protects an electric circuit. If a high current flows, we say that the fuse 'blows'. Which of the following words is a better description of what happens to the fuse? Explain your choice.
  The fuse:
  explodes   breaks   melts   blows up

Making more of maths

4 Reese was finding out about electrical resistance. Her teacher gave her several electrical components.
  Reese connected each one in turn to a battery and an ammeter. She made a bar chart to show her results.

  a Which component had the lowest resistance?
  b Make a list of the components, starting with the one with the lowest resistance.

Thinking skills

5 Devise an 'electric circuits' board game. Use snakes and ladders as the basic idea.
  What will be the snakes, and what will be the ladders? Explain your choices.
  Describe briefly how the game works.

Extension questions

6 Molly was measuring the current flowing through a bulb. She connected it to a single cell, and found that a current of 0.2 A flowed.
  Her teacher then gave her a second bulb, identical to the first one.
  Make predictions for the results of the following experiments. In each case, say what current you would expect to flow in the circuit, and why.
  a Molly connected the second bulb to a single cell.
  b Molly connected the two bulbs in parallel, and then connected them to a single cell.
  c Molly connected the two bulbs in series, and then connected them to a single cell.

7 Benjamin Franklin was an American scientist who did some pioneering experiments on electricity. Find out why he flew a kite in a thunderstorm. Why was this experiment dangerous, and what did he do to make it safer? How did this lead to the invention of lightning conductors?

UNIT REVIEW QUESTIONS

SAT-STYLE QUESTIONS

1. Mike set up the circuit shown. He had three bulbs to test. Each gave a different reading on the ammeter.

   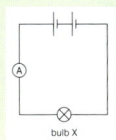

Bulb X	Ammeter reading = 0.4 A
Bulb Y	Ammeter reading = 0.2 A
Bulb Z	Ammeter reading = 0.6 A

   a Which bulb has the least resistance? (1)

   Mike then set up a circuit with two bulbs connected to the same battery as before.

   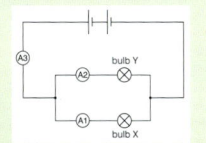

   b Are the bulbs connected in series or in parallel? (1)
   c What readings would you expect to see on the ammeters A1 and A2? (2)
   d What reading would you expect to see on the ammeter A3? (1)
   e Suggest two changes you could make to the circuit to increase the reading on ammeter A3. (2)

2. Pete's teacher gave him an electrical 'black box' to investigate. Pete connected the black box to a cell. He included an ammeter in the circuit to measure the current flowing through the black box.

   Pete added more cells, and recorded the current each time. The table shows his results.

Number of cells	Current flowing (A)
1	0.40
2	0.80
3	1.40
4	1.60

   a When the teacher looked at Pete's results, she suggested he might have made a mistake with one of them. Which one? (1)
   b Draw a graph of Pete's results. Draw a line through his correct results. Use your graph to make a prediction: what current would flow if the black box was connected to 5 cells? (3)

3. Pip went camping. She forgot to turn her new torch off when she went to sleep. It shone brightly for several hours, then grew dim and faded out.
   a Which graph might represent how the current flowing from the battery changed during the night? (1)

   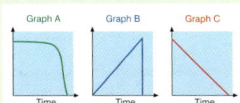

   b Which graph might represent how the energy stored in the battery changed during the night? (1)

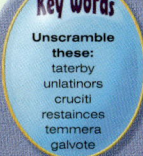

Key words
Unscramble these:
taterby
unlatinors
cruciti
restainces
temmera
galvote

Answers to SAT-style questions

1. a Z; it has the largest current. (1)
    b parallel (1)
    c A1 = 0.4 A, A2 = 0.2 A (2)
    d A3 = 0.6 A (1)
    e add another cell; add bulb z in parallel as well (2)
2. a the reading for three cells (1)
    b 2.0 A (3)
3. a Graph A (1)
    b Graph C (1)

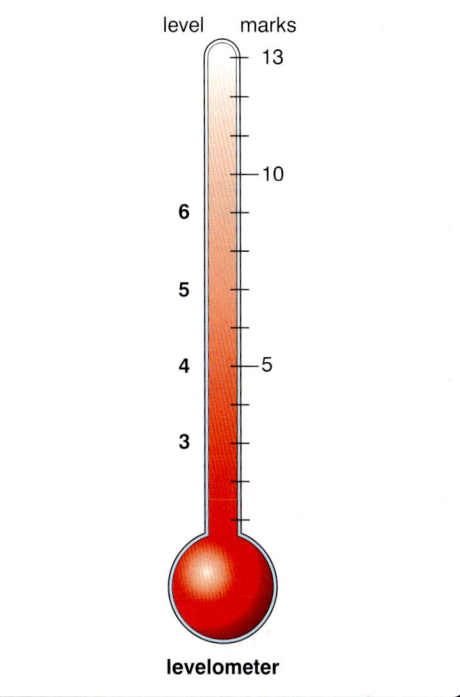

levelometer

7K Forces and their effects

National framework/QCA SoW references

National framework – Forces: Recognise forces have size and direction, identify the direction of forces, distinguish between mass and weight, use forces to consider floating, sinking, and friction, draw and interpret distance–time graphs.

QCA SoW (7K)

Where do we come across forces? (7K1)
Why do things float? (7K4)
How do different materials stretch? (7K2)
What is weight? (7K2, 7K3)
What does friction do? (7K5)
What affects how quickly a car stops? (7K6)
NC links: Sc1 2a, e, h, k; Sc4 2b, 2c, 2d.

Learning objectives

- To recognise situations where forces are acting.
- To use arrows to represent the direction of forces.
- To use a forcemeter to measure the sizes of forces.
- To know that weight is a force, measured in newtons.
- To know that weight is caused by gravity pulling downwards on an object.
- To know that mass tells how much matter is in an object, and is measured in kilograms.
- To identify the forces acting on an object in water.
- To find out what happens when forces balance.
- To find out how friction affects the movement of an object.
- To describe situations where friction is useful, and also how to reduce friction.
- To know the units used to measure speed.
- To know how the speed of a car affects its stopping distance.
- To draw and interpret distance–time graphs.

In scientific enquiry:

- To make appropriate measurements, and present results appropriately, to find out the relationship between force on a spring and extension of the spring.
- To measure forces and volumes accurately and to calculate the density of an object.
- To investigate the factors affecting the size of friction.

Answers to questions

What do you remember?

1 Upthrust.
2 Gravity.
3 Newton.
4 Air resistance is less.
5 Friction is greater.

Notes to support 'What do you remember?'

This unit builds on work from KS2. Pupils will have used a forcemeter to measure forces, identified surfaces with high and low friction, and shapes that move through water easily. They will have related weight to gravity, recognised that water pushes up on objects and learned that air resistance slows down falling objects. Before beginning this unit, it is helpful if they can:

- Explain that pushes and pulls can change the shape of an object, make it speed up, slow down or change direction.
- Measure distance and use a forcemeter to measure forces.
- State that forces have a direction.
- Recall experience of forces, such as magnetism, gravity, friction and air resistance.

Teaching strategy

This unit builds on the KS2 units 4E 'Friction' and 6E 'Forces in action'; it leads into units 9J 'Gravity and space', 9K 'Speeding up' and 9L 'Pressure and moments'.

Difficulties/Misconceptions

- Pupils may have difficulty identifying all the forces in situations where an object's shape or movement is affected by more than one force.
- Many pupils have difficulty distinguishing between mass and weight; rigorous use of correct terminology is needed here as the words are often used incorrectly in 'everyday' language.
- If pupils find ideas of friction difficult, thinking about reducing friction as 'making surfaces more slippery' can help.

Notes and tips

ICT can be used to help with recording, displaying and interpreting results from investigations into friction, extending springs and the relationship between mass and weight.

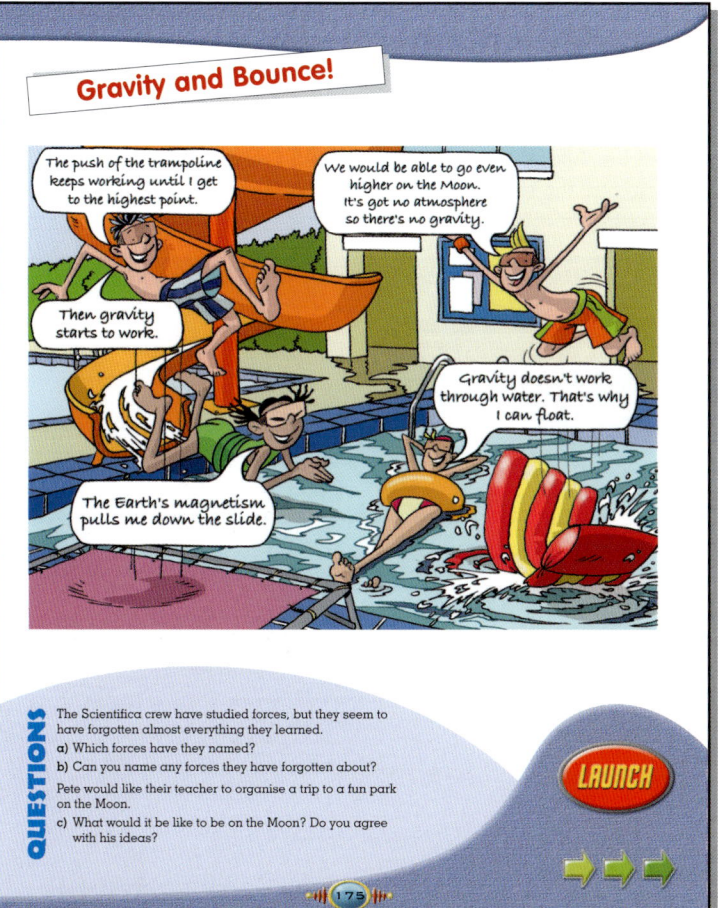

Learning styles

Visual
- **Imagining**.
- **Drawing**.
- **Drawing** graphs.

Auditory
- **Explaining**.

Kinaesthetic
- **Measuring** forces and upthrust.
- **Extending** springs.
- **Relating** mass and weight.
- **Investigating** friction.

Interpersonal
- **Collaborating**.

Launch activity notes – Gravity and bounce!

This activity can be used as the starting point to encourage pupils to revisit, and revise, their knowledge about forces. It gives a good opportunity to assess pupils' prior learning and adjust teaching accordingly.

- 'The Earth's magnetism pulls me down the slide.'

  Pupils will have learned at KS2 that magnets only affect other magnets and metals. Some pupils will know which metals and may have discussed rubber magnets or ceramic magnets.

- 'Gravity doesn't work through water. That's why I can float.'

  Pupils are very unlikely to have met this at KS2. Some pupils may spot that if this was true, everything they dropped in a swimming pool would float – encourage them to explain this to others.

- 'The push of the trampoline keeps working until I get to the highest point.'
  'Then gravity starts to work.'

You will spot the logical thinkers here, because they will question why gravity starts to work at different heights for different trampolines! This could lead into a discussion of which materials are 'bounciest'. Pupils will probably have considerable knowledge here based on a mixture of experience and intuition.

- 'We would be able to go even higher on the Moon. Yes, it's got no atmosphere so there's no gravity.'

  This could lead to many discussions about space and weightlessness, many of which are better left until the Year 9 topic on 'Gravity and space'. However, you could point out that if no atmosphere meant no gravity, mountaineers on top of Everest would begin to float around as the atmosphere is very thin there!

Answers

a) magnetism, gravity, push
b) friction, air resistance, upthrust
c) Jumping is easier – but because gravity is less, not because no atmosphere means no gravity.

7K1 Go on – force yourself!

National framework/QCA SoW references

National framework – Forces: Recognise that forces have size and direction and identify the directions of forces.

QCA SoW
7K Section 1: Where do we come across forces?
NC links: Sc1 2a.

Learning objectives

▸ To recognise situations where forces are acting.
▸ To use arrows to represent the direction of forces.
▸ To use a forcemeter to measure the size of forces.

Teaching strategy

Key points to highlight
- Forces affect things. Whenever something moves or changes shape, there is a force acting on it.
- Forces can be shown by arrows.
- A forcemeter measures the size of a force.
- The unit of force is the newton (symbol, N).

Difficulties/Misconceptions

Most pupils will be familiar with forces making things move, from Key Stage 2, and will confidently draw an arrow representing the direction of the push or pull. Some may have difficulty with situations where there are two or more forces making something change shape, such as pulling on opposite ends of a piece of chewing gum.

Skills

In the activities pupils are developing their visualisation and mental modelling skills, building for themselves a mental model of where forces must act to produce the results they see. Discussion and explanation among peer groups helps them do this.

Lesson starter suggestions

- **Forces all around us:** Pass round a range of objects (scrunched and un-scrunched paper, toys with springs, magnets, string with knots, etc.) for pupils to look at briefly and then pass on. In small groups, pupils to decide if forces act, or have acted, on the objects and how.
- **What do forces do?** Pupils to describe or draw three examples of pushes and three examples of pulls. What effect do these forces have on whatever they are pushing or pulling?
- **Forces are fun:** Show pictures of Newton's cradle executive toy, a child on a swing (adult pushing), a toy bow and arrow and a child on a pogo stick. Ask: 'All these toys use forces: can you find the forces? What do they do?'

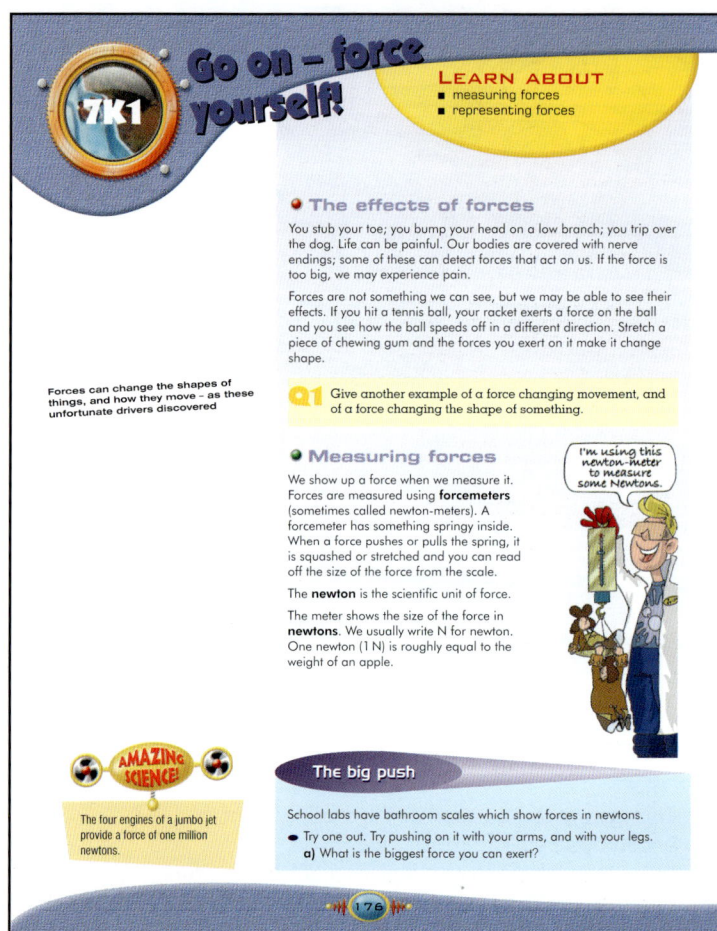

Activity and technicians' notes

The big push

Equipment required
- Bathroom scales marked in newtons

Teaching strategy contd.

Notes and tips
- This unit revises learning about forces from Key Stage 2, and extends it to a wide range of possibly unfamiliar situations. Most pupils will benefit from 'hands on' experience, so have plenty of objects/situations available where pupils can experience the forces you discuss.
- Some pupils may have heard the idea 'that if you push on something it pushes back'. This is true and they may look at these ideas if they take study of science to a much higher level. If you meet this idea here, tell them that we are surrounded by lots of balanced forces that push against each other and nothing happens. At the moment we are only interested in forces where we can see an effect of the force.

Extension ideas
Pupils design an experiment to investigate how the force between two magnets varies. They need to devise ways to ensure the changes they observe are caused only by the factor they are considering, not by gravity, for example.

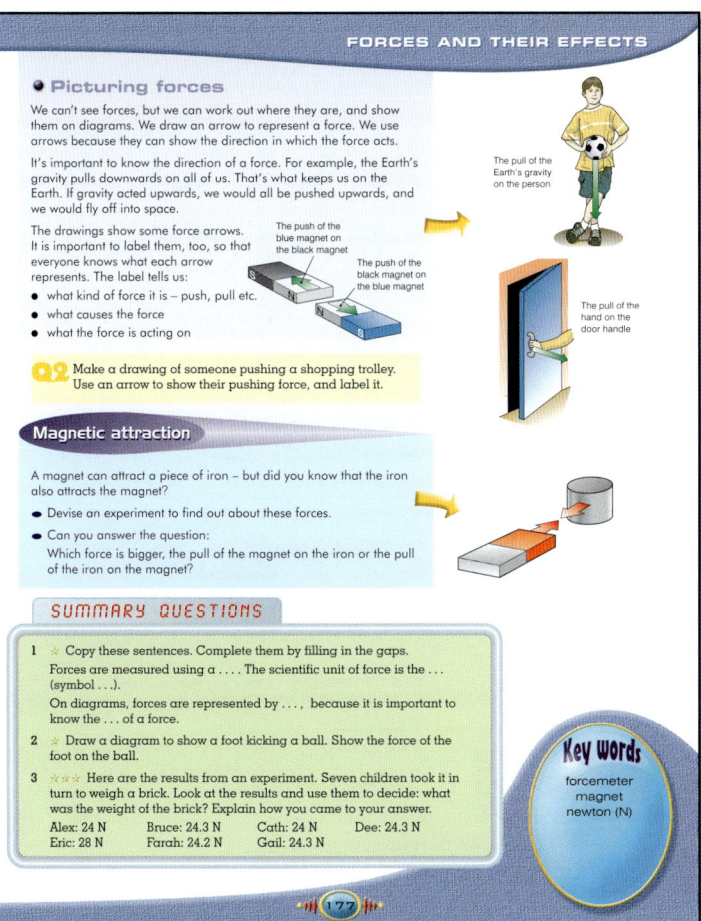

Suggestions for plenaries

- **Name that force:** In pairs, pupils take turns to name a force, their partner gives an example of where that force acts, and describes the effect it has.
- **What do forces do?** Pupils look at their drawings of pushes and pulls from the starter 'What do forces do?'. Ask them to draw arrows to show the directions of the forces.
- **Force anagrams:** Show drawings of situations where a range of forces act. Pupils un-jumble anagrams for the names of forces and put the name of the force with the correct picture.

Learning styles

Visual
- **Visualising** the forces acting in a range of situations.
- **Drawing** force diagrams.

Kinaesthetic
- **Practical** activities to measure forces.

Answers to questions

Text
Q1 Changing movement: e.g. a dog pulling on a lead to drag you forwards.

Changing shape: e.g. a potter moulding wet clay.

Q2 Arrow should be acting forwards on the shopping trolley. Labelled 'Person pushing'.

Summary
1. forcemeter, newton, N, arrows, direction
2. The arrow should act forward on the ball, in the direction the foot was moving in.
3. 24.2 N; average of six readings for weight. Eric's reading was ignored, as it is so different from the others that it was probably misread, and needs checking.

Activity and technicians' notes contd.

Magnetic attraction
Equipment required
- Magnets, iron bars, paper clips or small iron objects
- String
- Scissors
- Clamp stands
- Metre rulers, 30 cm rulers

Technician's notes
- Pupils may wish to clamp a metre ruler or 30 cm ruler in a horizontal position.

Learning outcomes

Following this, most pupils will:
▶ Describe situations in which forces act and identify the directions of the forces.

Some pupils will also:
▶ Explain that forces can combine to give a resultant force.

7K2 Bend and stretch

National framework/QCA SoW references

National framework – Forces: Recognise that forces have size and direction and identify the forces acting on an object.

QCA SoW

7K Section 3: How do different materials stretch?

7K Section 4: What is weight?
NC links: Sc1 2e, h; Sc4 2b.

Learning objectives

▸ To know that weight is a force, measured in newtons.
▸ To know that weight is caused by gravity pulling downwards on an object.
▸ To make appropriate measurements, and present results appropriately, to find out the relationship between force on a spring and extension of the spring.

Teaching strategy

Key points to highlight

- Springs can be used to measure forces. The larger the force, the further the spring stretches.
- Weight is the force on an object because the Earth's gravity pulls on it.
- How much a spring stretches depends on how stiff it is, and on the force pulling it.

Difficulties/Misconceptions

Many pupils think weight is the same as mass. They will learn the difference in section 7K3, but encouraging/insisting on correct scientific terminology here (weight is a force, measured in newtons) makes section 7K3 easier.

Skills

The activities give an opportunity to assess and develop pupils' scientific enquiry skills. Both can be done as full investigations, allowing pupils to plan, predict, and improve their observing, measuring and recording skills.

Lesson starter suggestions

- **Experiencing springs:** Pass round a range of different stiffness springs. Pupils to find out how easy they are to stretch or compress (warn them to only stretch or compress the spring a small amount to avoid damaging it). Can pupils see any pattern between the appearance of the spring and how easy it is to stretch or compress?
- **Springs around us:** Pupils to list as many examples of springs in things as they can. Ask: 'Are they stretched or compressed? What are they for?'
- **Forces acrostic:** Pupils to use the letters of the word FORCES to write six sentences, statements or facts showing what they know so far about forces. Ask if they can include anything about springs in any of their sentences.

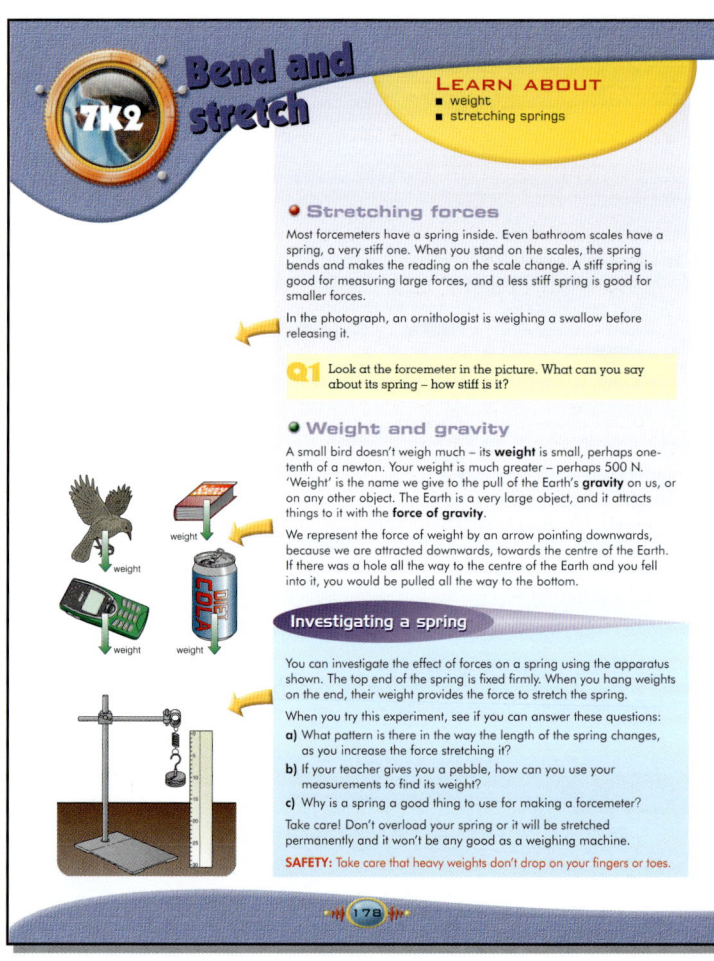

Activity and technicians' notes

Investigating a spring

Safety notes

Ensure feet and fingers are kept clear of dropped weights.

Equipment required

- Springs
- Clamp stands
- Metre rulers
- Graph paper
- Slotted masses (size and number of masses required will depend on stiffness of springs used – check beforehand)

Technician's notes

The total mass of the slotted masses provided for each group should be insufficient to overload, and permanently stretch, the spring.

Teaching strategy contd.

Notes and tips
The activity to bend a ruler requires pupils to make very precise measurements of distance. Pupils whose measurement skills are less precise could use longer rulers and heavier weights, so that the amount the ruler bends down at the end is increased. Pupils should use a ruler clamped vertically behind the end of their bending ruler, to measure how far the end of the bending ruler has moved.

Extension ideas
Pupils could devise their own code to keep secret Hooke's Law about springs. Challenge some pupils to consider why the spring doesn't go on stretching for ever when a weight is hung from it.

Learning styles

Visual
- **Drawing** graphs of extension and bending.

Kinaesthetic
- **Practical activities** to extend springs and bend beams.

Answers to questions

Text
Q1 The spring is very weak – suitable for measuring very small forces.
Q2 Use a beam/ruler that bends very easily when a force is applied to it.

Summary
1. weight, force, gravity
2. newtons, because weight is a force
3. Weight is a force and arrows are used to represent forces. The arrow points downwards because the force of gravity pulls things downwards.
4. Drop objects in water to show that some objects do fall through water, so gravity must be acting on them. It must be something else that makes some objects float.

Activity and technicians' notes contd.

Bending a ruler
Safety notes
Ensure feet and fingers are kept clear of dropped weights.

Equipment required
- Metre rulers, 30 cm rulers
- Clamp stands
- Slotted masses (size and number of masses required will depend on length and stiffness of rulers used – check beforehand)
- String (for attaching masses to rulers)
- Graph paper

Technician's notes
Boxes could be supplied to catch falling masses, if they slide off the ruler as it bends.

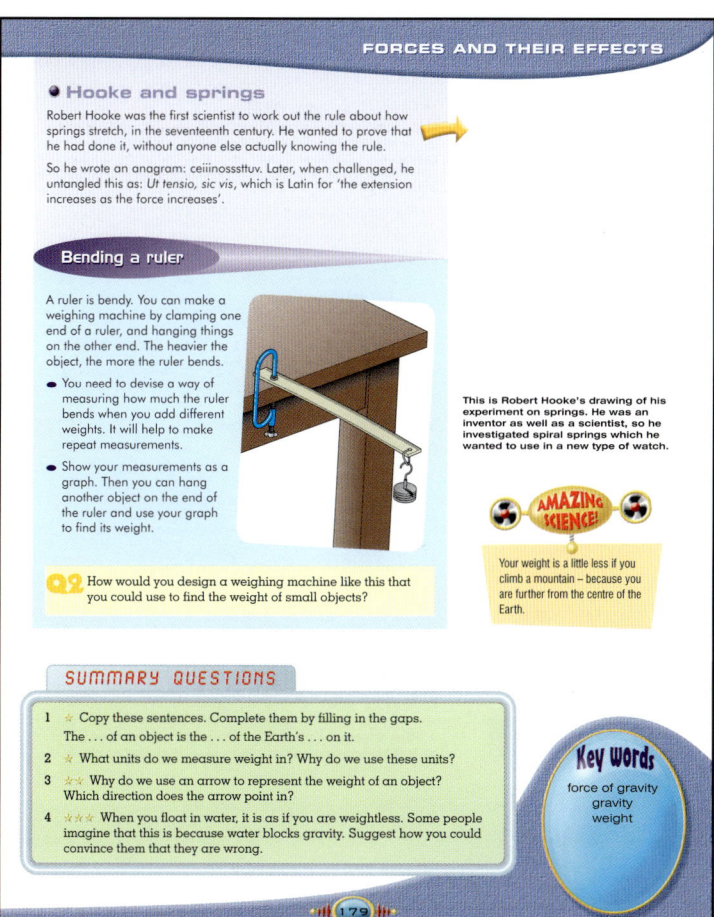

Suggestions for plenaries

- **Springs and toys:** Pupils to think of a toy with a spring in it, or design their own. They then draw it. Ask: 'What does the spring do? How easily would the toy break? What would break it? Why?'
- **Red, amber, green:** Pupils decide whether the statements about forces and weight are true (green card) or false (red card) or 'don't know' (amber card).
- **Anagrams:** Pupils un-jumble the anagrams for words, matching each word with its definition.

Learning outcomes

Following this, most pupils will:
▸ Make precise observations, repeat readings and use results to plot graphs.
▸ Identify the direction in which a force is acting.

Some pupils will also:
▸ Choose appropriate scales for graphs to show results effectively.
▸ State that weight is caused by the Earth's gravity.

7K3 Mass and weight

National framework/QCA SoW references

National framework – Forces: Use examples to distinguish between mass and weight.

QCA SoW

7K Section 4: What is weight?
NC links: Sc1 2k; Sc4 2b.

Learning objectives

- Mass tells us how much matter something is made of.
- Mass is measured in kilograms.
- Weight is a force caused by gravity pulling downwards on an object.
- Weight is measured in newtons, symbol N.

Teaching strategy

Key points to highlight

- Mass tells us how much matter there is in an object.
- Mass is measured in kilograms.
- Weight is the force on an object because gravity pulls on it.
- Weight is measured in newtons.
- The weight of an object on the Moon is less than the weight of the object on Earth, because the force of gravity is smaller on the Moon.

Difficulties/Misconceptions

Most pupils find the distinction between mass and weight confusing. Make sure they know that weight is a force, and the everyday use of the word 'weight' in sentences such as 'My weight is 50 kg,' is scientifically incorrect.

Skills

- In 'The mass-weight connection', pupils practise their graph plotting skills. Help them, where appropriate, to select suitable scales for the axes of their graphs – some pupils will need to be told what scales to use, or given prepared axes.
- They also practise visualisation skills in 'picturing' what would happen to the weight of various masses on the Moon.

Notes and tips

If you use the 'Weighty conundrum' starter, make sure you discuss with pupils all the numerous pitfalls:

- Firstly, it is incorrect to give a weight in kilograms – weight is a force and should be measured in newtons.
- Secondly; a force of 10 N pulling down on feathers is the same size as a force of 10 N

Lesson starter suggestions

- **Free sentences:** Ask pupils to write two sentences containing the word 'weight' and two sentences containing the word 'mass'. Discuss their sentences to establish what they know already.
- **Weighty conundrum:** Give pupils the old conundrum: 'Which weighs most, a kilogram of feathers or a kilogram of lead?' Can they work out the answer, and then spot all the reasons/inaccuracies?
- **Does weight matter?** Give pupils a range of situations where weight might affect something, e.g. a sky diver using a parachute, the speed a scrunched ball of paper falls, a heavy ball hurting more than a light one when you drop it on your foot. Pupils to decide which of these are to do with weight, and which to do with something else.

pulling down on lead – its just that for that size force you would have a small piece of lead and a very large bag of feathers! Throughout this section, encourage pupils to talk about masses and weights, insisting on them using correct terminology – the more they explain the difference to each other, the clearer it will become.

Extension ideas

Encourage pupils to distinguish between things that depend on the mass of the object, and those that depend on the weight, using this knowledge to consider how activities at a 'Moon hotel' might be different from activities on Earth.

Learning styles

Visual
- **Visualising/imagining** how objects would behave on the Moon.

Kinaesthetic
- **Practical activity** to find the connection between mass and weight.

Activity and technicians' notes

The mass-weight connection
Safety notes
Take care with heavy weights.

Equipment required
- Forcemeters
- A range of objects marked with their masses
- Graph paper

Technician's notes
Ensure masses are small enough not to overload the forcemeters. Use bathroom scales for very large masses.

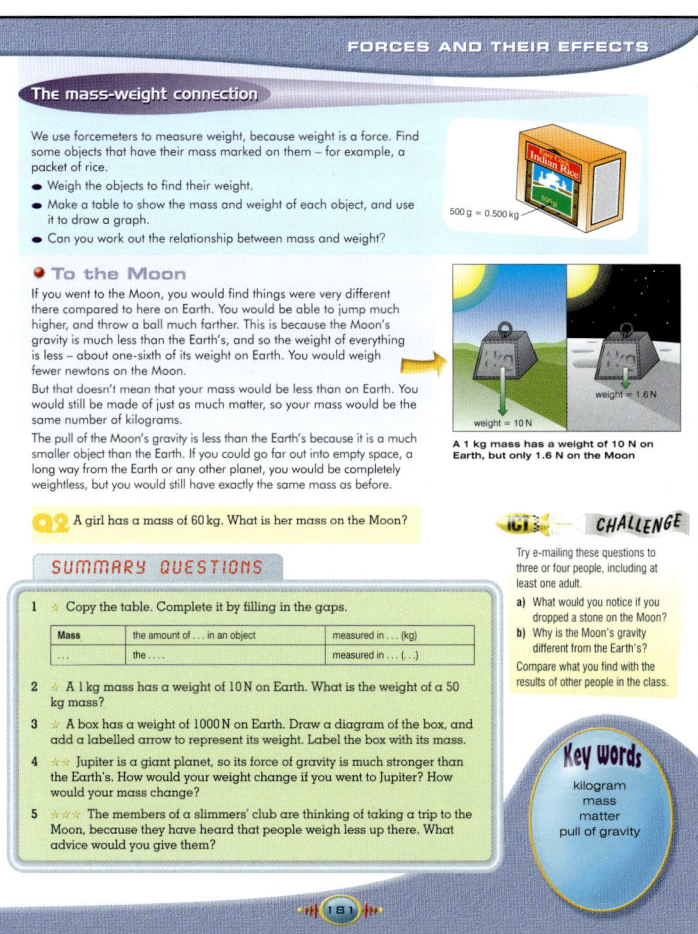

Learning outcomes

Following this, most pupils will:
▸ Use examples to distinguish between mass and weight.

Some pupils will also:
▸ State that weight is caused by the Earth's gravity, and describe how gravity is different on Earth and on the Moon.

Suggestions for plenaries

- **Connectives:** Pupils complete the sentence: 'The weight of an object can change . . .' in any way they like using one of the following connecting words: 'but, and, because, although however, if, to'.
- **Questions and answers:** Provide pupils with a set of answers, for which they have to write appropriate questions. Go over in class discussion.
- **Get it right!** Give pupils a series of sentences about mass and weight. Each sentence has a mistake that pupils have to find and correct.

Answers to questions

Text

Q1 A cricket ball or hockey ball has a much greater mass, so is harder to stop.

Q2 60 kg.

Summary

1

Mass	the amount of matter in an object	measured in kilograms (kg)
Weight	the force on an object caused by gravity	measured in newtons (N)

2 500 N (50 × 10)
3 Arrow points downwards; mass = 100 kg (1000 ÷ 10).
4 Weight would increase. Mass would stay the same.
5 They will weigh less on the Moon, but when they return their weight will return to what it was when the left (unless they stay there for a long time, of course!).

7K4 Floating and sinking

National framework/QCA SoW references

National framework – Forces: Describe situations where forces are balanced; use ideas of balanced and unbalanced forces to describe floating and sinking.

QCA SoW

7K Section 2: Why do things float?
NC links: Sc1 2k; Sc4 2c.

Learning objectives

- To find out the forces acting on an object in water.
- To find out what happens when the forces balance.
- To measure forces and volumes accurately to calculate the density of an object.

Teaching strategy

Key points to highlight

- Water, and any other liquid, pushes up on an object in it, with a force called 'upthrust'.
- The weight of an object in water stays the same.
- The upthrust 'cancels out' some of the weight, so a forcemeter reading will be lower.
- When an object is stationary, the forces on it are balanced.
- For a floating object, the weight and the upthrust are balanced.
- An object with a lower density than water will float.

Difficulties/Misconceptions

Density is an unfamiliar and abstract concept that many pupils find hard. Describing an object as 'light for its size' (low density) or 'heavy for its size' (high density) may help.

Skills

Ideas of upthrust and floating are difficult for many pupils. They will grasp these ideas more readily, and develop their literacy and communication skills, by explaining these ideas to each other.

Lesson starter suggestions

- **Anagram:** How many words can pupils make from the letters of the word FLOATING? Can they find the word that uses all the letters?
- **What floats?** Pupils work together to make suitable endings to these sentences; 'Things float if . . .' and 'You can make things float by . . .'. Use class discussion to establish present class understanding about floating.
- **Floating and sinking:** Give pupils a range of objects to group into those that float, those that sink, and those you don't know. Pupils explain why they have grouped objects as they have. Class discussion to determine class ideas about properties that make objects float or sink.

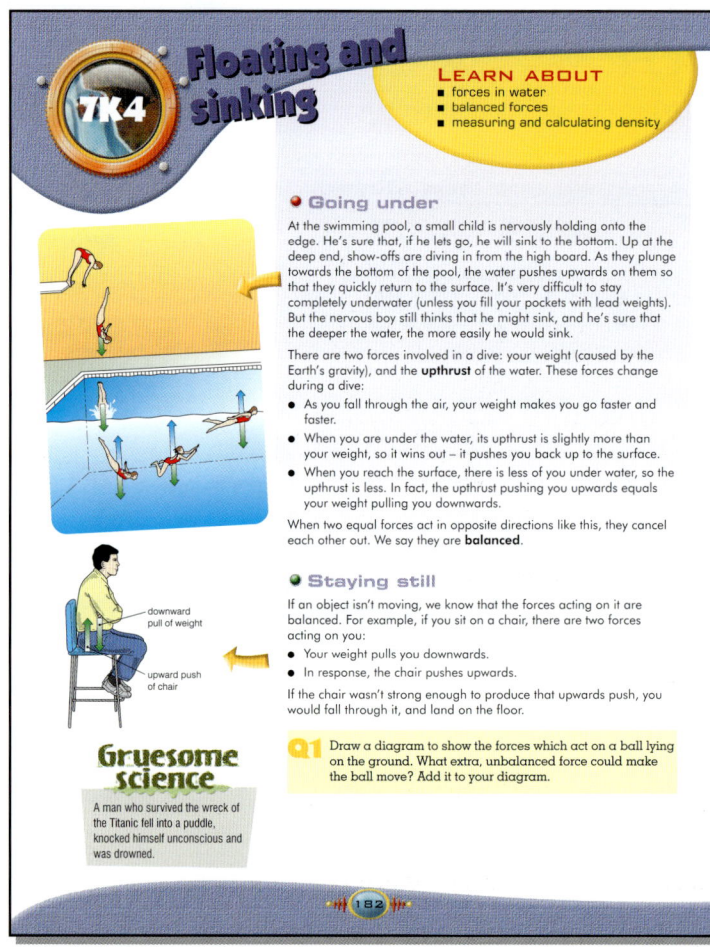

Answers to questions

Text

Q1 The arrow from the ball points downwards (weight). Force from the ground acts upwards on the ball. A sideways force would make the ball move along the ground.

Q2 The difference in weight readings tells how much of the weight has been 'cancelled out': the size of the force pushing against the weight.

Q3 Density = 1020 kg/m³; it will sink (density is greater than density of water).

Activity and technicians' notes

Measuring upthrust

Equipment required
- Forcemeters
- Range of sinking objects
- Bowl/beaker large enough to contain objects

Technician's notes
Ensure objects have small enough masses not to overload forcemeters.

Access required to:
- Water

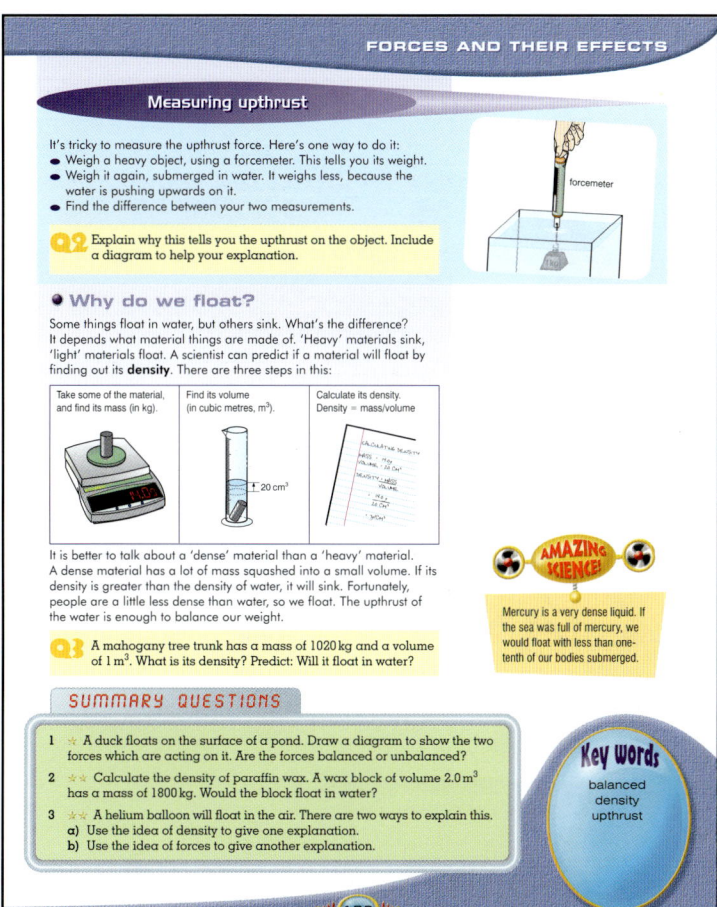

Teaching strategy contd.

Notes and tips
- Some pupils may benefit from pushing gently upwards on objects suspended from forcemeters, to see that the reading on the forcemeter goes down. This helps to make the direct connection with the water pushing up on an object in it.
- You may wish to discuss ideas of trapped air making the density of an object less, so making it more likely to float. Metal boats could be used as examples here.

Extension ideas
Pupils consider what floating in swimming pools tells them about their density. They could move on to consider ideas about displacement, and why different people displace different amounts of water and float at different depths.

Learning styles

Visual
- **Imagining** the forces they would feel in a swimming pool.
- **Imagining** various situations with balanced forces.
- **Drawing** force diagrams.

Kinaesthetic
- **Practical activity** to measure upthrust.

Answers to questions contd.

Summary
1. Downward force = weight of duck. Upward force = upthrust. These are balanced.
2. Density = (1800/2) = 900 kg/m^3. It will float (as density is less than 1000 kg/m^3).
3. a) The average density of the balloon and helium is less than the density of air, so it floats.
   b) Download force of weight is balanced by upthrust from air.

Learning outcomes

Following this, most pupils will:
- Make predictions about the upthrust on an object, and make observations and measurements to test their predictions.
- Describe floating as a situation in which forces are balanced.

Some pupils will also:
- Use ideas of balanced forces to explain results of floating experiments.

Suggestions for plenaries

- **Acrostic:** Pupils to use the letters of the word SINKING as the first letters in seven sentences, statements or facts about floating, sinking or forces. Go over in class discussion.
- **Calculating densities:** Give pupils a series of cards, each with a volume, a weight or a density. They have to match cards into sets of three.
- **True or false:** Give pupils a series of paired statements about floating and sinking. Pupils decide if each statement is true or false, and if one statement is an explanation of the other.

7K5 All about friction

National framework/QCA SoW references

National framework – Forces: Describe the effect of friction on the speed of an object, describe ways of reducing friction, describe situations where friction is useful.

QCA SoW

7K Section 5: What does friction do?
NC links: Sc1 2a; Sc4 2d.

Learning objectives

- How friction affects the movement of an object.
- To investigate the factors affecting the size of the friction force.
- To describe situations where friction is useful.
- To describe ways to reduce friction.

Teaching strategy

Key points to highlight

- Friction is a force that acts to oppose motion, to slow down surfaces moving over each other.
- Friction can be useful.
- Unwanted friction can be reduced by making surfaces smoother or using a lubricant.
- Air resistance and drag in water are partly due to friction.

Difficulties/Misconceptions

- Ensure pupils understand that the frictional force should always be drawn at the junction of the two surfaces in contact.
- Be very wary discussing details of why car and shoe treads work with all but the very ablest pupils. Putting a tread on a tyre neither increases nor decreases friction – the friction is independent of the area of contact (A-level work). A tyre tread channels water out from between tyre and road, so that the water does not act as a lubricant. A lubricant would make the friction much less on a wet road.

Skills

In the activity investigating friction, pupils practise their scientific enquiry skills of planning and carrying out a complete investigation, making decisions about which variables to control and how to control them.

Lesson starter suggestions

- **Different materials:** Pass round examples of shoes with soles made from different materials and/or gloves with different surfaces, e.g. goalkeepers gloves. Discuss when you would wear them, why they are good/bad.
- **Questions and answers:** Pupils to write three or four questions about sliding. They then try to answer another group's questions, finally contributing to class lists of 'Things we know about sliding' and 'Things we want to know about sliding'.
- **Playground fun:** Distribute pictures of children trying to climb up a slide, and sliding down a rope. Ask: 'Have you ever done these things? What happened?' Discuss why.

Activity and technicians' notes

Investigating friction

Equipment required

- Wooden planks
- Blocks of wood
- Slotted masses
- Metre rulers
- Stopwatches
- Protractors
- Sheets of different surfaces to cover wooden block with

Technician's notes

Some pupils may need a way to attach masses to their block of wood. Others may want blocks of wood with different surface areas in contact with the plank, though usually this can be achieved by turning the block on its side or on its end.

Teaching strategy contd.

Notes and tips
If pupils investigate the factors affecting the friction between two surfaces they will find that friction is affected by the type of surfaces in contact, and by the weight of the sliding object, but not by the area. This is often a surprise. It can be explained as follows:
- Increasing the weight pushes the surfaces together harder, so a bigger force is needed to make the 'bumps' move over each other.
- The area of actual contact between the surfaces is very small (approximately one ten-thousandth of the measured area). Decreasing the measured area in contact increases the pressure between the surfaces, squashing the 'bumps' and making them spread out until the actual area in contact is the same as it was before.

Extension ideas
- Pupils choose a sport and think of aspects where friction is necessary and where it is a nuisance.
- Some pupils could find out about the legal safety regulations for car tyres, why these regulations are important and why racing car tyres and 'normal' car tyres are different.

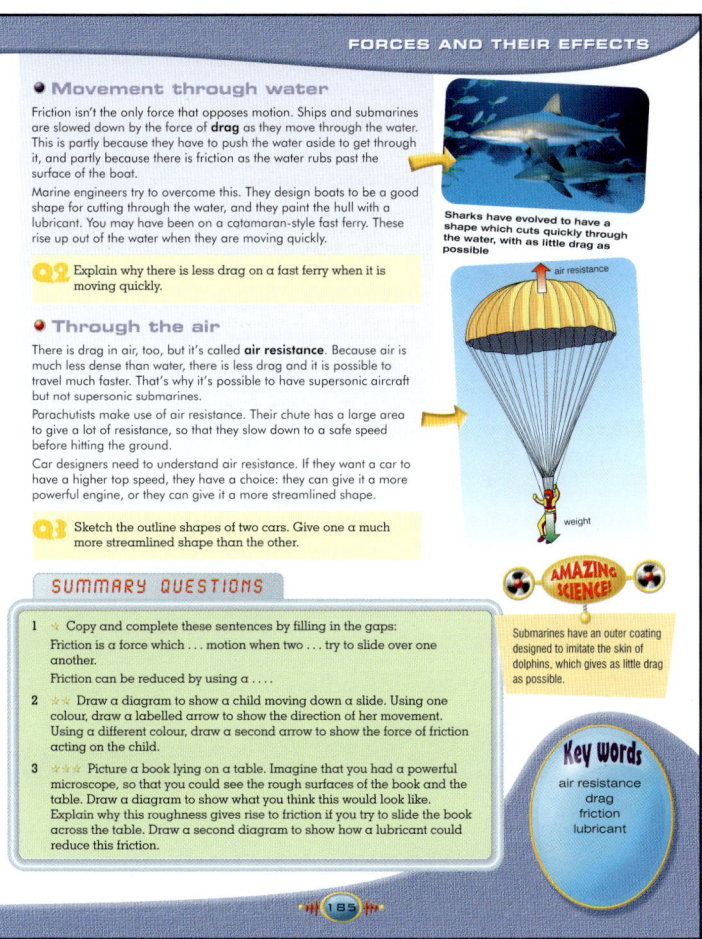

Learning styles

Visual
- **Imagining** a microscopic view of surfaces in contact.
- **Drawing** streamlined shapes and shapes that are not streamlined.

Kinaesthetic
- **Investigating** friction.

Answers to questions

Text
Q1 Water; slide would be a lot less slippery.
Q2 Ferry lifts in the water, so less of the hull is in contact with the water.
Q3 The most streamlined shape is a teardrop shape, with the blunt end at the front.

Summary
1. opposes, surfaces, lubricant
2. Friction acts in the opposite direction to the motion. It should be shown acting at the junction between the child and the slide.
3. Surfaces are bumpy. Bumps 'catch' on each other as the book moves. The lubricant moves the surfaces slightly further apart so less of the bumps 'catch'.

Suggestions for plenaries

- **Acrostic:** Use the letters of the word FRICTION as the first letters in eight sentences, statements or facts about friction. Go over them in a class discussion.
- **Questions and answers:** Can pupils now answer questions from the starter class list 'Things we want to know about sliding'? If not, help them decide how they can find the answers – possibly for homework.
- **Good and bad friction:** Give pupils a series of cards, each with an example where friction is useful or a nuisance. Pupils to sort the cards into two groups 'useful' and 'not useful', and then explain why.

Amazing science
The teflon 'non-stick' coating familiar from 'non-stick' saucepans was invented as a 'super smooth' coating for spacecraft.

Learning outcomes

Following this, most pupils will:
▸ Investigate friction, identifying key factors and control them.
▸ Describe situations where friction is useful, and some ways to reduce friction.

Some pupils will also:
▸ Explain how to ensure an investigation into friction is fair.
▸ Explain why friction exists between moving surfaces.

7K6 Graphs that tell stories

National framework/QCA SoW references

National framework – Forces: Describe the effect of forces on the speed of an object.

QCA SoW

7K Section 6: What affects how quickly a car stops?
NC links: Sc4 2c.

Learning objectives

▸ To know the units used to measure speed.
▸ To know how the speed of a car affects its stopping distance.
▸ To draw and interpret distance-time graphs for journeys.

Teaching strategy

Key points to highlight

- The units of speed are [a distance]/[a time].
- The stopping distance is the distance a vehicle travels while braking, before it stops.
- The stopping distance is increased by increasing speed and/or decreasing friction.
- Distance-time graphs can be used to show how far a moving object travels at different stages of its journey.

Difficulties/Misconceptions

For some pupils, speed may be the first derived quantity (one worked out from other quantities) that they have met. Concentrate on explanations of what speed is, e.g. the distance that an object will travel in one hour or in one second. The formal definition of speed is not needed at this stage, though you may wish to introduce it with some pupils.

Skills

The main skills pupils will practise in this section are graph interpretation skills. Some pupils may need to concentrate on reading scales and axes and becoming familiar with line graphs; other pupils may be ready to describe complicated graphs.

Lesson starter suggestions

- **Verbal quiz:** Ask the class to stand up. Ask each pupil a question about forces or friction. They are to sit down when they answer correctly. Differentiate questions for different pupils, so all have an equal chance of answering correctly.
- **Free writing:** Ask the pupils to write for one minute on 'speed and safety', concentrating on scientific ideas. Then they are to write for one minute on 'making cars stop', again concentrating on scientific ideas. Go over the ideas in a class discussion.
- **Describing journeys:** Give the pupils an example of a 'journey story'. Ask them to write their own story for a journey of their own choice, following the example given.

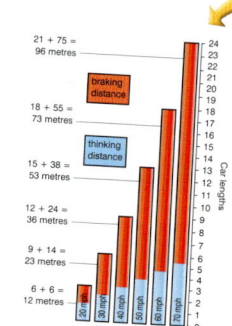

Teaching strategy contd.

Notes and tips
Pupils should work in groups for graph interpretation activities. They will benefit from explaining graphs of journeys to each other. You could check understanding by giving different graphs to each group, asking them to present an explanation of their graph to the rest of the class and to answer questions about it.

Extension ideas
Pupils could plot a distance-time graph using the information from a bus or train timetable, using the shape of the graph to decide when the bus or train was moving quickly or slowly. Some pupils may be ready for calculations using the formal definition of speed.

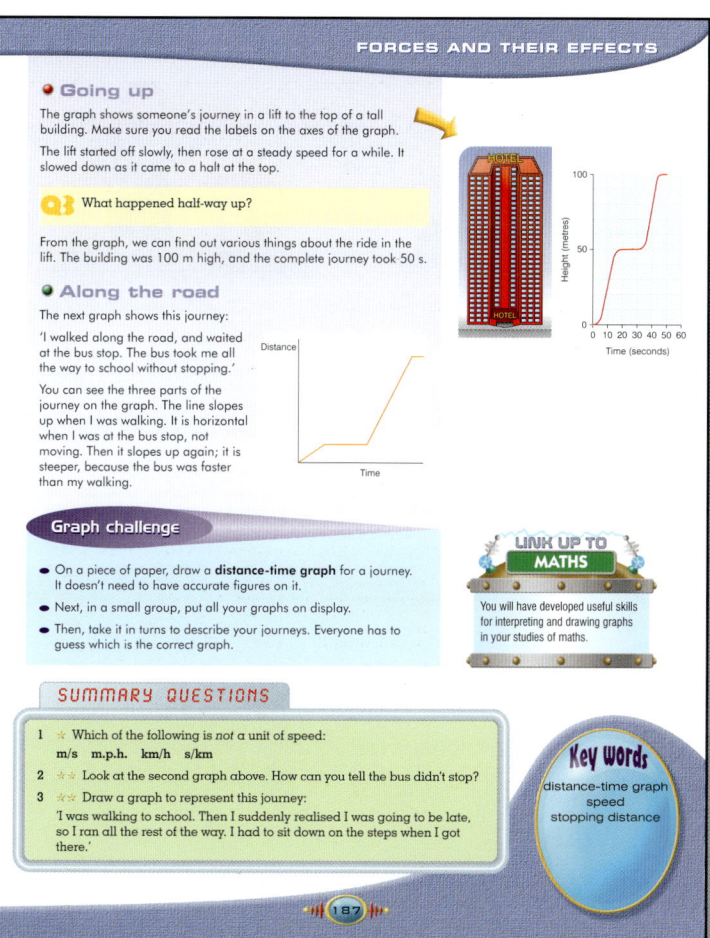

Learning styles

Visual
- **Drawing and interpreting** distance-time graphs.

Auditory
- **Explaining** journeys.

Interpersonal
- **Collaborating** to explain distance-time graphs.

Answers to questions

Text
Q1 Historically, in this country we have been using miles per hour much longer than kilometres per hour. Also, to avoid confusion: m/h might perhaps be confused with metres per hour.

Q2 a) 23 m

b) At 60 m.p.h., the stopping distance is more than twice as much (because the stopping distance is proportional to the square of the speed, not proportional to the speed).

Q3 It stopped to let someone in or out.

Summary
1. s/km
2. There is no horizontal line on the distance-time graph.
3. Graph with first section: shallow gradient; second section: steeper gradient; third section: horizontal line.

Activity and technicians' notes

Graph challenge
Equipment required
- Graph paper

Learning outcomes
Following this, most pupils will:
▸ Explain what is meant by the speed of an object.

Suggestions for plenaries

- **Concept map:** Make a class concept map to show what pupils have learned in this unit.
- **Graph stories:** In groups, pupils to match distance-time graphs, with the journey story for that graph.
- **Unit pairs:** Give pupils a series of cards, each with either the name of a property, such as speed, or the unit. Pupils to play 'pairs', matching the properties with the correct units.

7K Read all about it!

Teaching strategy

Archimedes and Galileo

Difficulties/Misconceptions

The concept of displacement to find density, and therefore composition of a crown, is one that some pupils may find quite complicated. However many pupils will be familiar with the story of Archimedes and his bath, so this is a good opportunity to revise this topic. Encourage pupils to retell the story in their own words, to assist understanding.

Skills

This section is a great opportunity for pupils to practise scientific enquiry skills. They could discuss the way in which Archimedes and Galileo worked – was it 'scientific'? How did Galileo learn from Archimedes, and what have we learned from both of them?

Notes and tips

- To practise ideas in this section, you could ask pupils to look at the cartoon characters in the summary and see if they can extend each comment to give some more information.
- They could even draw force diagrams for some of the situations mentioned.

Extension ideas

Pupils could find out more about Galileo or Archimedes. Or perhaps they could draw a cartoon to explain the work of one of these to pupils in a different class.

Questions

1. Why does the water level rise when you climb into a bath?
2. Why does gold take up less space than silver?
3. What did Galileo invent, after reading about Archimedes' work?
4. Why do goldsmiths today still find Galileo's invention useful?

Answers

1. your body displaces some water
2. gold is more dense than silver
3. weighing scales for finding out how much silver has been added to gold
4. modern gold jewellery is still a mixture of gold and silver, and goldsmiths need to know how much silver has been added

Archimedes and Galileo

You probably know the story of Archimedes in his bath. King Hiero had ordered a new gold crown, in the shape of a wreath of leaves. He suspected that the jeweller had cheated him by mixing silver with the gold. Could Archimedes find a way of checking the crown without damaging it?

Archimedes was in his bath when he thought of the solution. As everyone knows, when you get in the bath, the water level rises because your body displaces some of the water. Archimedes, seeing how he could put this to use, leapt from the bath and ran down the street shouting 'Eureka!'. That means 'I have it!'

Here is how the Roman historian Vitruvius described Archimedes' test of the crown.

The solution which occurred when he stepped into his bath and caused it to overflow was to put a weight of gold equal to the crown, and known to be pure, into a bowl which was filled with water to the brim. Then the gold would be removed and the king's crown put in, in its place. An alloy of lighter silver would increase the bulk of the crown and cause the bowl to overflow.

Archimedes was using the fact that gold is denser than silver, so it takes up less space. He found that the new crown was indeed a cheat, and the jeweller was punished.

Galileo lived 19 centuries after Archimedes, but he was able to read about his work. He investigated the force of **upthrust**, and realised he could go one step further than Archimedes. He invented a special type of weighing scales for finding out the amount of silver which had been added to gold. (Most jewellery is made from a mixture of gold and silver.)

The method was to hang a 'gold' object from the scales to weigh it in air. Then it was weighed again, hanging in water. The scales showed a smaller reading, because of the upthrust of the water. If there was more silver in the gold, there would be more upthrust, and so the reading would be smaller. Goldsmiths could quickly check the materials they were working with.

Archimedes and Galileo were both interested in the ideas of science, but they were also great inventors – the technologists of their times.

Danger – common errors

Balanced forces are tricky, because we often only look at forces that 'make something happen'. Pupils have met several examples of balanced forces in this unit:
- A mass hanging on a spring: the spring stretches until the weight downwards is balanced by the force of the spring trying to return to its original length.
- Floating: a floating object floats because its weight is balanced by the upthrust from the water.
- A car travelling at constant speed: the forward force from the engine of the car is balanced by the forces of friction and air resistance trying to slow down the car.

Pupils will meet this idea again in more detail in Unit 9K 'Speeding up'.

Learning styles

Interpersonal
- **Conversation and feedback** Archimedes and Galileo.
- **Conversation and feedback** about cartoon characters' discussions.

Learning outcomes

Following this unit, most pupils will:
- Describe situations in which forces act and identify the directions of the forces.
- Use examples to distinguish between mass and weight.
- Describe floating as a situation in which forces are balanced.
- Describe situations where friction is useful, and some ways to reduce friction.
- Explain what is meant by the speed of an object.

Some pupils will also:
- Explain that forces can combine to give a resultant force.
- State that weight is caused by the Earth's gravity, and describe how gravity is different on Earth and on the Moon.
- Explain why friction exists between moving surfaces.

In scientific enquiry:

Following this unit, most pupils will:
- Make precise observations, repeat readings and use results to plot graphs.
- Make predictions about the upthrust on an object, and make observations and measurements to test their predictions.
- Investigate friction, identifying key factors and controlling them.

Some pupils will also:
- Choose appropriate scales for graphs to show results effectively.
- Use ideas of balanced forces to explain results of floating experiments.
- Explain how to ensure an investigation into friction is fair.

7K Unit review

Answers to review questions

1
- **a** forces are balanced; rocket won't begin to move
- **b** upward push of rocket motor is greater than weight of rocket
- **c** there is no upward force, so weight will make rocket move downwards
- **d** Rocket could be made lighter (smaller weight) or upward force of rocket motor could be made larger.

2

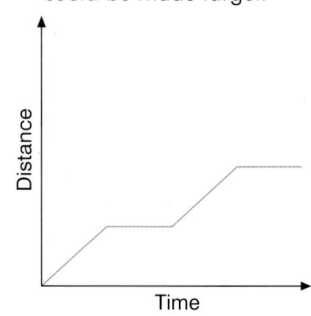

3 James Joule (1818–89) – work on heat and energy; energy is measured in joules
James Watt (1736–1819) – development of steam power; power is measured in watts
Sir Isaac Newton (1642–1727) described how gravity makes things move.

4

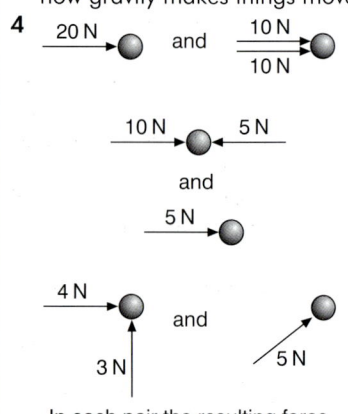

In each pair the resulting force is the same.

5

Mass in kilograms (kg)	Weight in newtons (N)
1	10
3	30
5	50

6 a B
  b C is more dense than D
  c F

7 weight of block in water is not zero; gravity is the same through water as through air, but weight is balanced by upthrust force from water.

8 Diagram should show bumps and irregularities on each surface. Lubricant moves surfaces further apart, so bumps don't 'stick' on each other.

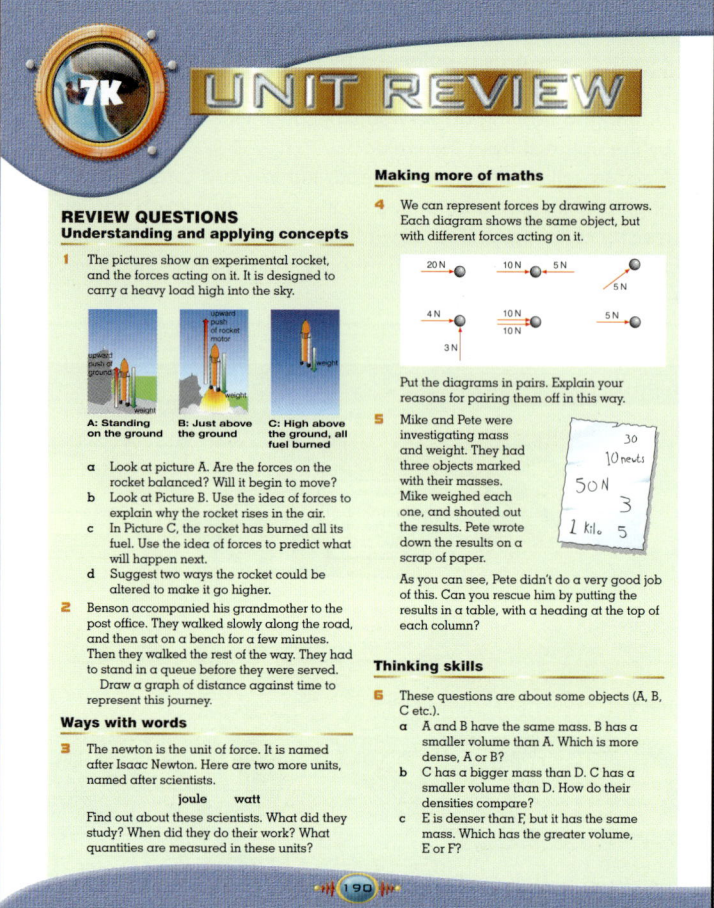

UNIT REVIEW QUESTIONS

Extension questions

7. You can use a newton-meter to weigh a block of wood. The block floats on water. Now if you use the newton-meter, its reading is zero.
   Does this mean that the weight of the block is zero when it is floating? Does this experiment prove that gravity doesn't work through water?

8. There is friction when one rough surface tries to slide over another. Imagine that you had a very powerful microscope, and that you could look at the two surfaces. Draw an illustration of what you think you would see.
   A lubricant such as oil or water can help to reduce friction. Draw another illustration to show your idea of how this works.

SAT-STYLE QUESTIONS

1. In an experiment to investigate how a spring stretches, Mike and Molly hung weights on the end of the spring and measured its length.
   Mike said, 'Every time we increase the weight, the spring will get longer.'
   Molly said, 'If we double the weight on the spring, it will get twice as long.'
   The table shows their results.

Weight (N)	Length of spring (mm)	Increase in length (mm)
0	40	0
1	46	6
2	52	12
3	58	
4	64	
5	70	

   a Study Mike's prediction, and look at the table of results. Was Mike's prediction supported by the results? (1)
   b Copy the table, and complete the final column. (1)
   c Draw a graph to show the results. (2)
   d Use your graph, or the table of results, to find out how much the spring stretched for every newton of load. (1)
   e Molly tried to make a better prediction than Mike. What she said is not quite right. Write down a better conclusion, based on the results they obtained. (1)

2. Emil is coming down a water slide.

   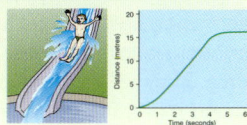

   a Name two forces acting on Emil when he is part way down the slide. (2)
   b The graph shows Emil's journey down the slide. Describe his motion between 1 and 4 seconds. (1)
   c Describe his motion between 5 and 6 seconds. (1)

3. Reese carried out an experiment to investigate floating and sinking. She had a wooden ruler. She weighed a lump of Plasticine, and then attached it to one end of the ruler. Then she floated the ruler in water, and recorded the length of the ruler sticking out of the water.

   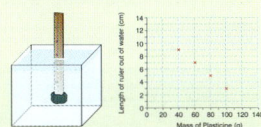

   Study the graph of Reese's results.
   a Use the graph to predict: what mass of Plasticine is needed to make the ruler sink? (1)
   b If there was no Plasticine, what length of the ruler would stick out of the water? (1)
   c Put these materials in order, starting with the most dense:
      water, wood, Plasticine. (3)
   d Reese says, 'As the ruler gets heavier, the upthrust of the water gets less, so eventually the ruler sinks.' Explain why Reese is wrong. (2)

   Key words
   Unscramble these:
   vitargy
   thewig
   argilokam
   aldebanc
   phruths
   dyehns

Answers to SAT-style questions

1 a Yes (1)
  b (1)

Weight (N)	Length of spring (mm)	Increase in length (mm)
0	40	0
1	46	6
2	52	12
3	58	18
4	64	24
5	70	30

c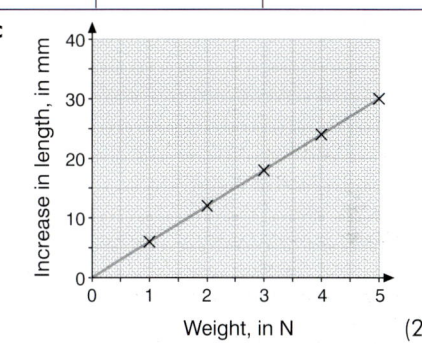
(2)

d Spring stretched 6 mm for every 1 N of load added. (1)
e If we double the weight on the spring it will stretch twice as much. (1)

2 a his weight and friction (2)
  b He is travelling down the slide at a steady speed. (1)
  c He has stopped at the bottom of the slide. (1)

3 a 130 g (1)
  b 13 cm (1)
  c plasticine, water, wood (3)
  d Upthrust stays the same. Ruler sinks because upthrust is no longer great enough to cancel out the increased weight. (2)

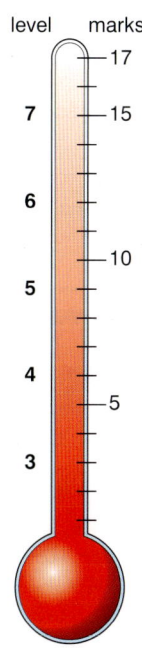

levelometer

7L The solar system – and beyond

National framework/QCA SoW

National framework – Energy: The Sun is a source of energy for the whole solar system, including the Earth.

QCA SoW (7L)

Causes of a year, a month, a day (7L1)
Seeing the Sun and Moon (7L1, 7L2, 7L4)
What causes the seasons on Earth? (7L3)
What the solar system consists of (7L2, 7L5)
What is beyond the solar system? (7L1, 7L6)
NC links: Sc1 1b; Sc1 2a, g, h, i, m; Sc4 4a, b, d, e.

Notes to support 'What do you remember?'

This unit builds on work from KS2. Pupils will have studied how we see objects when light reflects from them and enters our eye, and how shadows are formed when an object blocks the light from a light source. They will have modelled the spinning of the Earth to give day and night, the movement of the Earth around the Sun and the relative sizes and distances of the planets. Before beginning this unit, it is helpful if they can:
- Remember that the Earth, Sun and Moon are spherical.
- Describe how the Sun seems to move throughout the day, and how this affects shadows.
- Remember that day and night happen because the Earth turns.
- Remember that the Earth takes one year to go round the Sun, and the Moon takes one month to go round Earth.

Learning objectives

- To learn about the Sun and other stars as light sources.
- How we see non-luminous bodies, such as planets, by reflected light.
- How the movement of the Earth causes day and night, and the apparent movement of the Sun and stars.
- How the Moon orbits the Earth, and turns on its own axis.
- How sunlight reflecting from the Moon causes the phases of the Moon.
- To find out about stars, why they are only visible at night, and why they appear to move.

In scientific enquiry:

- To collect and interpret data about day length and temperature, and explain the seasonal variations.
- To find out about the various bodies in the solar system, using a range of secondary sources of information.
- To use a model to show how the Earth's tilt causes the seasons.
- To describe eclipses of the Sun and Moon and explain why they happen.

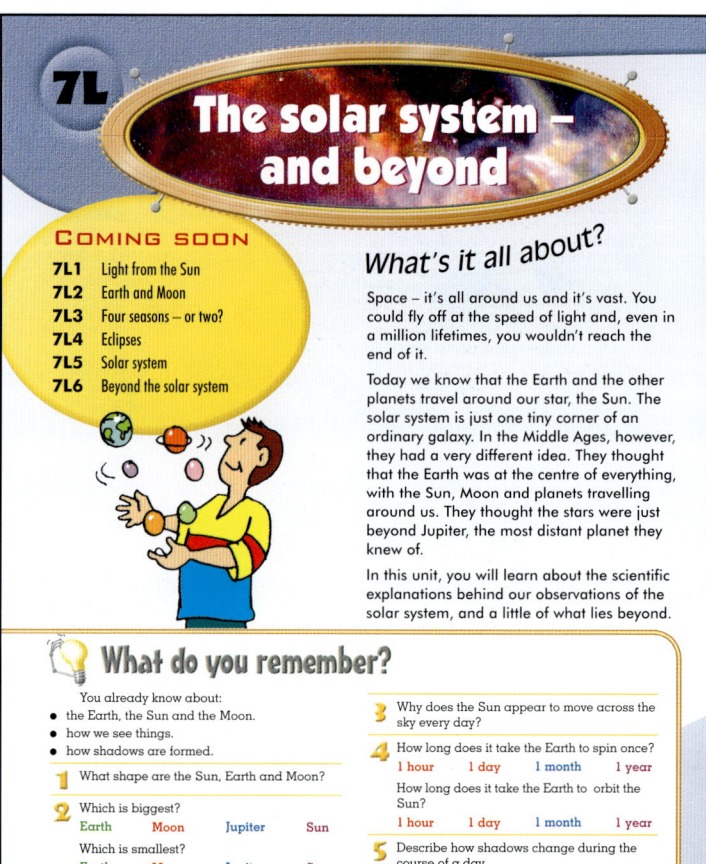

Answers to questions

What do you remember?

1. approx. spherical
2. Sun, Moon
   Order of size, largest to smallest:
   Sun, Jupiter, Earth, Moon.
   (Size, largest to smallest compared with Earth:
   Sun:109.00; Jupiter:11.20; Saturn: 9.50; Uranus:4.00; Neptune:3.88; Earth:1.00; Venus: 0.95; Mars:0.53; Mercury: 0.38; Moon:0.27; Pluto: 0.18)
3. The Earth turns on its axis
4. 1 day, 1 year
5. Start long, get shorter, then longer again. Direction changes as Sun 'moves'.

Teaching strategy

This unit builds on the KS2 unit 5E 'Earth, Sun and Moon' and leads into unit 9J 'Gravity and space'.

Difficulties/Misconceptions

- Many pupils struggle to develop a mental 'picture' of the solar system, and need repeated opportunities to model aspects of the solar system and to use video clips/computer animations.

- Pupils – and adults – often believe the tilt of the Earth varies as it orbits the Sun. Moving the Earth to the other side of its orbit around the 'Sun' to model 'winter' and 'summer' avoids encouraging this misconception.

Notes and tips

Computer animations showing how the orbits of Earth and Moon 'wobble' are very valuable, as this aspect of the solar system can be quite tricky to model in the classroom.

Learning styles

Visual
- **Observing**.
- **Imagining**.
- **Drawing**.
- **Interpreting**.

Auditory
- **Describing**.

Kinaesthetic
- **Modelling**.
- **Recording** using ICT.
- **Research** using ICT.

Intrapersonal
- **Report writing**.

Launch activity notes – Observing and explaining

This activity can be used as the starting point to encourage pupils to discuss their own ideas about 'how the solar system works'. It gives a good opportunity to assess pupils' prior learning and adjust teaching accordingly.

- 'The Moon is just the same size as the Sun.' 'We only see the Moon at night.'

  Pupils will have learned at KS2 that the Sun is much bigger than the Moon. They will also know that the Moon is much closer than the Sun. You may wish to discuss how the apparent equivalence in size supports this. At KS2 pupils will have learned that the Sun is a light source and the Moon isn't. Most of the time, the Moon is invisible in the daytime sky because the light from the Sun is so much brighter. Children may recall occasions (when it is foggy or very overcast) when they have seen the Moon in the daytime.

- 'It's dark at night because the Earth has gone behind the Sun.' 'In Australia, it's dark in the daytime.'

  Allowing pupils to offer their own explanations for night and day will illustrate prior knowledge and areas of confusion. These statements allow you to define 'daytime' as the time when the Sun is visible in the sky, and 'night-time' as the time when the Earth blocks out light coming from the Sun. Check pupils understand how shadows are formed: this will be essential to the understanding of eclipses.

- 'The planets look just like stars, but when I look through my binoculars, I can see they are much bigger than stars.' 'Stars only look smaller than planets because they are a long way away.'

  This final explanation is correct, of course. Use the apparent size of stars, to begin to develop an idea of how vast the Universe is, and how the solar system is only a small part of it.

7L1 Light from the Sun

National framework/QCA SoW references

National framework – Energy: The Sun is a source of energy for the Earth.

QCA SoW

7L Section 1: Causes of a year, a month, a day
7L Section 2: Seeing the Sun and Moon
7L Section 5: What is beyond the solar system?
NC links: Sc4 4a, d.

Learning objectives

- To explain the apparent movement of the Sun and why day and night happen.
- That the Sun and other stars are light sources. They are luminous.
- That the Moon, Earth and other planets are not luminous. We see them because light reflects from them.

Teaching strategy

Key points to highlight

- The Sun stays still; its apparent movement is caused by the Earth rotating.
- The direction of rotation of the Earth, and how this is related to the direction of the apparent movement of the Sun.
- The Sun is luminous; the Earth, Moon and planets are not.
- SAFETY: risk of eye damage/blindness from looking directly at the Sun.

Difficulties/Misconceptions

Some pupils may think very reflective objects are light sources. Modern space travel has enabled cameras on spacecraft to 'see' the dark side of planets – and they wouldn't have a dark side if they were light sources.

Lesson starter suggestions

- **What is the Sun like?** In pairs, pupils make brief notes to summarise present knowledge. They then write one or two questions they have about the Sun.
- **Modelling the Sun, Earth, Moon system:** Pupils choose suitable size balls to represent Sun, Earth, Moon. Can they model the positions/movement?
- **The Sun – good or bad?** Discuss what the Sun provides us with. How is it beneficial/harmful?

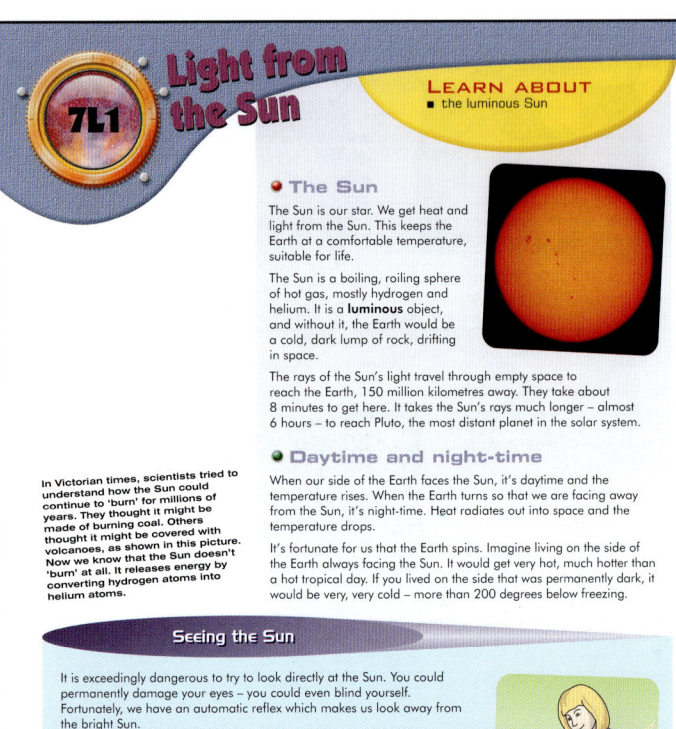

Activity and technicians' notes

Seeing the Sun

Safety notes

Stress strongly the dangers of looking directly at the Sun, with the aided or unaided eye.

Equipment required

- Binoculars
- 2 × stand and clamp
- White card screen

Technician's notes

- First clamp should hold binoculars securely, pointing directly at the Sun.
- Second clamp should hold card screen 'behind' binoculars, at right angles to Sun's rays.
- Adjust distance between eyepiece lenses and card screen to obtain a clear image of the Sun.

Access required to:

- A window looking directly at the Sun.
- A sunny day!

Teaching strategy

Skills

- The activity in this lesson is a teacher-led one, with pupils using their observation skills.
- They may also use prediction and planning skills to discuss ways to improve the image or make it larger.

Notes and tips

This unit provides an opportunity to review and revise learning from Key Stage 2. Some pupils may still find modelling the solar system difficult. Use models of the Sun (lamp if possible) and Earth (a globe) to illustrate how the rotation of the Earth causes night and day. Most pupils will find it helpful if they themselves move around to model the movement of Sun, Earth and Moon. Suitable materials for modelling the Sun, Earth, Moon system could include assorted sized spheres, coloured card circles, coloured clothing/hats for pupils to wear.

Extension ideas

- Pupils could consider why sundials do not have a vertical stick to form the shadow, and how the shadow changes during the day.
- They could also find out more about John Harrison's work to develop an accurate chronometer and how this assisted navigators.

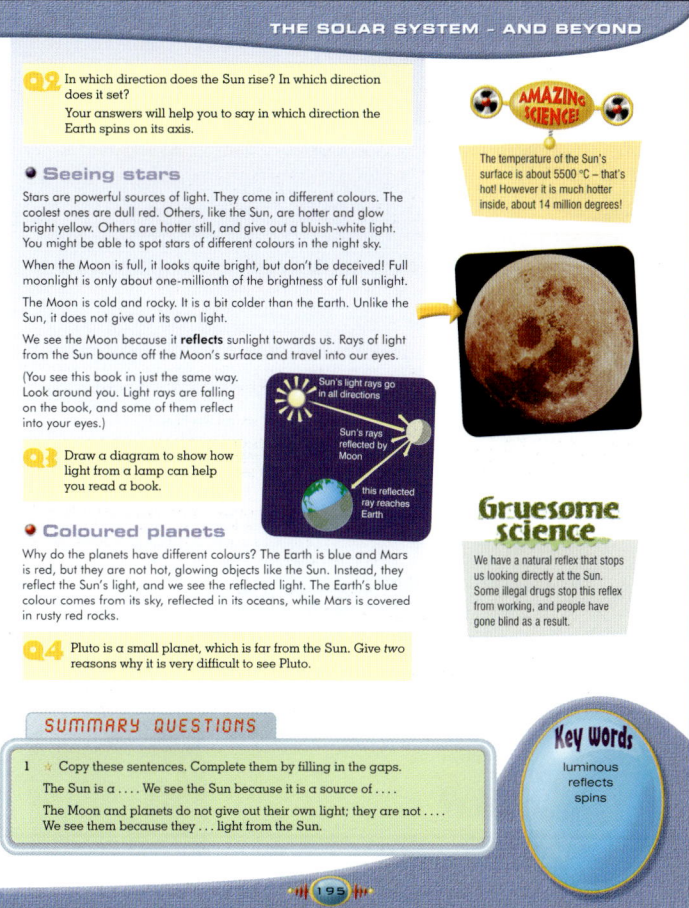

Learning styles

Visual

- **Observing** pictures of the Sun, and image of the Sun on the card screen.
- **Imagining** life on non-rotating Earth.
- **Drawing** ray diagrams of light from Sun and planets.

Answers to questions

Text

- **Q1** Look for sunspots moving across it.
- **Q2** It rises in the East and sets in the West. It turns towards the East, i.e. anticlockwise when viewed from above the North Pole!
- **Q3** Rays of light, with arrows, from lamp to book to eye. Angles of reflection equal to angles of incidence in each case.
- **Q4** Pluto is distant and very small. It is dim, as it only reflects a small amount of sunlight.

Summary

star, light, luminous, reflect

Learning outcomes

Following this, most pupils will:
▶ State that the Sun and other stars are light sources, but all the other objects in the solar system just reflect the Sun's light.

Some pupils will also:
▶ Compare the Sun with other stars.

Suggestions for plenaries

- **What is the Sun like?** Can pupils answer their 'Sun' questions from the starter activity? Encourage class cooperation to answer questions. Help pupils decide where to find answers for any questions they still cannot answer.
- **'Yes'/'No'/'Don't know' cards:** Pupils hold up the appropriate answer card for a series of statements about the Sun that you read out.
- **Fact and explanation:** Pupils link a series of facts about the Sun, with the correct explanations.

7L2 Earth and Moon

National framework/QCA SoW references

QCA SoW
7L Section 2: Seeing the Sun and Moon
7L Section 4: What the solar system consists of
NC links: Sc4 4d.

Lesson starter suggestions

- **What does the Moon look like?** Pupils to draw the appearance of the Moon: either close-up or in the sky. Discuss: 'Does it always look the same, how does it change?'
- **Modelling the orbit of the Moon about the Earth:** Use suitable size balls (Moon diameter is approx. ¼ Earth diameter) to model the orbit of the Moon about the Earth, including Moon's rotation on its own axis.
- **Living on the Moon:** Pupils to discuss, then write about what it would be like to live on the Moon.

Learning objectives

- The Moon is non-luminous.
- The Moon orbits the Earth, and turns on its own axis.
- How sunlight reflecting from the Moon causes the phases of the Moon that we see.

Teaching strategy

Key points to highlight

- The Moon is non-luminous: it reflects light from the Sun.
- Sometimes the Moon is out of sight because it is on the other side of the Earth. Sometimes it is 'in the sky' but too dim compared with the Sun to be visible.
- At any one time, half of the Moon is brightly lit by the Sun.
- As the Moon orbits the Earth, we see different amounts of the brightly lit side. This causes the phases of the Moon that we see.

Difficulties/Misconceptions

Most pupils are unable to picture what causes the phases of the Moon until they have seen a 'working model'. Seating the class in the centre of the room and asking them to observe an 'orbiting Moon' helps. Walk slowly around the group holding up a 'Moon' made from a ball painted half black, half white or yellow. Take care to keep the light half of the ball always pointing in the same direction (towards an imaginary 'Sun').

Skills

- The activity in this lesson allows pupils to practise the mathematical skill of visualising objects or systems in 3D.
- It also develops the verbal literacy skill of describing a system.
- This could be extended, possibly as a homework activity, to the skill of writing non-fiction explanations.

Notes and tips

- At Key Stage 2 some pupils may have modelled the Moon turning once on its own axis each month, keeping the same 'face' of the Moon towards us.
- Some pupils may wonder why the Moon orbits the Earth and not the Sun.
- Others may be aware that the tides are caused by the Moon.
- At this level, it is sufficient to know that the Moon orbits the Earth because it is much closer to the Earth, so the Earth has more effect on it than the Sun does. The tides do not appear on the National Curriculum for KS3 or KS4, but for able/interested pupils, some web sites give good animations of how the tides occur.

Extension ideas

Pupils could observe the full Moon (or pictures) closely and describe the detail they can see. Challenge: Were all the craters formed at the same time? How can you tell? Why aren't craters visible on Earth?

Learning styles

Auditory
- **Describing** how 'phases of the Earth' occur.

Visual
- **Using diagrams** to show how 'phases of the Earth' occur.

Kinaesthetic
- **Making models** to show how 'phases of the Earth' occur.

Answers to questions

Text

Q1 The Earth is a satellite of the Sun.

Summary

1 reflected, satellite, phases, orbit

Activity and technicians' notes

Travel to the Moon

Equipment required
- Different sized, different coloured balls for making models.

Learning outcomes

Following this, most pupils will:
▶ Relate phases of the Moon to a simple model of the Sun, Earth and Moon.
▶ State that the Moon is seen by reflected sunlight.

THE SOLAR SYSTEM – AND BEYOND

● Phases of the Moon

There is a third observation we have to explain. The shape of the Moon changes during a month. Sometimes we see a round, full Moon; sometimes we see a half Moon. Sometimes, when the Moon is 'new', we see a thin crescent Moon, close to the horizon at sunset. These are the **phases** of the Moon.

The Moon is cold – a few degrees colder than the Earth, because it has no atmosphere to keep it warm. It does not give out its own light, so we describe it as **non-luminous**. We see it because it reflects rays of sunlight. The diagram shows how the Sun's rays light up one side of the Moon, and are reflected towards the Earth.

We see the Moon because it reflects sunlight. When the Moon is opposite the Sun in our sky, we see a full Moon. When it is in the same direction as the Sun, we see only the crescent-shaped edge of the side that is illuminated.

● Bird's eye view

We are used to seeing the Moon from the Earth. However, to understand our observations completely, we need to leave the Earth and look down on it from above. The diagram shows what we could see from high above the North Pole:
- The Earth turns on its axis, once each day.
- The Moon orbits the Earth, once each month.
- The Earth and Moon each have one side lit up by the Sun.

You have to imagine standing on the Earth and looking towards the Moon. What would you see at each of its positions? The small diagrams in boxes show the views you would get as the Moon travels round the Earth.

The Earth and Moon, seen from above. The square boxes show the phase of the Moon at each position around its orbit.

Travel to the Moon

Astronauts have visited the Moon, and looked back at the Earth. They saw 'phases of the Earth', just as we see phases of the Moon.
- Imagine that you were visiting the Moon, and you met some puzzled Moon-creatures. You have to explain to them that your home planet doesn't really change shape.
- Devise a way of showing your explanation. You could use diagrams, or make a model using balls of different sizes to represent the Sun, Earth and Moon.

SUMMARY QUESTIONS

1. ☆ Copy these sentences. Complete them by filling in the gaps.
   The Moon is non-luminous; we see it by . . . light.
   The Moon travels around the Earth; it is our . . . .
   We see different . . . of the Moon, according to its position in its . . . .

Key words
non-luminous
orbit
phase
satellite

Suggestions for plenaries

- **Moon concept map:** Pupils draw a map/chart to illustrate their knowledge about the Moon. Discuss and combine to produce a class concept map.
- **Quiz loop:** Pupils write a question plus the answer on separate pieces of paper. Distribute round class. Ask first question. Pupil with the correct answer then asks their question and so on.
- **True or false:** Pupils identify statements that are true or false, then correct the false ones.

7L3 Four seasons – or two?

National framework/QCA SoW references

National framework – Energy: The Sun is a source of energy for the Earth.

QCA SoW

7L Section 3: What causes the seasons on Earth?
NC links: Sc1 2g, h, i, m; Sc4 4a, b.

Lesson starter suggestions

- **Seasons around the world:** Class discussion: Where have pupils 'holidayed', especially in our winter? What was the weather like? Summarise knowledge on the board.
- **Modelling:** Pupils in suitable clothing (or with suitably sized balls/cardboard cut-outs, etc.) model the movement of the Earth around the Sun (including tilt, and rotation on own axis).
- **Contrasting seasons:** Pupils work in pairs to record characteristics/differences for summer and winter. Can they define the characteristics/weather for spring and autumn?

Learning objectives

▶ To use a model to show how the Earth's axis is tilted as it orbits the Sun.
▶ To collect and interpret data about temperature and day length.
▶ To explain how the Earth's tilt causes seasonal variations in temperature and day length.

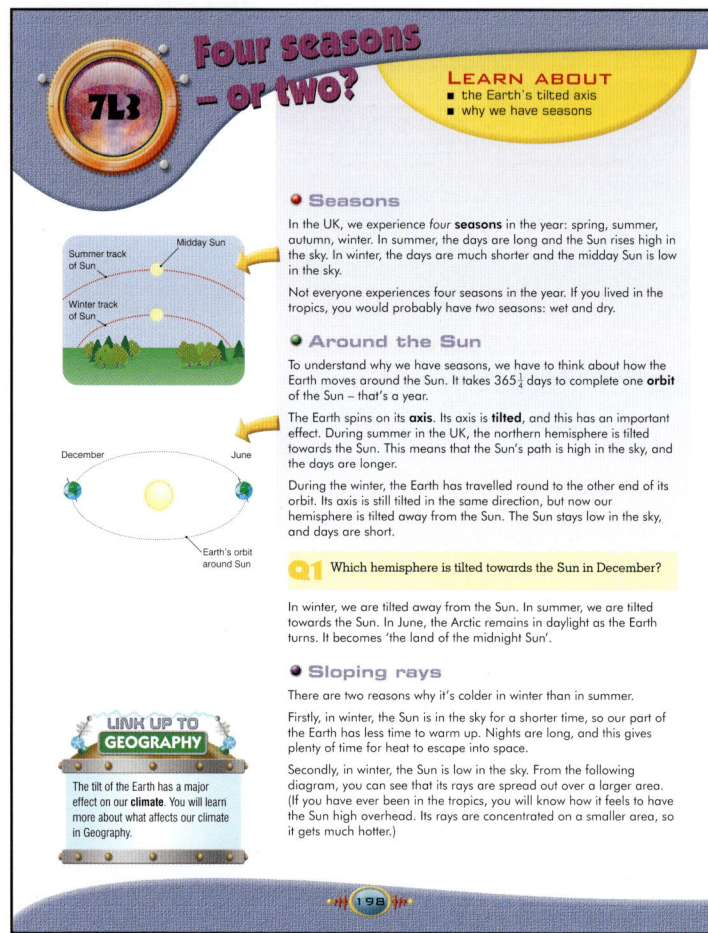

Teaching strategy

Key points to highlight

- The Sun's radiation is more spread out on some parts of the Earth's surface than on others.
- The more spread out the Sun's radiation, the cooler it is (the same amount of heat has to be shared over a larger area).
- The Earth's axis is tilted relative to the Earth's orbit about the Sun.
- In summer the UK 'leans towards' the Sun. The Sun's radiation is more concentrated and it is warmer.
- In winter the UK 'leans away' from the Sun. The Sun's radiation is more spread out and it is colder.

Difficulties/Misconceptions

- A common misconception is that the tilt of the Earth varies between summer and winter. Physical modelling of the system helps.
- Many pupils find it hard to picture simultaneously the rotation of the Earth on its own axis and the movement of the Earth around the Sun. Again, physical modelling using rotating pupils and/or globes helps.

Skills

- In this lesson, pupils again practise their 3D visualisation skills.
- Explaining aspects of the lesson to each other helps with literacy skills.
- This lesson can be linked to climate studies in geography.

Extension ideas

Ask pupils to think of alternative explanations for the seasons, such as the Earth orbiting the Sun in an ellipse, not a circle. Ask them to discuss evidence for and against their explanations.

Activity and technicians' notes

ICT challenge: Modelling the seasons

Safety notes

Ensure heat/light sources are not hot enough to cause burns.

Equipment required
- Large spheres/globes
- Bright lamp
- Data logger with light and heat sensors

Technician's notes

If spheres are used instead of globes, these should be marked with a 'North Pole', 'South Pole' and 'Equator'.

Access required to:
- Computers with data-logging software
- A room with lighting that can be dimmed (desirable but not essential)

Learning styles

Visual
- **Using a model** to observe how temperature and day length change.

Kinaesthetic
- **ICT challenge** to record temperature and light on a model 'Earth'.

ICT challenge

- Some pupils may gain more from this if it is done as a pupil-assisted, teacher demonstration activity.
- **It is very important to avoid the misconception that the tilt of the Earth actually varies.** Try using a naked light bulb as a heat source and holding the 'Earth', with a constant tilt, first on one side of it (for summer) then on the other side for winter.

Answers to questions

Text

Q1 the southern hemisphere

Q2 The Sun is higher in the sky in the summer.

Summary

1. orbit, axis, summer, winter
2. a) Shadows are shorter.
   b) The energy from the Sun's rays is concentrated on a smaller area, so the area gets hotter.

Suggestions for plenaries

- **Seasons around the world:** In a class discussion, ask pupils if they can now explain why countries near the Equator don't have the same seasons as Britain. Encourage pupils to contribute to a 'class explanation'.
- **'Yes'/'No'/'Don't know':** Ask pupils whether or not they agree with each of a series of statements about the seasons. Pupils to simultaneously hold up the appropriate answer card.
- **Connectives:** Pupils to complete the sentence: 'It is warmer in summer in Britain . . .' in any way they like, but they must continue the sentence with one of these words: 'and, because, although, but, however, therefore'. Emphasise that there are many possible correct answers. Pupils to share their answers with the class.

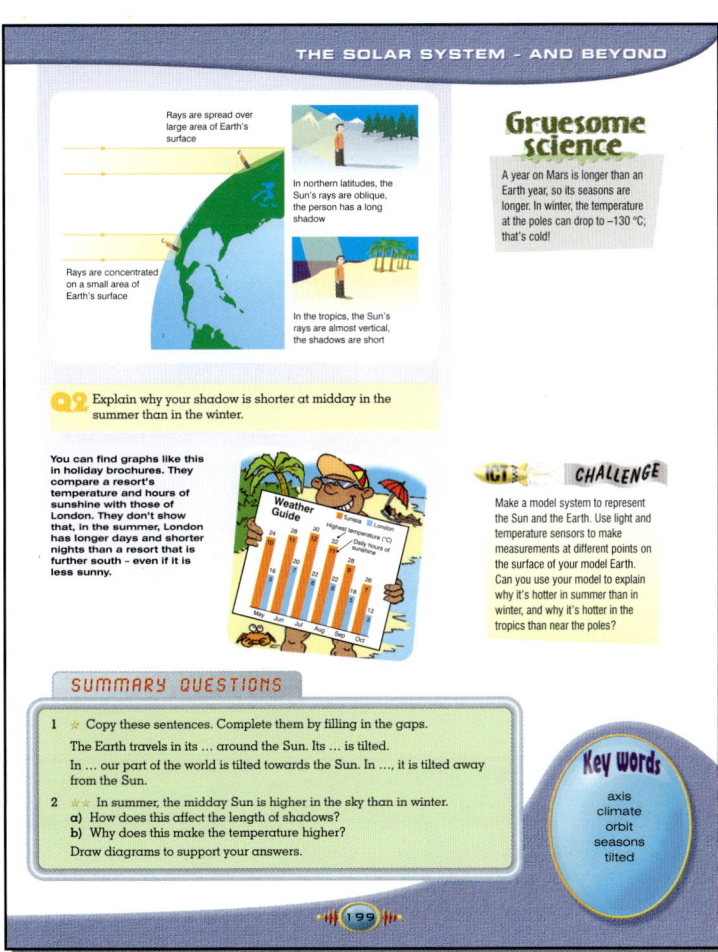

Learning outcomes

Following this, most pupils will:
- Interpret patterns in data, e.g. seasonal variations.
- Relate seasonal changes to a model of the Sun, Earth and Moon.

Some pupils will also:
- Use a model of the Sun, Earth and Moon to explain patterns in data, e.g. seasonal variations, and relate this to real observations.

7L4 Eclipses

National framework/QCA SoW references

QCA SoW

7L Section 2: Seeing the Sun and Moon
NC links: Sc1 1b; Sc1 2a; Sc4 4d.

Learning objectives

▸ To explain what an eclipse of the Sun (a solar eclipse) is like and why it happens.
▸ To explain what an eclipse of the Moon (a lunar eclipse) is like and why it happens.

Teaching strategy

Key points to highlight

- Both the Earth and Moon cast shadows.
- In a solar eclipse, the Moon casts a shadow on the Earth.
- In a lunar eclipse, the Earth casts a shadow on the Moon.
- Solar and lunar eclipses do not happen every month, because the orbit of the Moon around the Earth is not 'flat' compared with the orbit of the Earth around the Sun.
- Solar eclipses are rarer than lunar eclipses, because the Moon casts a smaller shadow than the Earth.

Difficulties/Misconceptions

Many pupils find the 3D aspect of the orbits of the Moon and Earth difficult. Instead they think of them moving in 'flat' circles as though they were moving on the surface of a plate. Modelling the orbits helps, particularly if you are able to use a light source to show the shadows cast.

Skills

- In this lesson pupils use scientific enquiry skills, considering how early scientists used creative thinking and experimental evidence to explain eclipses.
- In the activity, they use prediction and planning to solve an eclipse problem.

Lesson starter suggestions

- **Shadows:** Pupils to describe, using diagrams, writing, or both, how shadows are made. Can the Earth, Sun or Moon cast a shadow? Pupils to share ideas with a partner.
- **What is an eclipse?** The word 'eclipse' comes from the Greek word *ekleipo* meaning 'fail to appear'. Pupils to use this definition to describe what you think might happen in an eclipse.
- **Changing shadows:** Pupils to use diagrams to show how to change the size of the shadow a ball casts on a screen.

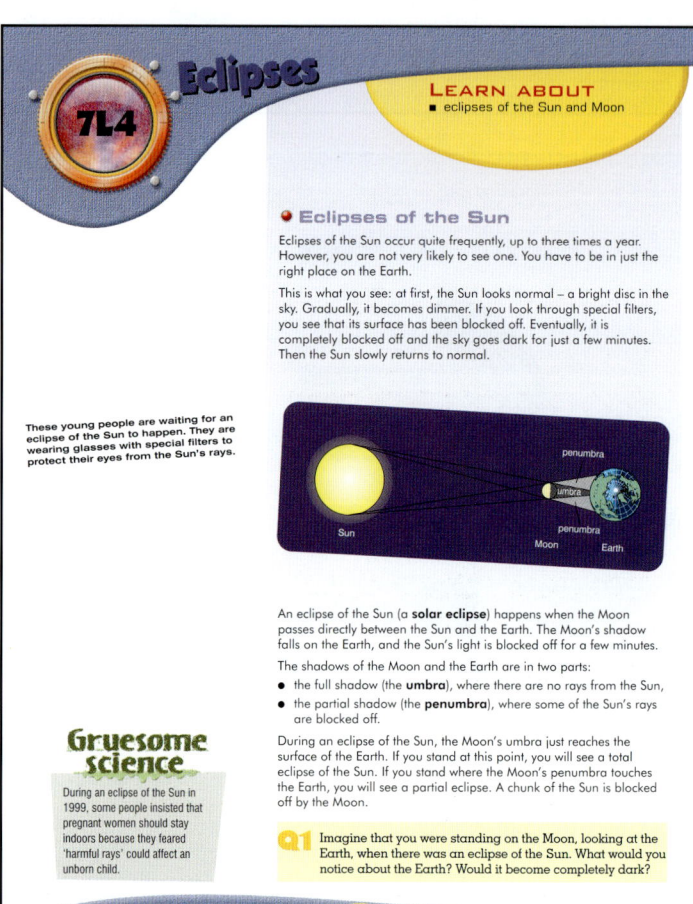

Activity and technicians' notes

Eclipse explanations

Equipment required
- Range of different sized balls
- Bright lamps to represent Sun

Technician's notes

It would be an advantage if light levels in the room could be dimmed.

Access required to:
- Electric power sockets, for lamps

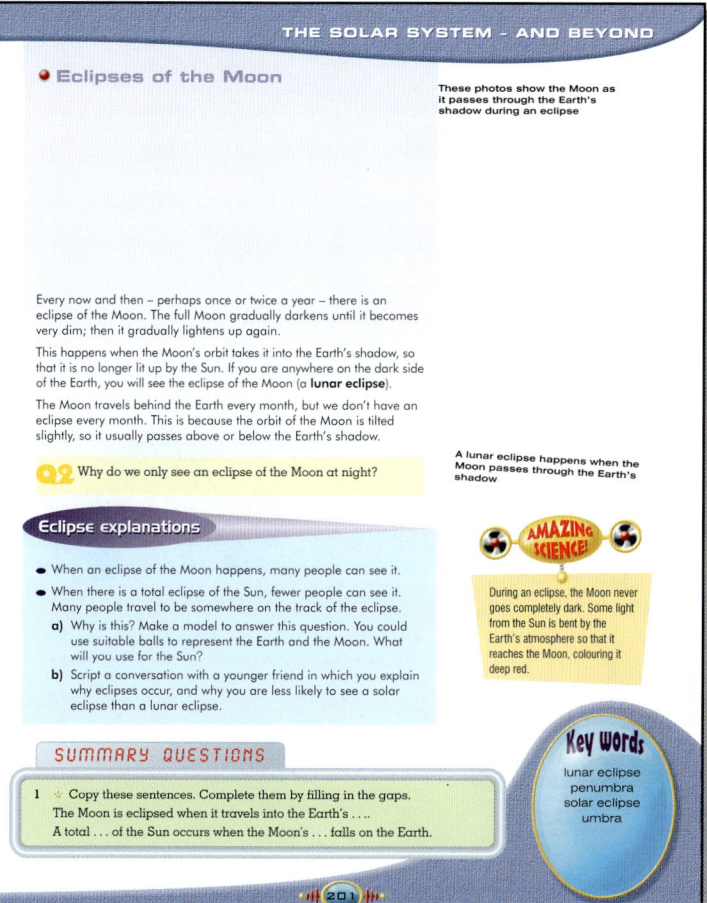

Suggestions for plenaries

- **Explaining eclipses:** Pupils work in pairs to jot down the sequence of events during a solar eclipse. They then share the explanations with the class. Encourage pupils to highlight good points in others' explanations. (This could be used as the starting point for a writing homework describing the experience of witnessing an eclipse.)
- **Eclipse legends:** Pupils work together to invent a short legend to explain a solar or lunar eclipse. The legend to be told to the class (aim to last 60 seconds maximum).
- **Fact and explanation:** Pupils link a series of facts about solar and lunar eclipses, with the correct explanations.

Teaching strategy contd.

Notes and tips

Before beginning this lesson, it is helpful if pupils are thoroughly familiar with Key Stage 2 work about how shadows are formed and the various ways of changing the size or shape of a shadow on a screen. A brief recap with a torch and a ball is generally all that is necessary.

Extension ideas

Pupils could find out about ways in which evidence from eclipses has been used to show that the Earth is a sphere (lunar eclipses) and that the distance of the Moon from Earth varies (solar eclipses).

Learning styles

Visual
- **Observing** animation of a solar eclipse.
- **Imagining** an eclipse of the Sun, seen from the Moon.
- **Observing** animation of a lunar eclipse.

Kinaesthetic
- **Modelling** eclipses of the Sun and Moon.

Intrapersonal
- **Writing** a script to explain why eclipses occur.

Answers to questions

Text

Q1 The Earth would not become completely dark. You would see a small round shadow of the Moon moving across the Earth's surface as the Earth rotated.

Q2 We have to be on the same side of the Earth as the Earth's shadow. That only happens at night.

Summary

1 shadow, eclipse, umbra

Learning outcomes

Following this, most pupils will:
- Describe and explain solar and lunar eclipses.
- Relate solar and lunar eclipses to a model of the Sun, Earth and Moon system.

7L5 Solar system

National framework/QCA SoW references

National framework – Energy: The Sun is a source of energy for the solar system.

QCA SoW

7L Section 4: What the solar system consists of
NC links: Sc4 4b, d, e.

Learning objectives

▸ To find out about the planets, asteroids, comets and satellites that make up the solar system.
▸ To use secondary sources to find out about the solar system.

Teaching strategy

Key points to highlight

- The planets are not the only bodies in the solar system. There are also asteroids, comets and many natural satellites of different planets.
- There are seven planets visible to the naked eye in the night sky. Uranus and Pluto are too distant to be seen without a telescope.
- As they get further from the Sun, the planets get colder.
- As they get further from the Sun, the planets take longer to complete one orbit, so one 'year' gets longer.

Difficulties/Misconceptions

The vast scale of the solar system is very hard to imagine. Help pupils with 'thought models' such as: 'If I set off from the Sun, travelling at 1million kilometres per hour it would take me six days to reach Earth and . . .'; or 'If the Sun was in the middle of a football pitch and the Earth was in goal, then Pluto would be . . .'.

Skills

This lesson provides good opportunities for pupils to practise research skills, using books, CD ROMs and the Internet to find information about the planets. Help them to structure their research by first planning questions they want to find answers to.

Lesson starter suggestions

- **Naming planets:** Ask: 'How many planets can you name in the solar system? Can you put them in order? Think of a mnemonic to help you remember the order. Do you know anything about any of the planets?'
- **Space luggage:** Imagine you are to tour the solar system. What equipment will you take with you? Why?
- **Planet anagrams:** Un-jumble the names of the planets. Then label the planets.

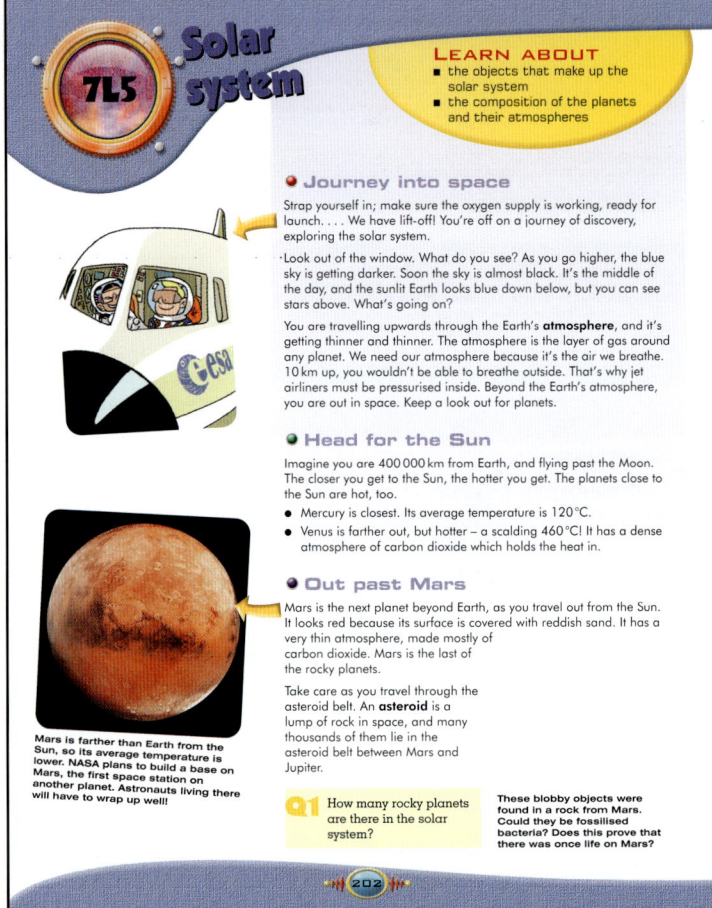

Notes and tips

- This lesson could be supported by a trip to a planetarium or space centre.
- Alternatively, a variety of virtual tours of the solar system can be found on the Internet.

Extension ideas

Pupils could think about exploring the solar system. Ask: 'Why would a printed map be little or no use? What can a computerised map or model show, that would make it much more useful?'

Learning styles

Visual
- **Observing** photo/PowerPoint presentation about the planets.
- **Imagining** a space journey through the solar system.

Kinaesthetic
- **ICT research** about different bodies in the solar system.

THE SOLAR SYSTEM – AND BEYOND

The gas giants

Jupiter and Saturn are two gas giant planets. They are made mostly of liquid hydrogen, but each has a core of solid rock, rather like the Earth. It's as if the Earth was trapped inside a thick, frozen atmosphere of hydrogen.

It's cold where these planets orbit, far from the Sun. That's why gases such as hydrogen and nitrogen are liquid or even solid here.

Beyond these gas giants are three more planets: Uranus, Neptune and Pluto. They are made of water, methane and carbon dioxide, all frozen solid.

Jupiter is famous for its Great Red Spot, a giant storm which has been raging in its atmosphere for hundreds of years

Many moons, cool comets

There are two types of satellite:
- **Natural satellites** are natural objects, such as our Moon; other planets also have moons orbiting them – Saturn holds the record at present, with over 30 known moons.
- **Artificial satellites** are spacecraft we have sent into orbit around the Earth, such as the ones that broadcast satellite TV programmes, and others that carry space telescopes.

Every now and then, a **comet** appears in the sky. It may be visible for a few days. You should see its long tail, pointing away from the Sun.

Planets have orbits that are almost circular. A comet's orbit is elongated – it is an ellipse.

A comet is a ball of dust and ice. As its orbit brings it close to the Sun, it starts to thaw. Material evaporates into space to make the tail. When it has passed the Sun and is disappearing back into the cold depths of the solar system, it refreezes.

This is not a giant pizza. It's Io, one of Jupiter's moons. It orbits Jupiter in less than two days.

Q2 Is a comet a satellite? Explain your answer.

 CHALLENGE

Choose one of the following:
- moons • asteroids • comets

Make an electronic scrapbook of photographs, collected from the Internet or from CD ROMs. You could scan some photos from books. Write a caption for each of your images. You could use PowerPoint to turn your scrapbook into a slideshow.

SUMMARY QUESTIONS

1. ☆ Copy these sentences. Complete them by filling in the gaps.
   The Moon is a natural … of the Earth.
   The … of a planet is almost circular; a … follows a more elongated path.
   We use … and … to find out more about the solar system.
2. ☆☆ a) Name the four rocky planets.
   b) Name two gas giant planets.
3. ☆☆ Why would it be difficult for people to live: a) on Mars, b) on Jupiter?

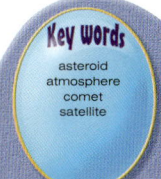
Key words
asteroid
atmosphere
comet
satellite

Learning outcomes

Following this, most pupils will:
▶ Present a report about an object in the solar system.
▶ Select information from secondary sources.
▶ Describe the positions of different planets in the solar system, and what they are like compared with Earth.

Some pupils will also:
▶ Use a variety of secondary sources to find out about the solar system.
▶ Observe and interpret patterns in data about planets in the solar system.

Answers to questions

Text
Q1 four: Mercury, Venus, Earth, Mars
Q2 Yes it is a satellite of the Sun, but with a very elongated elliptical orbit.

Summary
1. satellite, orbit, comet, telescopes spacecraft
2. a) Mercury, Venus, Earth, Mars
   b) Jupiter, Saturn.
3. Mars is cold, with a very thin atmosphere, composed mainly of carbon dioxide; Jupiter is even colder, composed mainly of liquid hydrogen.

ICT challenge

Technician's notes

Access needed to:
- Computers
- CD-ROM encyclopedias, Internet, Powerpoint software (optional)

Suggestions for plenaries

- **Spider diagram:** Pupils to work in pairs to select the three most important pieces of information about one planet, and contribute to a class spider diagram about the planets.
- **Moving house:** Pupils to choose which planet humans should colonise first. Ask: 'Why have you chosen this one? Explain your reasons.'
- **Muddled descriptions:** An alien visitor Al has muddled his descriptions of the planets in the solar system. Pupils to sort them out for him.

7L6 Beyond the solar system

National framework/QCA SoW references

QCA SoW
7L Section 5: What is beyond the solar system?
NC links: Sc4 4a, d.

Learning objectives

▸ To find out why we only see the stars at night.
▸ To find out why the stars we see in the night sky change during the night and during the year.

Teaching strategy

Key points to highlight
- We see stars because they are light sources.
- There are billions of stars – far more than we can see with the naked eye. Many are just like our Sun. (Pupils will learn about different types of stars and the evolution of stars at Key Stage 4.)
- The stars only seem to move in the sky because the Earth moves.
- Stars look dimmer if they are further away. (At Key Stage 4, pupils will learn that some stars really are brighter than others, but this can be ignored here.)
- As you travel north or south the stars that are visible change, so stars can be used to navigate.

Difficulties/Misconceptions
Some pupils may think all the stars are the same distance away. Talk about the different brightness of stars showing this is not so.

Skills
Pupils use literacy skills to interpret texts and to write explanations.

Notes and tips
- Remind pupils of how the Earth's movement makes the Sun appear to move. Extend this to the apparent movement of the stars.
- This lesson could also be supported by a trip to a planetarium or space centre.

Extension ideas
Pupils could find out about how the type of light from different stars tells astronomers about the stars. For example, dim stars are usually further away than brighter stars; red coloured stars are always cooler than our Sun, which is a yellow star.

Lesson starter suggestions

- **Star legends:** The ancient Egyptians believed that when their Pharaohs died, they returned to their home in the sky and 'pictures' of them could be seen in the stars. Ask: 'Do you know any other legends about stars?'
- **Star study:** Discuss in pairs or small groups: 'What do you know about the stars? Do you know how scientists try to find out about the stars? Is it important for scientists to learn about stars? Why, or why not?'
- **Star signs:** Ancient people pictured 12 signs of the zodiac; show pictures of what they saw. Does your 'star sign' affect you?

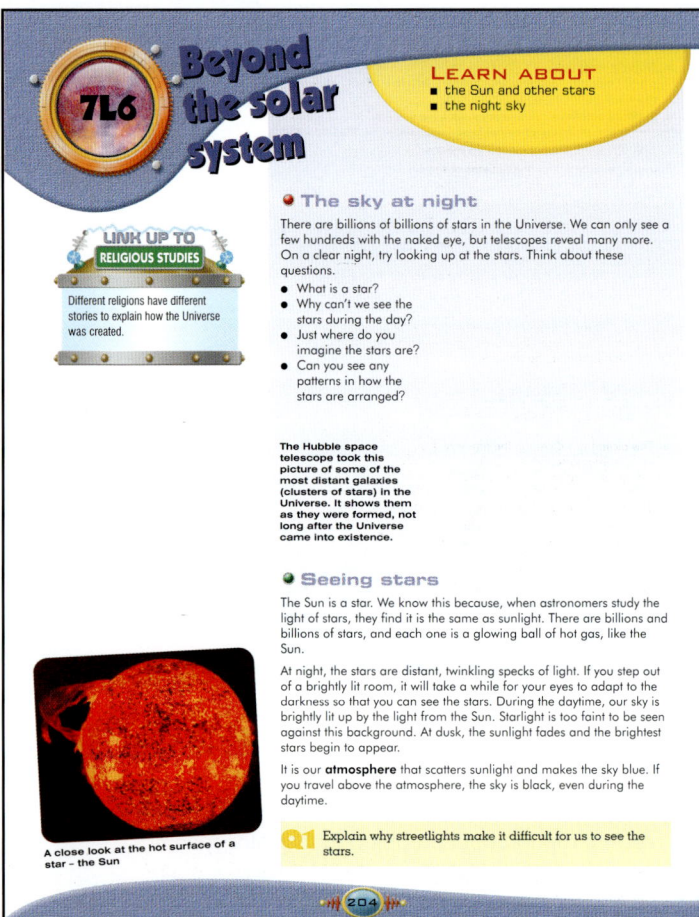

Answers to questions

Text

Q1 If the sky is brightly lit by light from streetlights, the dim stars do not show up.

Q2 In winter we are the opposite side of the Sun from summer, so at night-time we are looking in the opposite direction and so we see different stars.

Summary

1 planets, stars, Sun, day, night, Earth.
2 The Moon has no atmosphere, so light from the Sun is not scattered in all directions, as it is by the Earth's atmosphere. Therefore, on the Moon, the Sun looks bright but all the surrounding sky looks black.

Learning styles

Visual
- **Observing/imagining** the night sky.
- **Drawing diagrams** of stars visible in summer and winter.
- **Interpreting** early maps of the Universe.

Activity and technicians' notes

Far out

This is a paper-based activity, not requiring equipment.

Learning outcomes

Following this, most pupils will:
▶ Use a model of the solar system to explain why the stars we see change.
▶ State that the Sun and all other stars are light sources.

Some pupils will also:
▶ Describe how explanations about the solar system have changed with time.
▶ Compare our Sun with other stars.

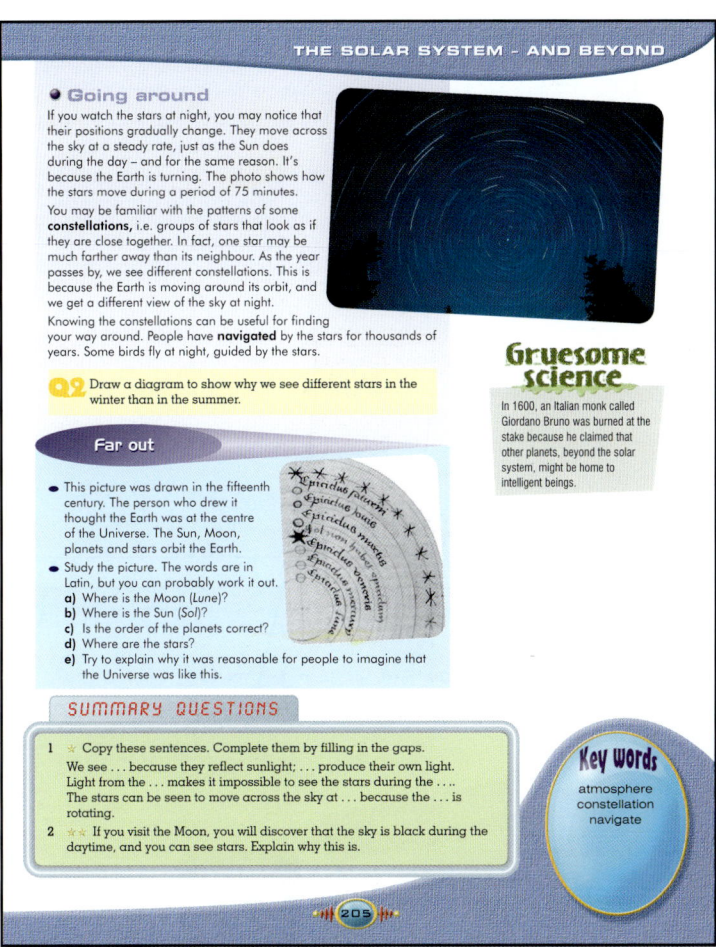

Suggestions for plenaries

- **Star science:** Pupils to imagine they are science correspondents for the 'Pharaoh Weekly'. They write a short article explaining to the ancient Egyptians why the stars they see in the sky keep changing.
- **Another Earth?** In 1600 Giordano Bruno said there might be other planets beyond the solar system, with intelligent beings on them. Ask: 'Do you think he was right or wrong? Would the planets be near stars? How could we find out? What could we look for?'
- **Star map:** Pupils to make a concept map showing information about stars, using different colours to show what is 'science knowledge' and what is other information.

7L Read all about it!

Teaching strategy

An eclipse/Eclipses now

Difficulties/Misconceptions

- Questions pupils may ask about eclipses include:
  Why is a solar eclipse visible in some places, but not everywhere?
  Why are solar eclipses so rare?
  Why does the 'path of the eclipse' move?
  Why does the Moon exactly cover the Sun?

- Use class modelling of eclipses, using a bright light and different sized balls, encouraging pupils to find the answers to these questions for themselves. They will find that:
  1. The Moon only casts a small shadow, so most of the Earth's surface remains in 'sunlight'
  2. The Moon's orbit about the Earth 'wobbles' – it is not often in the same plane as the Earth's orbit about the Sun, so most of the time the Moon's shadow does not fall on Earth.
  3. The Moon's shadow stays still (actually it moves very slightly because the Moon is orbiting Earth) but the Earth is turning on its own axis, so the shadow – the 'path of the eclipse' – seems to move across the Earth's surface.
  4. The Moon exactly covers the Sun because of the extraordinary coincidence that the Sun is 400 times the size of the Moon and also 400 times as far away. Sometimes an annular eclipse happens, when a bright ring of Sun is visible around the edge of the dark Moon. This happens when the Moon is slightly further from Earth (its orbit is not a perfect circle) and so looks slightly smaller.

Skills

This unit is very useful in developing pupils' skills at visualising things in three dimensions. However some pupils, often completely unrelated to general ability, find this extremely difficult and need repeated modelling to make this learning secure.

Notes and tips

Video footage of the 1999 solar eclipse, and computer animations of solar and lunar eclipses are very useful.

Questions

1. Give one thing in the account of an eclipse from 1061 that is unlikely to be true.
2. Why do astronomers today still find early accounts of eclipses useful?
3. Why are astronomers able to learn more about the Sun during an eclipse than they can normally?
4. Why does the Moon exactly cover the Sun during an eclipse?
5. When is it safe to take off your safety filters during an eclipse? When should you put them back on again?

Answers

1. birds falling while flying
2. They use them to check their calculations of when ancient eclipses happened to check they are calculating in the right way.
3. The edges of the Sun are visible when its most intense radiation has been blocked by the Moon.
4. The Sun is 400 times larger than the Moon but 400 times as far away.
5. when the Moon is completely covering the Sun; as soon as the Sun begins to emerge from behind the Moon

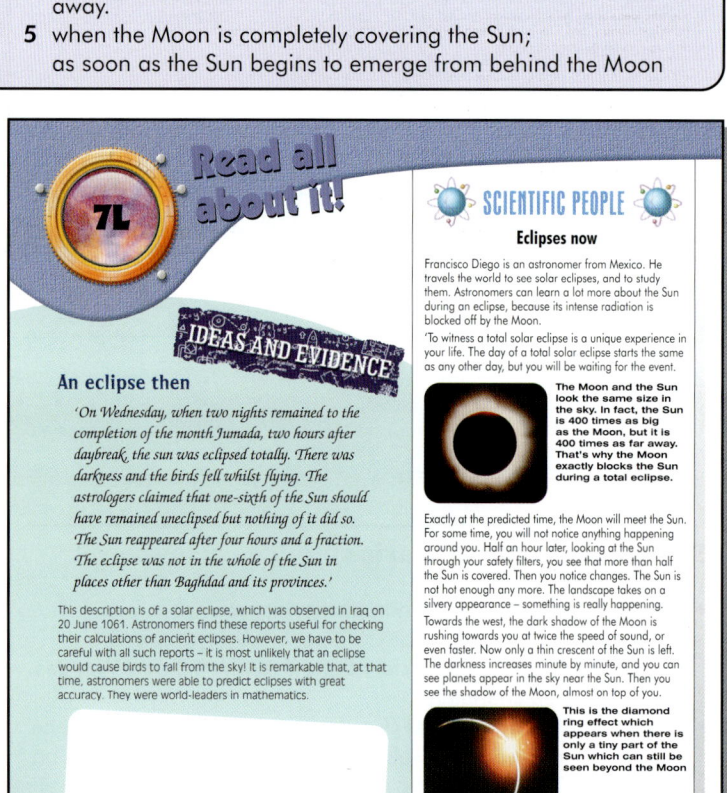

Danger – common errors

- Throughout this unit, modelling should have helped pupils view the solar system 'from outside'. Asking them to demonstrate and explain models may reinforce earlier learning from the unit.
- Some pupils may think that the 'squashed orbit' of Earth means that we are nearer the Sun in summer and that is why summer is hotter. Check for this misconception and correct it if necessary.

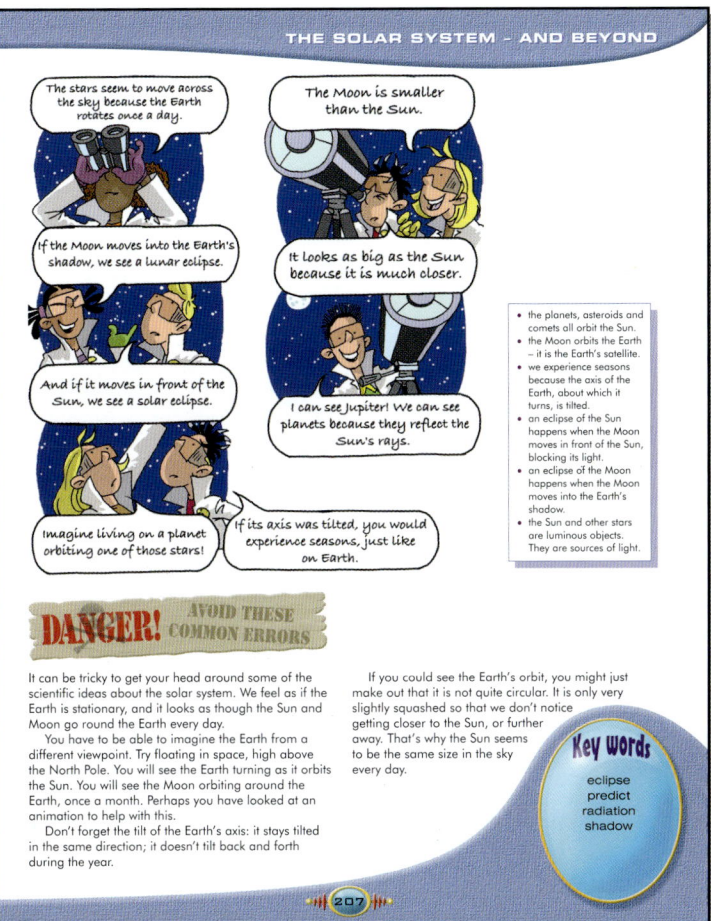

Learning styles

Visual
- **Observing** videos of 'Scientific people' and of solar eclipses.

Auditory
- **Listening** to video of 'Scientific people'.

Interpersonal
- **Conversation and feedback** about cartoon characters' discussions.

Learning outcomes

Following this unit, most pupils will:
- Relate phases of the Moon, seasonal changes, solar and lunar eclipses to a simple model of the Sun, Earth and Moon.
- Use a model of the solar system to explain why the stars we see change.
- Describe the positions of different planets in the solar system, and what they are like compared with Earth.
- State that the Sun and other stars are light sources, but all the other objects in the solar system just reflect the Sun's light.

Some pupils will also:
- Observe and interpret patterns in data about planets in the solar system.
- Compare the Sun with other stars.

In scientific enquiry:

Following this unit, most pupils will:
- Describe and explain solar and lunar eclipses.
- Interpret patterns in data, such as seasonal variations in temperature and day length.
- Present a report about an object in the solar system, selecting information from secondary sources.

Some pupils will also:
- Describe how explanations about the solar system have changed with time.
- Use a model of the Sun, Earth and Moon to explain patterns in data, such as seasonal changes, and relate this to real observations.
- Use a variety of secondary sources to find about the solar system.

7L Unit review

Answers to review questions

1 Stars move across the sky because the Earth is rotating on its own axis. Stars seem to rise in the East and set in the West, just as the Sun does.

2 If a planet's orbit were elliptical, it would be closer to the Sun at some times than others. It would have hot seasons and cold seasons, but the day length would remain the same throughout the seasons.

3 An example: The energy from the Sun can be used by solar cells, to generate electricity.

4 An example: My Very Easy Method Just Speeds Up Naming Planets

5 c They are looking at the Moon from slightly different positions.

  e If the Moon was further away from Earth, it would not cover the Sun completely during a solar eclipse, so we would see a bright ring of Sun around the dark central Moon.

6 Day follows night, night follows day. The Sun is a luminous object. The Earth turns on its axis.
  The Moon's appearance changes from day to the next. The Sun is a luminous object. The Moon is a non-luminous object. The Moon orbits Earth.
  The moon is occasionally eclipsed. The Sun is a luminous object. The Moon orbits the Earth. The Earth and Moon have shadows.

7 a–b Pupil drawing
  c Because of its position, Voyager 1 can only see part of the brightly lit side of each of the Earth and the Moon.

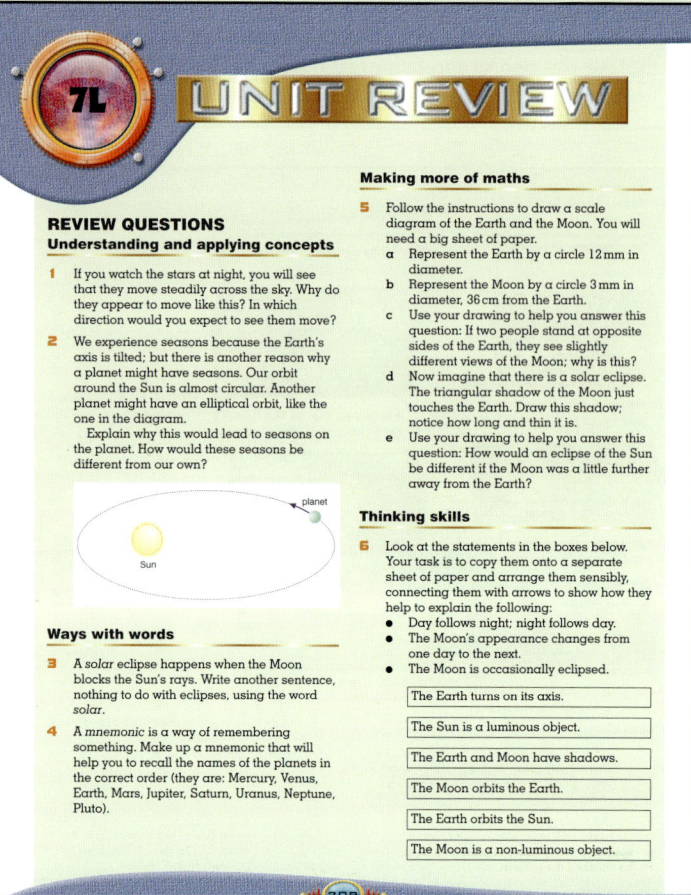

UNIT REVIEW QUESTIONS

Extension question

7 This is an unusual photograph. It shows both the Earth and the Moon, and was taken from the Voyager 1 spacecraft.
  a Imagine looking downwards on the Earth and Moon. Draw a diagram to show the relative positions of the Earth, Moon, Sun and spacecraft at the time this photograph was taken.
  b Show the parts of the sides of the Earth and Moon which are lit up by the Sun.
  c Explain how your diagram relates to the photograph. Why do both the Earth and the Moon appear as crescent shapes?

SAT-STYLE QUESTIONS

1 Every day, the Sun appears to travel across the sky. The picture shows the path of the Sun across the sky during a winter's day.

  a The picture shows the position of the Sun at mid-day. Explain how you can tell from the picture that it is mid-day. (1)
  b Copy the diagram and add arrows to show how the Sun moves along its path from dawn to dusk. (1)
  c Draw another line on the diagram to show the path of the Sun across the sky on a summer's day. (1)

2 During 2003, an eclipse of the Sun was visible from the north of Scotland. Mike wore his special safety glasses to protect his eyes when he watched the eclipse.

Mike said: 'You must wear safety glasses because the Sun is a luminous object. You don't need to wear them for an eclipse of the Moon, because the Moon is a non-luminous object.'
  a Mike said that the Sun is a luminous object. Was he correct? (1)
  b Mike was not correct in giving the reason why safety glasses must be worn when looking at the Sun. Give the correct reason. (1)
  c The diagram shows the positions of the Sun, Moon and Earth a few days before the eclipse. Copy the diagram and show how light from the Sun allows us to see the Moon. (2)

3 The drawing shows the Earth with the Sun's rays shining on it. Give reasons to support your answers to the questions that follow.

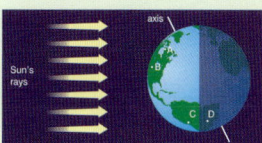

  a At which point is it night-time? (2)
  b At which point is the Sun highest in the sky? (2)
  c At which points is it winter? (3)

Key words
Unscramble these:
moulisun
clefter
shapes
broit
burma

Answers to SAT-style questions

1 a the Sun is at the highest point in the sky; the shadows are short (1)
  b the Sun moves from East to West (1)
  c the path goes higher in the sky, starts further East and ends further West (1)

2 a Yes (1)
  b The Sun is extremely bright and can damage your eyes. Less bright luminous objects are safe to look at. (1)
  c The light goes from the Sun to the Moon and reflects off to Earth. For reflection, the angle of incidence should be approximately equal to the angle of reflection. (2)

3 a D, it is on the unlit half of the Earth (2)
  b B, closest to midday (2)
  c C and D, on hemisphere tilted away from the Sun (3)

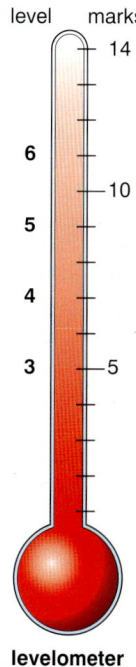

levelometer

Best practice in Sc1

Notes

These pages shows some of the Scientifica crew carrying out an investigation into dissolving when they were in Year 6. Many of your pupils will have done such an investigation.

The activity on page 212 is intended to encourage your pupils to consider the general procedure for carrying out an investigation, and to think critically about the investigation shown here.

You may need to point out that some of the ideas shown here are incorrect, and that it is their task to explain why.

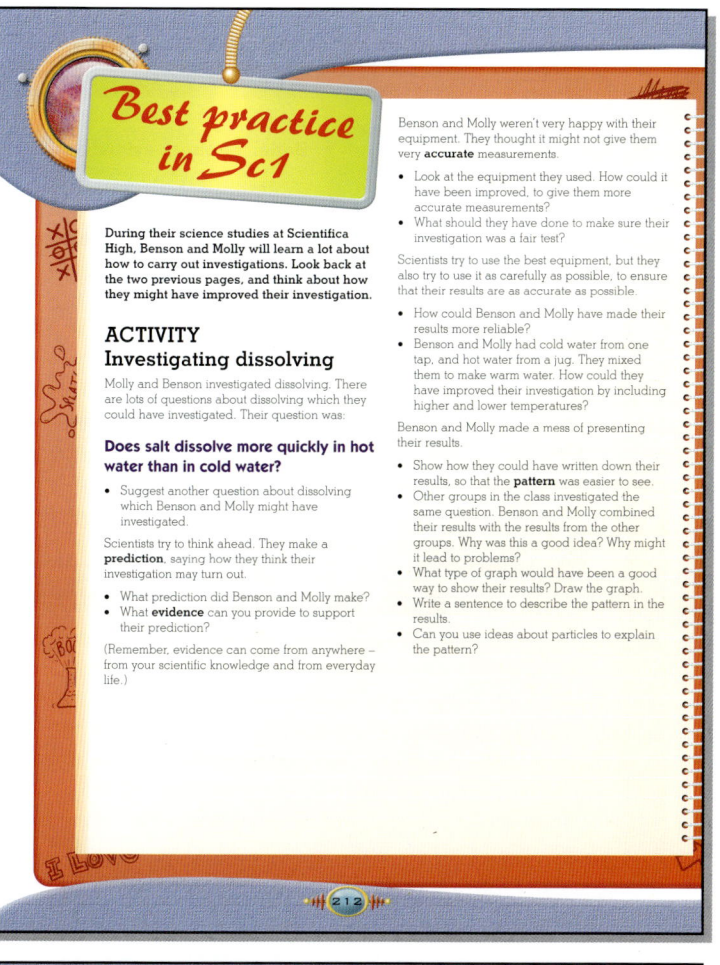

Best practice in Sc1

During their science studies at Scientifica High, Benson and Molly will learn a lot about how to carry out investigations. Look back at the two previous pages, and think about how they might have improved their investigation.

ACTIVITY
Investigating dissolving

Molly and Benson investigated dissolving. There are lots of questions about dissolving which they could have investigated. Their question was:

Does salt dissolve more quickly in hot water than in cold water?

- Suggest another question about dissolving which Benson and Molly might have investigated.

Scientists try to think ahead. They make a **prediction**, saying how they think their investigation may turn out.

- What prediction did Benson and Molly make?
- What **evidence** can you provide to support their prediction?

(Remember, evidence can come from anywhere – from your scientific knowledge and from everyday life.)

Benson and Molly weren't very happy with their equipment. They thought it might not give them very **accurate** measurements.

- Look at the equipment they used. How could it have been improved, to give them more accurate measurements?
- What should they have done to make sure their investigation was a fair test?

Scientists try to use the best equipment, but they also try to use it as carefully as possible, to ensure that their results are as accurate as possible.

- How could Benson and Molly have made their results more reliable?
- Benson and Molly had cold water from one tap, and hot water from a jug. They mixed them to make warm water. How could they have improved their investigation by including higher and lower temperatures?

Benson and Molly made a mess of presenting their results.

- Show how they could have written down their results, so that the **pattern** was easier to see.
- Other groups in the class investigated the same question. Benson and Molly combined their results with the results from the other groups. Why was this a good idea? Why might it lead to problems?
- What type of graph would have been a good way to show their results? Draw the graph.
- Write a sentence to describe the pattern in the results.
- Can you use ideas about particles to explain the pattern?

Notes

This activity is intended to lead your pupils through an evaluation of the investigation shown in the cartoon story on pages 210–11.

It could be easily incorporated into work on Unit 7H Solutions. Year 6 pupils often have a very limited understanding of the particle model of matter, so it is difficult for them to use scientific knowledge to explain dissolving.

Once you have discussed the ideas of predictions, patterns, accuracy, reliability, range of data, explanations etc., you might carry out a demonstration of a similar investigation, to show how these ideas might be put into practice. You can shortcut the need for repetitions by producing a set of results of the 'ones I prepared earlier' variety.

(Pupils always find it entertaining to see a demonstration of bad practice which they can criticise, but it is also very worthwhile giving a demonstration of good practice.)

SCIENTIFIC ENQUIRY

Pete and Reese are investigating burning. Look at how they carried out their investigation – are they better at investigating than when they were in Year 6?

On page 215, Pip and Mike are investigating how the temperature changes during the day. They are using electronic equipment to collect data, but they still have to think just as hard about what their results mean.

- It can help to have a trial run.
- Repeat readings can help you to judge how reliable your measurements are.
- Think about the strength of the evidence you have collected.

Key words
- accurate
- evidence
- pattern
- prediction
- reliable

Notes

This cartoon story shows Pete and Reese investigating burning. The emphasis is on obtaining better evidence by making repeat readings, and looking critically at data including graphs.

This activity could be incorporated with work on burning (Unit 7F Simple chemical reactions; Unit 7H Energy). Since the scientific content is not difficult to grasp, it could also be tackled at any point where you wish to discuss these aspects of investigative work.

Best practice in Sc1

Notes

This cartoon story shows Pip and Benson using an electronic data gathering system in the course of an experiment.

The accompanying activity encourages pupils to think about the advantages of electronic data logging. It also emphasises the need to think critically about the data collected – it is important not to simply accept it at face value.

This activity could be incorporated into work on night and day (Unit 7K: The solar system and beyond). It could also be tackled at any point where electronic data collection is being used.